Thomas Moore

British Wild Flowers

Thomas Moore

British Wild Flowers

ISBN/EAN: 9783741186134

Manufactured in Europe, USA, Canada, Australia, Japa

Cover: Foto ©Klaus-Uwe Gerhardt /pixelio.de

Manufactured and distributed by brebook publishing software
(www.brebook.com)

Thomas Moore

British Wild Flowers

BRITISH WILD FLOWERS.

FAMILIARLY DESCRIBED IN THE

FOUR SEASONS.

FOR THE USE OF BEGINNERS AND AMATEURS.

A NEW EDITION OF

'The Field Botanist's Companion.'

BY

THOMAS MOORE, F.L.S., F.R.H.S., ETC.

LONDON:

REEVE & CO., 5, HENRIETTA STREET, COVENT GARDEN.

1867.

PRINTED BY J. E. TAYLOR AND CO.,
LITTLE QUEEN STREET, LINCOLN'S INN FIELDS.

PREFACE.

—◆—

The present Work forms one of a projected series of Illustrated Volumes on British Natural History. It has not been prepared in consequence either of a lack of books on the same subject or of any deficiency in those which are accessible, but simply to bear its part towards the completion of a naturalist's library upon a plan as nearly as practicable uniform with Mr. Berkeley's 'Fungology,' with which the series commenced.

The book thus originated has been written with the aim of making it effective in training up students for the more advanced and technical Floras of Babington, Bentham, and Hooker. For this purpose, dissections of the parts of the flowers have been introduced among the figures; and the illustrated plants, which furnish a tolerably complete series as regards important features of structure, have for the same reason been rather fully described. By these means it is hoped that an insight into the structure and classification of plants will be acquired by those who may honour the book by making it their 'Companion;' and this insight, if attained, will assuredly enable them to use profitably the more technical volumes to which allusion has been made. That of Mr. Bentham has been herein most generally followed. Though it has been sought to use as few technical terms as possible, they have not been entirely avoided : that indeed being all but impracticable, without largely increasing its bulk, in a book devoted to matters of science, even when presented in a popularized form. Where, however, such terms have been used, the attempt has been made to soften them down and make them self-explanatory as much as possible. Beyond this, a Glossary has been provided to elucidate the rest.

We may here briefly point out how the book is intended to be used. First it will be evident, on scanning a few of its pages, pp. 22 to 42 for example, that the bouquet form of gathering wild flowers, in which, judging from one's correspondence, lady-botanists are most apt to indulge, is not the proper botanical form; that is to say, little sprigs of flowers, consisting mainly of flowers, and without leaves and fruits, are not the materials from which a proper knowledge of the plants or of their classification is to be learned. It will be seen on glancing over these pages, that not only are flowers with their stamens and their pistils required for examination, but there are carpels with their seeds to be sometimes looked into, and there are leaves with their ribs and veins to be closely scrutinized—these latter, indeed, being almost the first which the uninitiated botanist must inquire about. It is clear, therefore, that complete examples should be gathered, completeness being determined thus: all small plants should be entire, root and branch; while of larger ones portions as ample as may be manageable (if for an herbarium, nearly as long as the paper used, or such as can while fresh be readily folded to the length of the paper) must be selected, showing all the parts,—roots if conveniently obtainable; perfect leaves, both root-leaves and stem-leaves, if they differ at all, as they often do; and flowers including buds and old flowers with advanced fruits. Sometimes perfect full-grown fruits or seed-vessels are indispensable, and, as they are always desirable, they should always if possible be gathered. Furnished with such materials as these, and supposing ourselves occupied with them at any period of the pleasant springtide, let us look at p. 22 and p. 27, wherein the two great divisions of plants are indicated. These, it will be seen, consist of plants with parallel-veined leaves and those with net-veined leaves. It is not generally difficult to decide between these, for we may leave out of sight the very few exceptions to the general rule that occur. Supposing our plant is a Wallflower, it will be net-veined, and therefore exogenous. Now among the exogenous plants it will be seen that there are some polypetalous (the Thalamiflores and Calyciflores), some monopetalous (the Monopetals), and some apetalous or without petals (the Monochlamyds). It will at a glance be ascertained that this flower has petals as well as calyx, and more than one petal, so that it must be polypeta-

lous. Well then, to which of the polypetalous groups does it
belong—that with petals distinct and stamens hypogynous, or
that with petals distinct and stamens perigynous or epigynous?
It proves, when pulled to pieces, first gently tearing away the
calyx and then the petals, to have hypogynous stamens: so
that it is one of the Thalamiflores. Then comes the question,
are the carpels distinct or combined into an apocarpous ovary?
They are combined. After that the question, whether the
placentas are parietal or axile? They are parietal. Still again
another inquiry, whether the stamens are five or six in number? They are six, and tetradynamous. We should thus arrive at the fact that it belonged to group 3, or the Cruciferous
plants. Then turning to group 3 in p. 28, two decisions would
lead us to fix on the eighth genus, *Cheiranthus.* Passing on
to the eighth genus in p. 44, it would at last be found that
the plant was *Cheiranthus Cheiri.* By a similar use of the
tabulated information the names of the other plants would be
ascertained.

The book does not contain all our British plants, but only
those which have been deemed the most likely to be met with
either in home-walks or during more extended health-seeking
or pleasure-seeking trips. These have been divided under
the four seasons, an arrangement which it has been thought
would simplify the task of searching for the name of a flower
by excluding those which bloom during other seasons. Generally this would be the case, but instances may here and there

occur in which the flowering season does not very definitely fall within these limits; and there are some plants which bloom continuously or successionally. If a plant, therefore, should not be readily identified amongst those set down for the season in which it may be found in bloom, those of the preceding or succeeding seasons should also be scrutinized.

It is perhaps hardly necessary to add, that in using the Summaries, care should be taken to compare correctly with their correlatives the various terms and signs employed for subdividing the groups. Thus in p. 22, Exogenous plants are to be compared only with Endogenous plants (the words will be found printed in the same type) in p. 27; Thalamiflores, Calyciflores, Monopetals, and Monochlamyds, (also printed in correspondent type,) must be compared with each other; and the * in p. 23 compared with ** in p. 24: the signs which fall between * and ** in like manner being compared among themselves. The different signs are used for the purpose of classification in the consecutive order which is usual in printed books, namely, *, †, ‡, §, ||, a, 1, etc.

We should strongly recommend those who may take this book as a guide in acquiring a knowledge of our Wild Flowers, to gather as many of those which are figured as they may be able to collect, and to compare them closely both with the figures and with the lengthened descriptions, which latter will be found under the head "Illustrations," at pp. 1, 78, 348, and 386. They should especially separate the parts of the

a

flowers, so as to get a clear conception of the terms by which the several parts are known, as well as those by which their characteristic features, taken either separately or collectively, are indicated. It may be well further to point out, that it would be by no means labour lost to commit to memory the more important of the technical terms found in the Glossary. The advantage of knowing their meaning would at once be discovered on making real use of the book.

CHELSEA,
May, 1862.

EXPLANATION OF THE PLATES.

✦

PLATE I.

A.—Anemone nemorosa, *Linnæus :* p. 6.
   1. Cluster of young carpels.
B.—Caltha palustris, *Linnæus*   p. 7.
   1. Cluster of young carpels.
   2. One of the carpels, separate.
   3. The full-grown carpels or follicles.
C.—Cardamine pratensis, *Linnæus :* p. 7
   1. The stamens and pistil.
D.—Cheiranthus Cheiri, *Linnæus :* p. 8.
   1. The stamens and pistil.
   2. The ripe pod, with the valves separating.

PLATE II.

A.—Viola odorata, *Linnæus :* p. 8.
   1. The stamens and pistil.
   2. One of the spurred stamens, separate.
   3. The pistil with its curved style.

b

B.—Acer Pseudo-platanus, *Linnæus :* p. 9.

    1. One of the flowers, separate.

    2. The winged pair of fruits or samara.

C.—Oxalis Acetosella, *Linnæus :* p. 10.

    1. The stamens and pistils.

    2. The pistil, with its five styles.

D.—Ribes rubrum, *Linnæus :* p. 11.

    1. A flower, separate.

    2. One of the berries.

PLATE III.

A.—Saxifraga granulata, *Linnæus :* p. 11.

    1. Two of the clustered tubers.

    2. A flower, with the calyx and corolla removed.

    3. The calyx surrounding the ovary, with its two styles.

B.—Leontodon Taraxacum, *Linnæus :* p. 12.

    1. One of the florets, separate.

    2. One of the fruits or achenes, with the pappus expanded.

C.—Bellis perennis, *Linnæus :* p. 13.

    1. One of the ray florets, separate.

    2. One of the disk florets, separate.

D.—Vinca minor, *Linnæus :* p. 14.

    1. A portion of the corolla-tube, showing the attachment of the stamens to its inner surface.

    2. A flower, with the corolla and two of the calyx-segments removed, showing the ovaries, style, and pulley-shaped stigma.

PLATE IV.

A.—Menyanthes trifoliata, *Linnæus :* p. 14.

    1. A corolla laid open, showing the stamens attached to the inner fringed surface.

2. The pistil.
3. One of the capsules.

B.—Primula vulgaris, *Hudson :* p. 1.
1. The pistil.

C.—Daphne Mezereum, *Linnæus :* p. 15.
1. One of the perianths laid open, showing the stamens attached to the inner surface.
2. The pistil.
3. A section of one of the berries.

D.—Euphorbia amygdaloides, *Linnæus :* p. 16.
1. One of the small flower-heads, with crescent-shaped glands, several erect male flowers, and a recurved female flower.
2. One of the male flowers separated, consisting of a scale and stamen only.
3. The scale of the male flower.
4. The tricoccous stalked ovary, with its three-cleft style, forming the female flower, which has no perianth.

PLATE V.

A.—Ulmus montana, *Smith :* p. 16.
1. A perianth with stamens.
2. The pistil.
3. The winged seed.

B.—Salix Caprea, *Linnæus :* p. 17.
1. A branch, with the male catkins.
2. One of the male flowers, separate.
3. A female catkin.
4. One of the female flowers, separate.
5. A seed.

C.—Orchis maculata, *Linnæus :* p. 19.
(*The name is misprinted "mascula" on the plate.*)
1. A flower, showing the three recurved sepals, two convergent petals, and three-lobed lip.

b 2

2. The column and cells of the single perfect anther.

3. One of the pollen-masses.

D.—Cypripedium Calceolus, *Linnæus :* p. 19.

    1. Front view of the column, with its abortive central stamen, and two lateral perfect ones.

    2. Back view of the same.

PLATE VI.

A.—Crocus vernus, *Willdenow :* p. 4.

    1. The base of the flower-tube, laid open in the upper part, showing the attachment of the stamens.

    2. One of the stamens.

    3. The three-cleft stigma.

B.—Galanthus nivalis, *Linnæus :* p. 3.

    1. A stamen.

    2. The pistil, showing the ovary at the base.

C.—Scilla verna, *Hudson :* p. 20.

    1. One of the flowers, separate.

    2. A segment of the star-shaped perianth, with its stamen.

    3. The pistil.

D.—Hyacinthus non-scriptus, *Linnæus :* p. 20.

    1. Portion of a flower, showing the pistil, and the insertion of the stamens.

    2. A transverse section of the ovary.

PLATE VII.

A.—Berberis vulgaris, *Linnæus :* p. 80.

    1. A flower, separate.

    2. A petal, with its two glands.

    3. A stamen, showing the valves of its anthers.

    4. The pistil.

    5. One of the fruits.

B.—**Nymphæa alba,** *Linnæus :* p. 81.

    1, 2. Different forms of stamens.

    3. The pistil, with its sessile radiating stigma.

C.—**Papaver Rhœas,** *Linnæus :* p. 82.

    1. A stamen.

    2. The pistil, with its sessile radiating stigma.

    3. The ripe capsule, showing the apertures for the escape of the seeds.

D.—**Reseda lutea,** *Linnæus :* p. 82.

    1. One of the flowers, separate.

    2. One of the upper petals.

    3. Other forms of petals.

    4. The ovary, accompanied by a stamen.

PLATE VIII.

A.—**Helianthemum vulgare,** *Gærtner :* p. 83.

    1. Back view of the calyx.

    2. The pistil.

B.—**Drosera rotundifolia,** *Linnæus* p. 93.

    1. One of the gland-fringed leaves.

    2. The stamens and pistil.

C.—**Polygala vulgaris,** *Linnæus :* p. 83.

    1. A flower, seen from beneath.

    2. Side view of a flower.

    3. The pistil.

D.—**Frankenia lævis,** *Linnæus :* p. 84.

    1. A portion of the stem, showing the opposite leaves, and axillary clusters of smaller leaves.

    2. A flower.

    3. One of its petals.

    4. The pistil.

PLATE IX.

A.—Dianthus plumarius, *Linnæus:* p. 84.
1. One of the clawed petals.
2. The pistil, with its two curved styles.

B.—Lychnis Githago, *Lamarck:* p. 85.
1. A petal with adherent stamens.
2. The pistil, with five styles.

C.—Malva sylvestris, *Linnæus:* p. 85.
1. The disk-shaped fruit, composed of several contiguous carpels, and surrounded by the calyx.
2. The staminal column with the stigmas protruding from the centre.
3. One of the stamens, separate.
4. The pistil separated from the column of stamens.

D.—Tilia europæa, *Linnæus:* p. 86.
1. One of the flowers, separate.

PLATE X.

A.—Hypericum pulchrum, *Linnæus:* p. 86.
1. One of the oblique petals.
2. One of the parcels of stamens.
3. The calyx surrounding the pistil, the petals and stamens being removed.

B.—Geranium pratense, *Linnæus:* p. 87.
1. The stamens and pistil.

C.—Linum usitatissimum, *Linnæus:* p. 87.
1. The stamens and pistil.
2. The capsule surrounded by the calyx.
3. One of the seeds.

D.—Impatiens Noli-me-tangere, *Linnæus:* p. 89.
1. A flower with most of the petals removed.
2. A young capsule.

PLATE XI.

A.—Euonymus europæus, *Linnæus* : p. 90.
  1. A flower.
  2. A ripe fruit.

B.—Rhamnus catharticus, *Linnæus* : p. 90.
  1. A staminate or barren flower.
  2. A pistillate or fertile flower.
  3. The ripe fruit.

C.—Tamarix anglica, *Webb* : p. 88.
  1. Portion of a branchlet, showing the close scale-like leaves.
  2. A flower separated from the spike.
  3. The stamens and pistil.

D.—Rosa canina, *Linnæus* : p. 78.
  1. A vertical section of the flower, the petals being removed, showing the ovaries attached to the inside of the calyx-tube, with the stigmas just protruding from the orifice.
  2. A stamen.
  3. One of the ovaries.
  4. The ripe fruit.

PLATE XII.

A.—Lythrum Salicaria, *Linnæus* : p. 91.
  1. A vertical section of a flower.
  2. A capsule, bursting open.

B.—Hippuris vulgaris, *Linnæus* : p. 94.
  1. A flower, consisting of the ovary and style with one stamen.
  2. The stamen separate.

C.—Epilobium hirsutum, *Linnæus* : p. 94.
  1. The top of the ovary, with the calyx, stamens, and style.
  2. One of the seeds.

D.—Lathyrus pratensis, *Linnæus :* p. 91.

    1. The standard or vexillum forming the dorsal petal.

    2. The lateral petals or wings.

    3. The lower petal or keel.

    4. One of the flowers complete.

    5. The pod or legume.

E.—Bryonia dioica, *Jacquin :* p. 95.

    1. A male or staminate flower.

    2. A female or pistillate flower.

PLATE XIII.

A.—Œnanthe crocata, *Linnæus :* p. 96.

    1. A flower separated from the umbel.

    2. One of the cylindrical fruits.

    3. A transverse section of the fruit, showing the two car-
        pels, and the ridges and vittæ.

B.—Pastinaca sativa, *Linnæus :* p. 97.

    1. A flower detached from the umbel.

    2. One of the flattened fruits.

    3. A transverse section of the fruit.

C.—Sedum acre, *Linnæus :* p. 92.

    1. A leaf, showing the spur behind the point of attach-
        ment.

    2. The ripe follicles.

    3. A flower.

    4. The follicles or seed-vessels, before bursting.

D.—Cornus sanguinea, *Linnæus :* p. 98.

    1. One of the flowers.

    2. A fruit.

E.—Montia fontana, *Linnæus :* p. 92.

    1. The calyx or flower-cup.

    2, 3. Different views of the corolla and stamens.

    4. The pistil.

PLATE XIV.

A.—Lonicera Periclymenum, *Linnæus* : p. 98.
   1. One of the tubular two-lipped flowers.
   2. A cluster of the berries.

B.—Galium verum, *Linnæus* : p. 99.
   1. A flower complete.
   2. The pistil separated.

C.—Centranthus ruber, *De Candolle* : p. 99.
   1. One of the curious spurred flowers.
   2. The ripe fruit, with its feathery pappus unrolled.

D.—Dipsacus sylvestris, *Linnæus* : p. 100.
   1. A flower with its bract.
   2. A pistil.

PLATE XV.

A.—Carduus nutans, *Linnæus* : p. 101.
   1. One of the florets, separated from the capitule or head.

B.—Campanula rotundifolia, *Linnæus* : p. 102.
   1. The anthers and pistil, showing the inferior ovary.

C.—Erica cinerea, *Linnæus* : p. 103.
   1. One of the pitcher-shaped flowers.
   2. A stamen, showing the appendage at the base of the anther-cells, and the pores by which the pollen escapes.
   3. The pistil.

D.—Ligustrum vulgare, *Linnæus* : p. 104.
   1. A flower.
   2. A sprig bearing berries.

PLATE XVI.

A.—Villarsia nymphæoides, *Ventenat* : p. 104.
   1. A portion of the throat of the corolla, showing the attachment of the stamens.
   2. The pistil.

B.—**Polemonium cœruleum,** *Linnæus :* p. 105.

    1. A portion of the throat of the corolla, showing the insertion of the stamens.

    2. The pistil, surrounded by an annular disk.

    3. A capsule.

C.—**Convolvulus arvensis,** *Linnæus :* p. 106.

    1. One of the stamens.

    2. The pistil, surrounded by an annular disk.

D.—Solanum Dulcamara, *Linnæus :* p. 109.

    1. A flower, showing the connivent anthers.

    2. One of the anthers separate, showing the terminal pores.

    3. The pistil.

PLATE XVII.

A.—Digitalis purpurea, *Linnæus :* p. 111.

    1. The didynamous stamens, separated from the corolla.

    2. The pistil.

    3. A transverse section of the ovary

B.—Orobanche minor, *Sutton* p. 112.

    1. A flower separated from the spike, with its bract.

    2. The corolla laid open, showing the attachment of the stamens.

    3. One of the sepals.

    4. The pistil with a bract and sepals, the corolla removed.

C.—**Verbena officinalis,** *Linnæus :* p. 114.

    1. A flower, separate.

    2. The corolla laid open, showing the attachment of the stamens, and surrounding the pistil.

D.—Salvia pratensis, *Linnæus :* p. 113.

    1. A flower, separate.

    2. The stamens, showing the short filament, with the elongated connective bearing a perfect anther on its long arm and an abortive one on its short arm.

3. Another view of the lower part of the connective.
4. The pistil, showing the four-lobed ovary.

PLATE XVIII.

A.—Myosotis palustris, *Withering*: p. 107.
1. The corolla laid open, showing the stamens and the scaly appendages of the throat.
2. The pistil with its four-lobed ovary.

B.—Pinguicula vulgaris, *Linnæus*: p. 114.
1. A flower with the corolla removed.
2. The two-valved one-celled capsule.

C.—Armeria maritima, *Willdenow*: p. 109. '
1. A flower removed from the head.
2. One of the petals with a stamen attached to its base.
3. The pistil, with five hairy styles.

D.—Plantago media, *Linnæus*: p. 110.
1. A flower, separate.
2. The pistil.

PLATE XIX.

A.—Polygonum Bistorta, *Linnæus*: p. 115.
1. A flower, separate.
2. The pistil with its three styles.

B.—Aristolochia Clematitis, *Linnæus*: p. 116.
1. A flower complete.
2. The pistil, showing the rayed stigma and epigynous stamens.
3. A capsule.

C.—Pinus sylvestris, *Linnæus*: p. 117.
1. One of the pairs of leaves, with its sheath.
2. One of the male catkins, with its bract.
3. An anther, with its two adnate cells, and scale-like connective.

4. One of the scales of the female catkin, with its two naked ovules.

5. Inner view of a similar scale when mature, with its two winged seeds.

D.—Tamus communis, *Linnæus :* p. 118.
1. One of the male flowers with stamens.
2. One of the female flowers with inferior ovary and three-branched style.
3. The style, separate.
4. A ripe berry.

PLATE XX.

A.—Paris quadrifolia, *Linnæus :* p. 119.
1. One of the awl-shaped stamens.
2. The pistil.
3. A ripe berry, with the persistent perianth.

B.—Hydrocharis Morsus-ranæ, *Linnæus :* p. 119.
1. The stamens removed from the flower.
2. One of the stamens, separate.
3. The pistils, with six two-cleft stigmas.

C.—Ophrys apifera, *Hudson :* p. 126.
1. Front view of the lip.
2. Side view of the lip.
3. The column, showing the anther-case and pollen-masses.
4. One of the pollen-masses separate.

D.—Iris Pseud-acorus, *Linnæus :* p. 125.
1. A stamen.

E.—Convallaria majalis, *Linnæus :* p. 127.
1. A flower with its pedicel and bract.
2. Vertical section of a flower, showing three of the stamens and the pistil.
3. A ripe berry.

PLATE XXI.

A.—Butomus umbellatus, *Linnæus :* p. 125.
   1. A flower with the perianth removed.
   2. One of the carpels.
   3. A transverse section of the carpels.

B.—Typha latifolia, *Linnæus :* p. 121.
   1. A male flower, separate.
   2. A female flower, separate.

C.—Acorus Calamus, *Linnæus :* p. 123.
   1. A flower with its green scales and stamens, and broad sessile stigma.
   2. One of the scales with its accompanying stamen, separate.
   3. An ovary.

D.—Potamogeton natans, *Linnæus :* p. 123.
   1. One of the flowers.
   2. A stamen, separate.
   3. One of the carpels.

E.—Lemna trisulca, *Linnæus :* p. 122.
   1. A frond with its branches and root.
   2. A flower.

F.—Juncus effusus, *Linnæus :* p. 124.
   1. A portion of the cylindrical stem.
   2. A flower.
   3. A capsule with the dry persistent perianth.

G.—Carex riparia, *Curtis :* p. 129.
   1. A flower of the male spikelet with its glume.
   2. A flower of the female spikelet with its glume.

PLATE XXII.

A.—Arundo Phragmites, *Linnæus :* p. 131.
   1. One of the flowers, separate.

B.—Bromus mollis, *Linnæus :* p. 130.

    1. One of the flowers, separate.

    2. The pistil with its feathery styles.

C.—Parnassia palustris, *Linnæus :* p. 349.

    1. One of the fringed glands or nectaries.

    2. The pistil with its sessile stigmas.

D.—Ulex nanus, *Forster :* p. 349.

    1. A thorny branchlet with its flower.

    2. The standard or dorsal petal.

    3. The keel or lower combined petal.

E.—Hedera Helix, *Linnæus :* p. 350.

    1. A flower separated from the umbel.

PLATE XXIII.

A.—Scabiosa succisa, *Linnæus :* p. 351.

    1. A flower separated from the head.

    2. The pistil.

B.—Arbutus Unedo, *Linnæus :* p. 352.

    1. A stamen, showing its pores and awns.

    2. The pistil.

    3. Some of the ripe berries.

    4. A transverse section of a berry.

C.—Gentiana Pneumonanthe, *Linnæus :* p. 353.

    1. A portion of the corolla, showing the attachment of the
       stamens to its inner surface.

    2. The pistil.

D.—Mentha Pulegium, *Linnæus :* p. 354.

    1. A flower separated from the verticillaster or whorl-like
       collection of flowers.

    2. The pistil.

E.—Chenopodium polyspermum, *Linnæus .* p. 354.

    1. A flower, separate.

    2. The fruit enclosed by the persistent perigone.

PLATE XXIV.

A.—Colchicum autumnale, *Linnæus :* p. 355.
   1. A stamen.
   2. The pistil invested by the base of the perianth-tube.
   3. The three-celled capsule.

B.—Eriocaulon septangulare, *Withering :* p. 356.
   1. A leaf.
   2. One of the male flowers.
   3. One of the female flowers.
   4. The pistil with its subulate stigmas.

C.—Ilex Aquifolium, *Linnæus :* p. 386.
   1. An abortive flower.
   2. A perfect flower.
   3. A ripe berry.
   4. A transverse section of the berry.

D.—Viscum album, *Linnæus :* p. 387.
   1. A cluster of male flowers.
   2. A cluster of female flowers.
   3. A ripe berry.

SPRING FLOWERS.

—◆——

"Bring flowers.

They speak of hope to the fainting heart,
With a voice of promise they come and part;
They sleep in dust through the winter hours,
They break forth in glory; bring flowers, bright flowers."

Mrs. Hemans.

ILLUSTRATIONS.

ESPECIALLY deserving of precedence amongst Spring Flowers is the Primrose,* first or primal flower of the year, which at any time after the winter solstice may be seen peeping forth from many a sheltered bank, on the look-out, as it were, for sunnier days, or else may be found snugly nestling in some sheltered copse or hedgerow, blossoming unheeded, the herald of approaching spring. Even in more rigorous seasons, when the wintry blasts are severe and prolonged, this earliest flower is found soon arousing from its winter sleep. Seldom indeed does the first month of the year pass away without the echoing cry through the dreary town, "Primroses, all a-growin', all a-blowin': buy my pretty primroses." And though it is only when spring has come at last, and in good earnest,

* *Primula vulgaris*—Plate 4 B.

B

that the thick tufts of modest blossoms show "the rathe
primrose" in its fullest beauty, we do well to be thankful for
the earnest which it gives us of the Flora of the new-born year.

This well-known favourite flower, besides illustrating the
Primulaceous family, will afford us a general botanical lesson,
ere we pass on to notice other heralds of the spring. Gather
one of the flowers which are snugly nestled amongst the broad
and wrinkled leaves, and at the end of its slender stalk will be
seen a narrow green five-pointed five-angled funnel, which is
the calyx or flower-cup, the outermost of the series of parts
which constitute the flower, and which in most flowers, being
green, may be readily distinguished. Within this stands the
corolla, the yellow attractive part of the flower. In this case,
if gently pulled, it will be found to come away all in one,
and hence it is called monopetalous, or consisting of one petal,
the parts of which it is constituted having, as it were, cohered
to form this one piece. This is the condition in which the
corolla is found throughout the MONOPETALS, one of the larger
groups in which plants are classified. Sometimes these Monopetals are very irregular in form, but in the Primrose we
have an example of one which is perfectly regular. Let us
see of what it consists:—first, there is a long slender tube,
which is straight; then there is a broad flat expanded part or
limb, and that consists of five lobes or segments of similar
size and form, and spreading equally, so that we may infer
that the corolla is here formed of five equal coherent parts.
We have thus a corolla which is perfectly regular or symmetrical in plan. The particular form a corolla assumes
has, in most cases, a particular name; that of the Primrose
is called hypocrateriform, or salver-shaped, but other regular
monopetalous forms will be found by-and-by, in summer, in
the funnel-shaped corollas of the Convolvulus, or the bell-

shaped corollas of the Campanula. We will not now stay to examine the Primrose further, but pass on with the remark, that on the inside of the corolla-tube are fixed five stamens, and just visible in its mouth is the round-headed stigma on its long slender stalk, looking very much like a pin dropped into the tube. The Primrose illustrates one form under which the large Dicotyledonous group, to which we shall have to recur, is developed.

At a very early period, too, comes the Snowdrop,* one of the Amaryllidaceous family, a doubtful wilding perhaps, but here and there established in meadows and pastures, in seemingly wild localities: always welcome as " the early herald of the infant year," or, as Mrs. Barbauld calls it, " the first pale blossom of the unripened year," its pendent bells rivalling in purity the snow-flakes which not unfrequently fall around them. The grassy leaves and pendulous flowers of the Snowdrop are familiar to every one ; the three white outer concave segments of the latter form the sepaline divisions of the perianth, and represent the calyx, and the three inner, which are smaller and tipped with green, form the petaline divisions representing the corolla. In the inside are six stamens ; while the ovary or immature seed-vessel is formed entirely beneath the other parts of the flower, that is, below the actual base of the parts though in reality uppermost as the flower hangs : hence it is called inferior.

We have in the Snowdrop an excellent illustration of another large group or class of plants—the MONOCOTYLEDONS, so named because their seeds are furnished with only one instead of the two cotyledons or lobes which are found in the larger proportion of the flowering plants, hence called Dicotyledons. Take one of them for examination. There is first a

* *Galanthus nivalis—* Plate 6 B.

bulbous stem or base to the leaves and flower-stalks: this is
frequent among Monocotyledons, but not characteristic. Then
the leaves are ribbed with veins all running side by side
lengthwise, a peculiar feature by which the group may in all
ordinary cases be recognized. Next the flowers consist of six
divisions, which is a nearly certain mark of a Monocotyledon.
There is no separate calyx and corolla in the Snowdrop as in
the Primrose, but the two will be found blended together, all
the parts having become corolla-like. When thus combined,
the calyx and corolla form what is called a perianth; three of
the segments, which will be found to be exterior, represent the
calyx, and are hence called sepaline divisions, and three are
interior, representing the petals, and are hence called petaline
divisions. These features—the straight-veined leaves, and the
parts of the flowers arranged in threes or multiples of three—
are generally distinctive of the large and important class of
MONOCOTYLEDONS, also called ENDOGENS, from the internal
manner of accretion in their stems.

The Crocus,* too, is one of Spring's earliest harbingers,
starting up almost as if by magic from the scarce-thawed
earth, and making it resplendent with the richest colours almost before the snow has vanished from the surface. The
Spring Crocuses, though blooming at so early a period, present
little other difference compared with those kinds which bloom
in autumn. This favourite flower is a well-known representative in gardens of the Iridaceous family, and is so far naturalized in meadows and pastures in some parts of England, as to
claim admission amongst our field plants, though perhaps not
a true aborigine. It has a kind of solid bulb called a corm,
and produces grassy leaves. Its large funnel- or vase-shaped
six-lobed flowers, expanding in the sunshine, purple in those

* *Crocus vernus*—Plate 6 A.

which occur in the wild state, differ from those of the Snow-drop in being erect instead of pendent, but like the latter plant, the Crocus belongs to the great family of Monocotyledons, and to that series in which there are six coloured leaves to the perianth or flower, these being combined at the base into a long slender tube; above they are scarcely distinguishable into an outer and an inner series, each consisting of three leaves, representing the sepals and petals which were found to exist in the Primrose, where however they occur in a state of cohesion, the parts it will be recollected being united into a tubular calyx and a monopetalous corolla.

The Crocus flowers have three stamens, and a stigma which is dilated and fringed at the top. The long slender tube of the perianth, which is in fact the stalk-like portion that at length becomes visible, is a good deal hidden by the leaves and sheathing membranes which emerge with it from the ground; and the ovary, or young seed-vessel, is buried amongst the bases of the leaves.

And now having briefly adverted to these earliest of the early of Flora's offerings, which besides have afforded illustrations of the groups of Monopetalous and Monocotyledonous plants, we will proceed to glance in something like order at a few other examples representative of the Vernal Flora, which have been selected as the subjects of our illustrations. We commence with the DICOTYLEDONS, called also EXOGENS, from the external manner of accretion in their stem, a large group, known generally by their net-veined leaves, and sharing with the Monocotyledons and Cryptogams the whole Vegetable Kingdom.

At a very early period of the year, in moist woods and pastures, the surface of the ground will be found whitened with

a multitude of small starry blossoms of a small Ranunculaceous plant. These are the blossoms of the Wood Anemone,*
a dwarf herb, which has fleshy underground stems, from which
spring up three-parted leaves and white cup-like flowers, below
which latter an involucre or guard of three leafy parts resembling the true leaves is placed.

Here, then, we have an illustration of a very different kind
of flower from any of those which have been previously noticed.
We have in fact one of the group of Polypetalous (that is,
many-petaled) Exogens or Dicotyledons. Passing over all but
the flower itself, what do we find? There is first a single row
of what look like petals and appear to form a corolla, and
within these is a large tuft of small yellowish bodies, which
are the stamens and pistils. The petal-like bodies are however in reality a coloured calyx, divided into many (about six)
separate pieces or sepals, standing in place of petals, which
are entirely wanting. It is because such flowers have several
distinct and separate parts to form their floral envelopes that
they are called polypetalous, and our subject represents one
condition of a large group, in which however both calyx and
corolla are generally present. The rule is, that when only one
floral envelope is found—the calyx and corolla are called
floral envelopes—it is regarded as a calyx, whether it be green
or coloured. In our Wood Anemone the pistils will be found
to be numerous and distinct, and they consequently grow up
into a group of distinct fruits or carpels, which contain each
one seed. This little spring flower can only be seen in perfection when the atmosphere is dry, for in humid weather and at
night the petal-like calyx closes up.

A purple-flowered variety, with smaller flowers, generally
formed of eight, rarely of six, narrow-ovate sepals, of a uni-

* *Anemone nemorosa*—Plate 1 A.

form deep purple, has been lately found at Pinner, and also at Chislehurst.

Growing in wet open places, and amongst the earliest of wild flowers, is another Ranunculaceous plant, petal-less like the foregoing, namely, the Marsh Marigold,* a specious-looking stout-growing perennial, with bold roundish leaves, hollowed at the base in what is called a heart-shaped form, and whose bright golden flowers have much the structure of those of the Wood Anemone, but are larger and more conspicuous from being elevated on a tall branching stem. They have a varying number of about five or six coloured sepals and no real petals, a tuft of numerous stamens, and a variable number of carpels or fruits, each one containing several seeds. Somewhat resembling this, and one of the same group, but dwarfer, and having both calyx and corolla present, so as to form a complete regular polypetalous or many-petaled flower, which for the purpose of comparison it may be useful to examine in connection with the Marsh Marigold, is the Lesser Celandine (*Ranunculus Ficaria*), found abundantly in moist waste places, and easily recognized by its glossy-looking yellow star-like flowers, and its white-mottled angular-lobed leaves.

The Ladies'-smock,† during the months of early spring, imparts its own blush to the surface of moist low-lying meadow land, among the herbage of which it grows up. This plant, also a Polypetalous Exogen, sometimes called Bitter-cress and Cuckoo-flower, is a dwarf herb, growing erect to about a foot in height, and having pinnate leaves; the flowers are large and showy, and will serve to illustrate the structure of a considerable polypetalous regular-flowered group or Order, known as Cruciferous plants, or Cross-bearers, from the cir-

---

* *Caltha palustris*—Plate 1 B.
† *Cardamine pratensis*—Plate 1 C.

cumstance of their flowers having four equal petals arranged
in opposite pairs so as to form a cross. The group may be
known by this circumstance, and by having six stamens, two
shorter than the rest. This Cruciferous Order, besides being
an extensive one, is important, containing, amongst other
subjects of utility, the whole Cabbage family.

To the same Order belongs the Wallflower,* " grey ruins'
golden crown," a flower well known in every garden, and
prized for its delicious fragrance, found here and there in a
wild or semi-wild state on walls and old buildings, or in rocky
situations, generally near habitations. In the Wallflower we
have a plant of subshrubby growth, furnished with simple
leaves, and its yellow or reddish-bronzy flowers are succeeded
by what are called siliquose pods containing the seeds, as is
also the case with the Ladies'-smock. In both these plants
the inflorescence or collection of flowers forms a kind of corymb in the earlier stages, lengthening out by degrees into a
more or less elongated raceme. This flower has been made
the emblem of friendship in adversity, because, though Time,
the rude and sacrilegious despoiler of consecrated places, may
waste and overthrow the structures of the past, and leave them
uninhabited, the Wallflower, "mantling o'er the battlement,"
still lends a melancholy grace "to haunts of old renown."

And then, who does not know the Violet,† the very emblem
and personification of sweetness—sweeter, as Shakspere says,
than "Cytherea's breath"? This lovely plant is common on
banks and under hedges, a dwarf herb, with heart-shaped
leaves and polypetalous flowers, which, "kissed by the breath
of heaven, seem coloured by the skies." They are of a distinct type from any of the foregoing, and have a separate

* *Cheiranthus Cheiri*—Plate 1 D.
† *Viola odorata*—Plate 2 A.

calyx and corolla, the former consisting of five green sepals, the latter of five coloured usually purple or white petals, which form an irregular flower, two of the petals being placed to form the upper half, and of the remaining three which form the lower part, the lowest is extended backwards at the base, producing a kind of spur. The Violet has therefore an irregular polypetalous spurred corolla. There is also a peculiarity in the anthers, which are five in number, and are more or less closely joined in a sort of ring around the ovary, the two lowermost of the five being, like the lower petal, lengthened into a spur. These spurred polypetalous flowers serve to distinguish the Violets from all other British plants except the Balsams, which are known by having only three sepals and three petals, all coloured.

The Sycamore* illustrates another family of the same group—the Polypetalous Exogens, namely, the Aceraceous or Maple family. This is a well-known tree, very extensively planted in this country, seeding readily and springing up from self-sown seeds so freely that it may be regarded as naturalized here, though the mountains of central Europe and western Asia have been its ancient home. This tree, which flowers early in spring, puts forth broad palmately-lobed leaves, and bears rather inconspicuous flowers in pendent clusters, which look not unlike immature bunches of small grapes. The individual flowers consist of five small green sepals, five small green petals, and about eight stamens inserted on a thickened disk around the ovary. The fruit of these plants consists of two carpels, each extended into a wing at top; they are popularly called keys, but in technical language such a fruit is called a samara.

Along with the Wood Anemone, already adverted to, may

* *Acer Pseudo-platanus*—Plate 2 B.

be found the dwarf and unpretentious Wood Sorrel,* a lovely
little wilding, representing the family of Oxalidaceous plants.
This has peculiar knotty fleshy stems, and trifoliate leaves
like those of the clover. It has also polypetalous (here five-
petaled) regular flowers, in which both calyx and corolla will
be found, the calyx consisting of five green sepals, the corolla
of five equal obovate white petals ; within these are ten stamens, half of which are as long again as the others, and five
separate styles surmounting the ovary. This elegant little
plant is maintained by some antiquaries to be that which furnished St. Patrick with his illustration of the doctrine of the
Trinity, though others contend for the Trefoil or Clover, which
is now more commonly adopted as the Irish Shamrock.

The foregoing plants (excepting the Primrose, the Snowdrop, and the Crocus) all belong to a primary division of the
Dicotyledons or Exogens, called THALAMIFLORES, and they
have these distinguishing marks in common :—

(a) The petals are distinct from the calyx, and from each
    other : very seldom absent.

(b) The stamens are hypogynous, that is, they have their
    point of attachment below the ovary, which latter
    is the young seed-vessel.

We have next to consider one or two examples of another
great subdivision of the Dicotyledons, called CALYCIFLORES,
which has these peculiarities :—

(a) The petals are usually distinct.

(b) The stamens are perigynous, that is, appearing to grow on
    one of the organs surrounding the ovary, either calyx
    or corolla ; or epigynous, that is, apparently growing
    from the summit of the ovary itself.

* *Oxalis Acetosella*—Plate 2 C.

Of this group we find in very early spring an illustration in the well-known Red Currant* of our gardens, a member of the Grossulariaceous family, which, though a cultivated plant, is frequently found in a wild state, both in Scotland and in the north of England. This, as is well known, is a dwarfish branching shrub, bearing palmately-lobed leaves, and racemes of small greenish flowers, which latter consist of a calyx adherent to the ovary and divided into five sepals, a corolla consisting of as many small scale-like petals placed at the base of the segments of the calyx, five perigynous stamens, and an inferior ovary, which becomes a succulent berry, varying in colour.

Here also may be referred the Meadow Saxifrage,† a common but very pretty species, representing the Saxifragaceous family, and which in early spring is found, sometimes very abundantly, in meadow and pasture land. It is a perennial herb, producing underground a number of small fleshy tubers from which its stems arise. The lower leaves are kidney-shaped and lobed, and the stem grows upright six inches to a foot in height, bearing towards the top a few large white flowers, which have a calyx adherent to the ovary to about its middle, then separating into five lobes, five perigynous petals, ten perigynous stamens, and a two-celled ovary with two distinct styles.

Another of the great divisions of the Dicotyledons, the MONOPETALS, or Monopetalous flowers, has been already adverted to in referring to the Primrose, but we have other illustrations to offer. The distinguishing features are :—

(a) The petals united, at least at the base, into a single. piece.

---

* *Ribes rubrum*—Plate 2 D.

† *Saxifraga granulata*—Plate 3 A.

(b) This monopetalous corolla either epigynous, bearing the stamens; or altogether distinct from the stamens; or hypogynous, bearing the stamens.

Of the first of these subdivisions we have some early-flowering examples in the family of Composites, or Compound flowers. Among them is the common, well-known, golden-flowered Dandelion,* to be found everywhere, and combining in itself the characters of a gay spring flower, a troublesome weed, and a valuable remedial agent. This plant has a thick, fleshy tap-root, from the crown of which spreads a tuft of oblong run-cinately pinnatifid leaves, among which spring up numerous hollow peduncles, bearing, not a large yellow flower as is the vulgar notion, but a head of numerous yellow flowers of peculiar character. It is the fact of their bearing numerous flowers in one head, and so as to seem to constitute a single flower only, that has procured for the large family to which the Dandelion belongs, the name of Composites, or Compound flowers. Let us examine one of these a little more closely. At the top of the stalk are two or three rows of crowded green scaly leaves, of which the innermost are erect, and the outer ones recurved; these constitute the involucre or guard-leaves, which in the Composite plants are always found surrounding the flower-heads, and the parts of which are called bracts or scales. The top of the stalk is expanded into a broadish flattened surface, which the involucral leaves fence round, and on this surface, which is called the receptacle, the flowers are closely packed side by side. The Dandelion belongs to a group of Composites in which the flowers are all alike—all ligulate or strap-shaped, and hence the group has been named Ligulates. Take up one of the flowers,—they are here properly called florets, or little flowers,—and let us see of what it consists. At the base is a

---

* *Leontodon Taraxacum*—Plate 3 B.

compressed reversed ovate fruit, roughish on the upper part; then comes a little stalk, and a cylinder of long fine hairs called the pappus, from out of which at top issues the long yellow limb of the floret. By-and-by, as it gets older, the little stalk will lengthen into a long slender shaft, and the cylinder of hairs will expand like the rays of an umbrella, and in this way will float away the seeds. But the corolla: this is attached just above the point where the rays of the pappus diverge, and consists of a slender tube which some distance up is split on one side, and so forms the flat strap-shaped ligulate limb. Out of this tube emerges the two stigmatic branches, just beneath which the linear anthers are united into a sheath surrounding the upper part of the style. This cohesion of the anthers into a tube enclosing the style is one of the marks of a Composite flower.

But we have another illustration, which will serve to explain a different set of these Composite flowers, called the Corymbifers from their heads being generally in corymbs—the Daisy,* Burns's "bonny gem," which is almost ever and everywhere in blossom. The Daisy is a dwarf perennial herb, with a spreading rosulate tuft of obovate or spathulate leaves, from among which arise the numerous slender stalks, bearing not however a corymb, but a single or solitary head of flowers. Here, the involucre consists of about two rows of green bracts, and within them, at the outer edge of the receptacle, is a row of ligulate female florets, with a very short tubular portion at their base, from whence issues the two-branched style, these having no stamens. These are the ray florets. The other part of the receptacle is filled with small tubular funnel-shaped equal florets, which are hermaphrodite, having both stigmas and anthers. These small tubular disk florets form the yellow

* *Bellis perennis*—Plate 3 C.

centre of the Daisy flower-head. The manner in which the
poet Burns apostrophized this "modest crimson-tipped flower,"
while following the plough, and lamenting over the destruction he was causing it and could not avoid, is singularly pa-
thetic—"the share upturns thy bed, and low thou lies."

Of the subdivision of Monopetals in which the stamens are
distinct from the corolla, examples will be found in the spring-
blooming *Vaccinium* or Whortleberry family.

The division in which the corolla is hypogynous, bearing
the stamens: in other words, in which the flowers are perigynous, is well illustrated by the Primrose, already adverted
to in our opening page. Another illustration is afforded by
the Lesser Periwinkle,* one of the Apocynaceous or Dog-
bane family, found occasionally in hedges and woody banks,
and commencing to flower in spring—ay, sometimes very
early in the year. This is a perennial, with long trailing stems,
clothed with opposite ovate-oblong leaves, and producing also
short erect flowering-branches, which bear solitary axillary
flowers; these have a free calyx with five narrow deep divisions; a monopetalous corolla, in which the tube is almost
campanulate, and the limb consists of five flat spreading segments, having a lateral twist; five stamens enclosed in the
tube of the corolla; and two ovaries, distinct at the base, but
connected at top by a single style, terminating in an oblong
stigma, contracted in the middle. The twisted corolla and
pulley-shaped stigma are special marks of the Apocynaceous
family, to which the Periwinkle belongs.

The Buckbean,† an aquatic plant, found in bogs and shallow
pools, and famous for its tonic properties, comes into flower
soon after the Periwinkle, and affords another example of the

---

* *Vinca minor*—Plate 3 D
† *Menyanthes trifoliata*—Plate 4 A.

same subdivision, belonging to another natural family, that of the Gentians. It has stoutish creeping or floating stems, with strong coarse roots, and forming at the end a tuft of leaves, consisting of three obovate or oblong leaflets, set on a long stalk, which is sheathing at the base. The flowers come in erect racemes, and each consists of a short calyx with green lobes, and a bell-shaped corolla, deeply five-lobed, the lobes spreading or even reflexed, white tinged with red, the inside elegantly fringed. The flowers have five stamens joined to the tube of the corolla, and the fruit is a capsule opening in two valves.

We must pass on to another great division of Dicotyledons in which the structure is of a simpler character. These are the MONOCHLAMYDS, having Monochlamydeous or one-coated flowers. They form a considerable group, comprising some few showy families, and including many others in which the blossoms are quite inconspicuous. Here there is normally not more than one floral envelope (sometimes none) to the stamens or pistil, which are the essential organs of the flower. The envelope when present is called the perianth or perigone, and is in reality a green or coloured calyx, the corolla being constantly wanting.

To this series belongs the gay but poisonous Mezereon shrub,* one of the Thymelaceous family, a very early-blooming plant, found in woods and thickets. The branches of this dwarf deciduous shrub are clothed with the little clusters of flowers, "blushing wreaths investing every spray," while yet the young leaves are undeveloped. The latter appear later, and are oblong or lanceolate in form. The flowers have a short broadish tube to the perianth, a limb of four spreading lobes, and eight

---

* *Daphne Mezereum*—Plate 4 C.

stamens which are attached in two rows to the inner face of the tube. The flowers are purple or in some plants white, very sweet-scented, and succeeded by a "vesture gay" of red or yellow berries, containing a single seed; this was present in the form of a pendulous ovule in the urn-shaped ovary, which stood free within the base of the perianth.

Of the same Monochlamydeous division, and growing in similar habitats, but of much more frequent occurrence, is the Wood Spurge,* representing the varied and extensive Euphorbiaceous family. This plant is in early spring conspicuous on account of its umbels of light yellowish-green bracts. It has almost woody stems, of a reddish colour, bearing narrow-oblong leaves, above which the umbel of five or six principal branches is produced. These bear floral leaves or bracts in pairs, of a yellowish-green colour, and between each pair a small green body, apparently a flower but really a flower-head, consisting of a small cup-shaped involucre resembling a perianth, having four minute teeth, and alternating with them as many horizontal yellowish glands, which are here crescent-shaped; within these are several stamens each bearing a pointed filament and a minute scale at its base, thus showing them to be distinct male flowers; while in the centre is a single female flower on a recurved stalk, consisting of a three-celled ovary and a three-cleft style. This ovary grows into a fruit of three carpels, called cocci, whence the fruit is called tricoccous.

Of a different character, still Monochlamydeous, is the Wych Elm,† the common wild Elm of Scotland, Ireland, and the north and west of England, a representative of the Ulmaceous family. Here we have a deciduous large-grow-

---

* *Euphorbia amygdaloides*—Plate 4 D.

† *Ulmus montana*—Plate 5 A.

ing tree, whose branches are, during summer, clothed with broadly-ovate pointed leaves. In early spring however, before the leaves appear, the buds along the twigs burst open, and each developes a dense cluster of reddish flowers surrounded by brownish bracts that soon fall away. The flowers are mostly hermaphrodite, and consist of a bell-shaped perianth with from four to six teeth and as many stamens. The ovary is flat, with two short diverging styles, and is succeeded by a flat thin leaf-like or winged seed.

The common Sallow,* one of the Amentiferæ or Catkin-bearing family, is of a different character, though part of the same great Monochlamydeous division of Dicotyledons. The large family of Sallows and Willows consists of trees and shrubs, with unisexual flowers growing in catkins in early spring on the leafless twigs, the male catkins being produced on distinct plants from those which bear the females. The common Sallow itself is a tall shrub with broadish-ovate or oblong greyish downy leaves. The catkins of the male plants are cylindrically oblong, an inch long or more, formed of overlapping silky-hairy scales, but no perianth; in the axil of each scale are placed two stamens, which are longer than the scale itself, so that the fully developed catkins are rather conspicuous from the crowded prominent yellow anthers. The female catkins are longer and narrower, and have in the axils of the scales, instead of a pair of stamens, a silky ovary, which tapers into a longish beak and is terminated by the forked stigmas. Notwithstanding the bright gleam of vegetable beauty which at this early season the Sallow affords in its favourite haunts by the streamlet's margin, in moist coppice woods, or overhanging a watery ditch, it may have probably escaped the attention of many who may been attracted by these golden ' palms' of

* *Salix Caprea*, Plate 5 B.

c

Easter-tide, that no fruit is ever borne by these specious catkins. Yet it is so. Near at hand however will be found other bushes with other catkins, without the alluring hue of gold, and these on closer inspection will be seen to consist of the pale-green silky ovaries or young fruits, surmounted by the forked stigmas, intended to catch the dust that flies off from the catkins of the golden hue, which dust is borne to the ovary-bearing or female plant by the agency of insects or on the wings of the wind. Thus even the rude blast, annoying though it sometimes may appear to be, has its appointed office to perform in Nature's laboratory, one of which is to carry the fertilizing powder or pollen from plant to plant, and thus to secure the fulfilment of the appointed law by which each herb and tree bears seed after its kind.

The remaining principal division of the Flowering Plants, is that which is called Monocotyledons, the chief peculiarities of which have been already pointed out in referring to the Snowdrop and Crocus. The two flowers just named belong to the regular-flowered natural families in which the ovary is inferior, or developed beneath the other parts of the flower, which thus appear to grow from the top of the ovary. Our spring flowers however afford us some illustrations of the Orchideous family, a peculiar series of Monocotyledons in which the flowers are remarkably irregular, and also of the Liliaceous family, a natural group of regular-flowered Monocotyledons in which the ovary is superior, the other parts of the flower being developed from beneath it, so that it is enclosed by them.

Let us examine more closely the specimens of the Orchidaceous family, first taking the Spotted Palmate Orchis of our meadows.* The flowers in this family are of very singular structure. As in most other Monocotyledons, they are made upon a trimerous plan: that is, the parts are in threes; and, as in the majority of the petaloid division of Monocotyledons, the perianth is six-leaved: that is, twice three organs are brought together in close association, but here they acquire great diversity of character. In the Spotted Palmate Orchis, which has an upright stem, furnished below with spotted simple narrow elongate leaves, and terminated by an erect spike of spotted pink flowers, the three outer parts or sepals are nearly alike, and of a narrowish or lanceolate form, while the inner series of three consists of two convergent petals, which resemble the sepals but are shorter, and a lip which is much larger, three-lobed, entirely different from the rest. In the centre, opposite to the lip, is another part called the column, which is a fleshy body formed by the combination of one stamen with the pistil. Theoretically three stamens should be present, but in the Orchis two of these are constantly abortive, and the central one only is developed. The anthers in the whole family are very peculiar in structure. This plant like many other orchids has a pair of fleshy tubers or tuberous roots, one of which becomes wasted by the development of the current year's growth, while the other is forming for the succeeding season.

The Lady's Slipper,† another of the Orchids, shows some variations of structure from that just noticed. It is larger-growing, with ovate pointed ribbed leaves. The flower-stems are a foot and a half high, supporting one or rarely two large handsome flowers, of which the upper or dorsal sepal, opposite the lip, is broadly lanceolate, and there is a similar one formed

* *Orchis maculata*—Plate 5 C (misnamed *mascula*).

† *Cypripedium Calceolus*—Plate 5 D.

by the combination of the two lateral sepals behind or beneath
the lip, while the two petals are narrower, spreading right
and left: these all being of a deep brownish-purple. The lip
is a large inflated pouch-shaped body, and is compared to a
a slipper; it is yellow, variegated with purple. The column
is broadish, and much shorter than the petals; and in this
case the central stamen is abortive, and the two lateral ones
are perfect.

The Spring Squill,* a regular-flowered Monocotyledon, and
one of the Liliaceous family, is a delicate little bulbous plant,
found in sandy wastes and pastures, especially near the sea.
It has narrow-linear leaves, and a flower-scape bearing at top
a short raceme of small blue starry flowers. The perianth of
these flowers is composed of six nearly equal spreading segments, all coloured alike, and within them close to their base
are inserted six stamens, one opposite each segment, the centre
being occupied by the ovary. This is a superior ovary, the
other parts of the flower being developed exterior to and
beneath it.

Another example of a spring-flowering Liliaceous plant, also
regular-flowered, is afforded by the Bluebell,† the "shade-
loving hyacinth" of the poets, a pretty bulbous plant, very
abundant in Britain, in woods, hedgerows, and other shady
places. This plant has broadish-linear leaves, and a flower-
stem about a foot high, bearing a terminal one-sided or nodding raceme of pendent flowers, each flower having a small
leafy bract at the base of its pedicel or stalk. The perianth
is tubulose, the segments united for a short distance at the
base, spreading only at the top where they become recurved.
It is usually of a deep blue, though in some cases white, or

---

* *Scilla verna*—Plate 6 C.
† *Hyacinthus non-scriptus*—Plate 6 D.

more rarely pinkish. It is sometimes referred to the *Scilla* family, but a comparison with the foregoing will show considerable difference between them. This flower is occasionally called Harebell: a name which is more appropriately given to the little *Campanula rotundifolia*. The Hyacinth of the ancients,—which was named in commemoration of a comely youth, Hyacinthus, who was accidentally killed by Apollo, and which bore certain lines or marks on its petals, in allusion to the absence of which our native species was called *nonscriptus*,—was made by the Greeks the emblem of Death; and hence an American poet has taken it as the symbol of sorrow: "the deep blue tincture that robed it seemed the gloomiest garb of sorrow." Another poet has called it the "sublime Queen of the mid-May."

And with this well-known and favourite flower we close our sketch of the illustrative examples of the flowers of the spring season. In the succeeding pages we shall give a more complete and formal, as well as classified enumeration of the principal kinds of wild plants which blossom during the early portion of the year. The examination of these, as day by day they become unfolded, will, we hope, prove an agreeable and instructive pastime to many readers, and give them an increased zest for the search after those which are yielded by the seasons that follow.

> Who loves not Spring's voluptuous hours,
> The carnival of birds and flowers?
> Yet who would choose, however dear,
> That Spring should revel all the year?—*Montgomery.*

SUMMARY OF SPRING FLOWERS.

[I.—GROUPS AND ORDERS.]

EXOGENOUS PLANTS or DICOTYLEDONS.

Leaves with netted veins. *Flowers* usually quinary—the parts in fives, or quaternary—the parts in fours. *Embryos* with two (rarely more) cotyledons; hence dicotyledonous. This group includes the Thalamiflores, Calyciflores, Monopetals, and Monochlamyds.

Thalamiflores: Polypetalous dichlamydeous plants, with petals distinct (*i. e.* separable) from the calyx, and stamens hypogynous; it includes the Orders numbered 1 to 9.

> * *Carpels more or less distinct (i. e. apocarpous), sometimes solitary with one lateral placenta.*

1. **Ranunculaceous plants**—herbs; stamens indefinite, usually numerous; anthers opening by two longitudinal clefts. [For the genera in the several groups, see page 28 and sequel.]
2. **Berberidaceous plants**—shrubs; stamens equal in number to the petals; anthers opening by recurved valves; carpel solitary.

> ** *Carpels combined into an undivided (i.e. syncarpous) ovary which has two or more placentas.*
>
> † *Seeds attached to the sides of the carpels (i. e. placentas parietal).*
>
> ‡ *Stamens 6, tetradynamous (4 long and 2 short), distinct.*

3. **Cruciferous plants**—herbs; corolla regular; petals four.

‡‡ *Stamens 5, cohering in a ring around the ovary.*

4. Violaceous plants—herbs; corolla irregular; petals five.

†† *Seeds attached in the axis or centre of the carpels (i. e. placentas axile).*
‡ *Flowers symmetrical, i. e. the parts equal or proportional in number.*

5. Caryophyllaceous plants—herbs; leaves undivided, without stipules; flowers regular; stamens definite.
6. Geraniaceous plants—herbs; leaves lobed, with stipules; flowers regular; styles and carpels combined around a long beaked axis.
7. Oxalidaceous plants—herbs; leaves in British species trifoliate; flowers regular; styles distinct.

‡‡ *Flowers unsymmetrical, i. e. the parts neither equal nor proportional in number.*

8. Aceraceous plants—trees; flowers regular; stamens distinct, definite; fruit consisting of two winged nuts.
9. Polygalaceous plants—herbs; flowers very irregular; stamens combined in two parcels.

Calyciflores: Polypetalous dichlamydeous plants with the petals usually distinct, and the stamens perigynous or epigynous; Orders 10 to 17.

* *Stamens perigynous.*
† *Carpels more or less distinct or single.*
‡ *Ovary superior, the calyx distinct from the carpels.*

10. Leguminous plants—herbs or shrubs; flowers very irregular, papilionaceous; the fruit a single carpel forming legume; stamens 10, all or 9 of them united.
11. Rosaceous plants—herbs or trees; flowers regular, rarely without petals; stamens indefinite; fruit one-seeded nuts or drupes, or follicles containing several seeds.

‡‡ *Ovary inferior, the calyx adhering to the carpels.*

12. **Pomaceous plants**—trees or shrubs, the fruit a pome.
13. **Saxifragaceous plants**—herbs, the fruit a capsule.

†† *Carpels combined into an undivided ovary, which has more than one placenta.*
‡ *Ovary superior, the calyx distinct from the carpels.*

14. **Rhamnaceous plants**—shrubs or trees; calyx-lobes four or or five, valvate; stamens opposite the petals.
15. **Celastraceous plants**—shrubs or trees; sepals four or five, imbricate; stamens alternate with the petals.
16. **Portulacaceous plants**—herbs; sepals two, imbricate; petals five, cohering at the base.

** *Stamens epigynous.*

17. **Grossulariaceous plants**—shrubs; flowers regular; ovary inferior, one-celled.

Monopetals: Dichlamydeous plants, with the petals united (from the base more or less upwards) into a single piece; Orders 18 to 30.

* *Stamens hypogynous.*

18. **Ericaceous plants**—shrubs; stamens free, equal to or twice as many as the lobes of the corolla, the anthers with an appendage, and opening by two pores.

** *Stamens epigynous.*
† *Stamens free, or distinct from the corolla.*

19. **Vacciniaceous plants**—shrubs; stamens twice as many as the lobes of the corolla, the anthers opening by two pores.

†† *Stamens epipetalous, or affixed to the corolla.*
‡ *Carpels two- or more-celled.*

20. **Caprifoliaceous plants**—shrubs; leaves opposite, without stipules; cells of the carpels many-seeded.

21. **Galiaceous plants**—herbs; leaves and leaf-like stipules forming radiate whorls around the square stem; cells of the carpels one-seeded.

‡‡ *Carpels one-celled, one-seeded.*

22. **Valerianaceous plants**—herbs; stamens fewer than the lobes of the corolla; anthers distinct.
23. **Composite plants**—herbs; stamens equalling in number the divisions of the corolla (*i. e.* isomerous); anthers united into a tube around the style (*i. e.* syngenesious).

*** *Stamens perigynous.*
  † *Seeds attached to a free central placenta.*

24. **Primulaceous plants**—herbs; stamens equalling in number and opposite to the lobes of the corolla.

  †† *Seeds attached to the sides of the carpels, or in the axial angle of the cells.*

  ‡ *Corolla regular or nearly so.*

25. **Apocynaceous plants**—herbs; corolla contorted, the stamens isomerous, alternating with its lobes; ovaries two, distinct, cohering by their stigma.
26. **Gentianaceous plants**—herbs; stamens isomerous, alternating, ovary of 1–2 many-seeded cells; placentas parietal.
27. **Boraginaceous plants**—herbs; stamens isomerous, alternating; ovary four-lobed, with one ovule in each lobe.

  ‡‡ *Corolla irregular.*

28. **Scrophulariaceous plants**—green leafy herbs; ovary two-celled, the cells many-seeded; placentas axile.
29. **Orobanchaceous plants**—brown leafless herbs; ovary two-celled, the cells many-seeded; placentas parietal.
30. **Labiate plants**—green leafy herbs; flowers unsymmetrical; ovary four-lobed, with one ovule in each lobe.

Monochlamyds: Perianth single (*i. e.* consisting of a calyx only), or altogether wanting, and replaced by scaly bracts; Orders 31 to 40.

 * *Stamens and pistils combined in the same flower.*

 † *Stamens perigynous.*

31. **Thymelaceous plants**—shrubs; perianth coloured; carpels solitary, simple, becoming a berry or drupe.
32. **Ulmaceous plants**—trees; carpels blended into a two-celled ovary, becoming a membranaceous winged fruit.

 †† *Stamens epigynous.*

33. **Aristolochiaceous plants**—herbs; perianth coloured, superior; stamens inserted on the perianth.

 ** *Flowers diclinous,* i. e. *the staminate and pistillate ones separate.*

 † *Flowers having a calyx (i. e. monochlamydeous).*

34. **Elæagnaceous plants**—shrubs or small trees with scurfy leaves; carpels solitary, simple.
35. **Euphorbiaceous plants**—herbs or shrubs; carpels combined into a three-celled ovary, each cell containing one or two pendulous ovules.
36. **Empetraceous plants**—small shrubs; carpels combined into a 6–9-celled ovary, each cell containing one erect ovule.

 †† *Flowers naked, the calyx being replaced by scaly bracts (i. e. achlamydeous).*

37. **Amentaceous plants**—trees or shrubs; carpels superior, naked.
38. **Corylaceous plants**—trees or shrubs; carpels inferior, seated in an involucre (cupuliferous).
39. **Coniferous plants**—trees or shrubs, with resinous acerose persistent leaves; ovules naked, becoming nuts enclosed within the hardened scales of a woody cone.

40. **Taxaceous plants**—trees or shrubs, with linear persistent leaves; ovules solitary, seated in a succulent cup.

ENDOGENOUS PLANTS or MONOCOTYLEDONS.

Leaves with parallel veins. *Flowers* usually ternary—the parts in threes. *Embryos* with one cotyledon; hence monocotyledonous. This group includes the Orders numbered from 41 to 48.

> * *Flowers imperfect, naked* i.e. *without either perianth or glumes, or consisting of scales.*

41. **Araceous plants**—herbs; inflorescence spathaceous *i. e.* emerging from a spathe; flowers unisexual.

> ** *Flowers perfect, with a petal-like whorled perianth.*
> † *Ovary inferior.*

42. **Iridaceous plants**—herbs; perianth six-leaved; stamens three, distinct.

43. **Amaryllidaceous plants**—herbs; perianth six-leaved; stamens six, distinct.

44. **Orchidaceous plants**—herbs; perianth six-leaved irregular; stamens gynandrous (*i. e.* combined with the style).

> †† *Ovary superior.*

45. **Liliaceous plants**—herbs; perianth six-leaved (sometimes combined) regular.

> *** *Flowers perfect, with a dry calyx-like whorled perianth.*

46. **Juncaceous plants**—herbs; perianth regular, six-leaved, brown.

> **** *Flowers glumaceous, i.e. formed of imbricated chaffy scales or glumes.*

47. **Cyperaceous plants**—herbs; leaves with entire sheaths; bracts one to each flower.

48. **Graminaceous plants**—herbs; leaves with sheaths split on the side opposite the blades; bracts two to each flower.

[II.—GENERA OR FAMILIES.]

1. Ranunculaceous Plants. RANUNCULACEÆ.

* *Carpels several, short, one-seeded; (flowers regular.)*
   † *Sepals 4–5 or more, often petal-like, but no real petals.*

(1) **Anemone**—flowers with a three-leaved involucre. [For the species of the several genera, see page 42 and sequel.]

   †† *Petals 5 or more, usually more conspicuous than the sepals.*

(2) **Myosurus**—petals small, with a tubular claw.
(3) **Ranunculus**—petals with a small scale or hollow near the base.

   ** *Carpels several, many-seeded; (flowers regular.)*

(4) **Caltha**—sepals five, large; petals none.
(5) **Helleborus**—sepals five, large; petals small, tubular.

   *** *Carpels solitary, many-seeded.*

(6) **Actæa**—flowers nearly regular, small; fruit a berry.

2. Berberidaceous Plants. BERBERIDACEÆ.

(7) **Berberis**—sepals, petals, and stamens six each.

3. Cruciferous Plants. CRUCIFERÆ.

* *Pod siliquose, much (at least 3 or 4 times) longer than broad.*
   † *Calyx with 2 evident sacs or pouches at the base.*

(8) **Cheiranthus**—pod compressed; stigmas two-lobed or capitate.

   †† *Calyx equal at the base.*

(9) **Barbarea**—pods linear, four-angled; the valves with a prominent rib.

(10) **Arabis**—pods linear, compressed; valves nearly flat, with a prominent rib.

(11) **Cardamine**—pods linear, compressed; valves nerveless.

(12) **Alliaria**—pods linear, terete, the valves convex 3-ribbed; stalk of the seed flattened and winged.

(13) **Dentaria**—pods narrow-lanceolate, tapering, compressed; valves flat, nerveless.

** *Pod siliculose, less than three times as long as broad.*

† *Pod (or pouch) dorsally compressed, the valves parallel to the broad partition.*

(14) **Draba**—pouch oval or oblong, compressed; cells many-seeded; filaments simple.

(15) **Alyssum**—pouch roundish or oval, compressed; cells two-seeded; filaments toothed or winged near the base.

†† *Pod (or pouch) laterally compressed at right-angles to the narrow partition, the valves keeled.*

(16) **Hutchinsia**—pouch elliptical, entire; cells two-seeded.

(17) **Teesdalia**—pouch emarginate, the valves narrowly winged; cells two-seeded; filaments with a scale-like appendage near the base.

(18) **Thlaspi**—pouch emarginate; the valves broadly winged; cells many-seeded; filaments simple.

4. Violaceous Plants. VIOLACEÆ.

(19) **Viola**—lower petal and filaments of the two lower stamens spurred.

5. Caryophyllaceous Plants. CARYOPHYLLACEÆ.

* *Sepals united into a cylindrical tube.*

(20) **Lychnis**—styles five, rarely four; capsule opening in 5-10 teeth at top.

** Sepals distinct or cohering only at the base; (leaves without stipules).
† Capsule opening in 4–5 valves.

(21) **Sagina**—sepals and petals four or five, the latter entire, sometimes wanting; styles four or five.

†† Capsule opening in 8–10 teeth.
‡ Petals entire.

(22) **Arenaria**—sepals and petals five each; styles three, rarely four.

(23) **Mœnchia**—sepals and petals four each; styles four.

‡‡ Petals toothed or jagged, not cleft.

(24) **Holosteum**—sepals and petals five each; styles three, rarely four.

‡‡‡ Petals bifid or cleft.

(25) **Cerastium**—sepals and petals four or five; styles five, rarely four or three; capsule opening at top with short teeth.

(26) **Stellaria**—sepals and petals five; styles three, rarely five; capsule opening halfway down.

6. Geraniaceous Plants. GERANIACEÆ.

(27) **Geranium**—stamens ten, monadelphous; carpels one-seeded.

7. Oxalidaceous Plants. OXALIDACEÆ.

(28) **Oxalis**—stamens ten; ovary five-celled, the cells containing several seeds.

8. Aceraceous Plants. ACERACEÆ.

(29) **Acer**—flowers polygamous; calyx five-lobed; stamens seven or nine, rarely five.

9. Polygalaceous Plants. POLYGALACEÆ.

(30) **Polygala**—two inner sepals wing-like; petals three to five adhering to the tube of the stamens, the lower one keeled.

10. Leguminous Plants. LEGUMINOSÆ.

* *Stamens monadelphous (i. e. in one parcel).*

(31) **Ulex**—calyx elongated deeply two-lipped yellow, the upper lip with two, the lower with three small teeth at the tips.

(32) **Genista**—calyx short two-lipped, the upper lip larger two-parted, the lower deeply three-toothed.

(33) **Sarothamnus**—calyx short two-lipped, the upper with two, the lower with three small teeth at the tips.

** *Stamens diadelphous (i. e. in two parcels).*

† *Pod (or legume) continuous, one-celled.*

(34) **Trifolium**—legume few(1–4)-seeded, scarcely longer than the calyx; leaves trifoliolate.

(35) **Vicia**—legume many-seeded, two-valved; leaves pinnate, tendrilled.

†† *Pod (or legume)* divided by transverse articulations into one-seeded cells.

(36) **Arthrolobium**—flowers umbellate, without bracts; legume terete, scarcely contracted at the joints.

(37) **Hippocrepis**—flowers umbellate, without bracts; keel acuminated; legume much flattened, of numerous joints each curved like a crescent or horseshoe, or having one margin plane the other concave.

(38) **Ornithopus**—flowers umbellate, bracteated; keel obtuse; legume compressed, the joints equally contracted on both sides.

11. Rosaceous Plants. ROSACEÆ.

* *Petals four or five.*

† *Fruit a solitary drupe; (calyx deciduous).*

(39) **Prunus**—ovary of one free carpel containing two ovules, becoming a succulent fruit enclosing a hard stone containing usually one seed.

†† *Fruit consisting of numerous small nuts or drupes; (calyx persistent).*

(40) **Fragaria**—fruit dry, on a succulent receptacle.

(41) **Potentilla**—fruit and receptacle both dry

** *Petals none.*

(42) **Alchemilla**—calyx double, eight-parted the four outside divisions smaller; stamens four or fewer.

12. Pomaceous Plants. POMACEÆ.

* *Calyx-segments large, foliaceous.*

(43) **Mespilus**—petals large roundish; fruit turbinate, with the upper ends of the carpels exposed; disk dilated.

** *Calyx-segments small.*

(44) **Cratægus**—petals large roundish; fruit round or oval, with 1–5 bony 1–2-seeded cells or carpels.

(45) **Pyrus**—petals large roundish; fruit fleshy, with five distinct cartilaginous 1–2-seeded cells.

(46) **Cotoneaster**—petals small erect; fruit turbinate, its cells forming distinct nuts cohering to the inside of the fleshy calyx.

13. Saxifragaceous Plants. SAXIFRAGACEÆ.

* *Petals five.*

(47) **Saxifraga**—stamens ten all antheriferous; styles two; ovary two-celled, superior or partially inferior.

** *Petals none.*

(48) **Chrysosplenium**—ovary adherent to near the top, bearing a one-celled capsule.

14. Rhamnaceous Plants. RHAMNACEÆ.

(49) **Rhamnus**—fruit a small berry or drupe, enclosing three or four small one-seeded nuts.

15. Celastraceous Plants. CELASTRACEÆ.

(50) **Euonymus**—capsule with four (rarely 3–5) angles, enclosing as many cells, the valves opening along the middle of each cell; seeds solitary in the cells, enveloped in a coloured fleshy arillus.

16. Portulacaceous Plants. PORTULACACEÆ.

(51) **Montia**—flowers minute, the five petals slightly united, split open in front.

17. Grossulariaceous Plants. GROSSULARIACEÆ.

(52) **Ribes**—fruit a berry filled with juicy pulp, in which the seeds are suspended by long stalks.

18. Ericaceous Plants. ERICACEÆ.

(53) **Arctostaphylos**—calyx and corolla inferior; ovary five-celled, with one ovule in each cell, becoming a globular berry.

19. Vacciniaceous Plants. VACCINIACEÆ.

(54) **Vaccinium**—calyx-tube adherent to the ovary; corolla superior; ovary five-celled, with several ovules in each cell, becoming a globular berry.

D

20. Caprifoliaceous Plants. CAPRIFOLIACEÆ.

(55) **Adoxa**—calyx two- or three-toothed; corolla with short tube and four or five spreading divisions; fruit a berry.

21. Galiaceous Plants. GALIACEÆ.

(56) **Asperula**—corolla-tube as long or longer than the lobes; calyx blended with the ovary without a visible border.

22. Valerianaceous Plants. VALERIANACEÆ.

(57) **Valerianella**—fruit crowned by a small cup-shaped or toothed border.

23. Composite Plants. COMPOSITÆ.

  * *Disk florets tubular and perfect (i. e. both staminate and pistillate), those of the ray ligulate or filiform and either pistillate or neuter.*

  † *Pappus none; (heads rayed.)*

(58) **Bellis**—ray-florets ligulate, conspicuous; involucral scales nearly equal in length; achenes compressed.

  †† *Pappus pilose; (heads discoid.)*

(59) **Senecio**—involucre cylindrical or conical, of one row of equal scales, with several small ones at the base; achenes terete,

  ††† *Pappus (of disk-florets) pilose; (heads rayed.)*
  § *Florets of the ray with pilose pappus.*

(60) **Tussilago**—ray-florets in many rows; involucres nearly simple; achenes terete.

  §§ *Florets of the ray without a pappus.*

(61) **Doronicum**—scales of the involucre of two or three rows, nearly equal; achenes terete.

** *Florets all perfect (i. e. both staminate and pistillate) and ligulate.*

(62) Leontodon—pappus pilose, filiform, soft, deciduous, beak long; involucre of fruit reflexed.

24. Primulaceous Plants. PRIMULACEÆ.

(63) Primula—leaves all radical; flowers solitary or umbellate, on radical peduncles; ovary superior.

25. Apocynaceous Plants. APOCYNACEÆ.

(64) Vinca—corolla hypocrateriform, with a cylindrical tube, and flat spreading limb of oblique segments.

26. Gentianaceous Plants. GENTIANACEÆ.

* *Terrestrial plants with opposite leaves.*

(65) Gentiana—calyx not divided below the middle; stamens five; style two-lobed, persistent.

** *Aquatic plants, with alternate leaves.*

(66) Menyanthes—leaves trifoliolate; corolla fringed within.

27. Boraginaceous Plants. BORAGINACEÆ.

* *Corolla without scales at the mouth of the tube.*

(67) Pulmonaria—calyx tubular, the lobes not reaching to the middle; stamens included in the tube of the funnel-shaped corolla.

** *Corolla having the mouth of the tube more or less closed by subulate scales.*

(68) Symphytum—corolla tubular-campanulate, with five small teeth; filaments entire.

D 2

28. Scrophulariaceous Plants. SCROPHULARIACEÆ.

(69) Scrophularia—corolla small, nearly globular, with five unequal lobes; stamens four, often with rudiment of a fifth.

29. Orobanchaceous Plants. OROBANCHACEÆ.

(70) Lathræa—calyx bell-shaped, with four broad short teeth; upper lip of the corolla helmet-shaped.

30. Labiate Plants. LABIATÆ.

* *Two upper stamens longer than the two lower.*

(71) Nepeta—calyx tubular, fifteen-ribbed; upper lip of the long-tubed corolla straight, notched.

** *Two upper stamens shorter than or only equaling to the lower.*
† *Corolla distinctly two-lipped.*

(72) Lamium—calyx regularly five-lobed; stamens longer than the corolla-tube.

†† *Corolla apparently one-lipped.*

(73) Ajuga—upper lip of corolla minute, tooth-like, entire or notched.

31. Thymelaceous Plants. THYMELACEÆ.

(74) Daphne—perianth four-cleft, often coloured; stamens eight.

32. Ulmaceous Plants. ULMACEÆ.

(75) Ulmus—perianth bell-shaped, with four to six short lobes, and as many stamens.

33. Aristolochiaceous Plants. ARISTOLOCHIACEÆ.

(76) Asarum—perianth bell-shaped, regular, three-cleft; stamens twelve.

34. Elæagnaceous Plants. ELÆAGNACEÆ.

(77) **Hippophae**—perianth of male flowers with two small segments and four stamens, of the females tubular, minutely two-lobed.

35. Euphorbiaceous Plants. EUPHORBIACEÆ.

(78) **Buxus**—perianth small, of four segments, in the male flowers with four stamens, in the females with three styles.

36. Empetraceous Plants. EMPETRACEÆ.

(79) **Empetrum**—perianth of six scales, in two rows, with external imbricating similar bracts, the males with three stamens, the females with six or more radiating stigmas on a very short style.

37. Amentaceous Plants. AMENTACEÆ.

* *Scales of the male catkins broad, imbricated ; anthers longer than their filaments.*

† *Male and female catkins short, sessile, erect.*

(80) **Myrica**—male catkins with spreading concave shining scales, the female ones shorter, with long protruding styles.

†† *Male catkins cylindrical, usually pendulous.*

(81) **Alnus**—flowers three distinct with four stamens under each scale of male catkins ; female catkins small, ovoid.

(82) **Betula**—no distinct flowers ; stamens six to twelve under each scale ; female catkins short, cylindrical.

** *Scales of the male catkins narrow-linear or divided ; anthers small, on slender filaments.*

(83) **Salix**—male and female catkins cylindrical, compact, the scales entire ; stamens two, rarely three to five, with one or two gland-like inner scales.

(84) **Populus**—male and female catkins cylindrical, compact, the scales jagged; stamens several, within an oblique cup-shaped inner scale.

38. Corylaceous Plants. CORYLACEÆ.

* *Male flowers in globose catkins; (anthers 2-celled.)*

(85) **Fagus**—female flowers two or three together in the centre of a scaly four-lobed prickly involucre; stigmas three, fili-form.

** *Male flowers in long cylindrical catkins.*
† *Anthers two-celled.*

(86) **Quercus**—female flowers solitary, within cup-shaped scaly at length indurated involucres; stigmas three, oblong.

†† *Anthers one-celled.*

(87) **Corylus**—female flowers crowded in short bud-like catkins; involucre of one piece, at length enlarged, leafy and jagged, containing a single nut; stigmas two, filiform.

(88) **Carpinus**—female flowers in a lax catkin; involucre of two distinct at length enlarged leaves, containing two nuts; stigmas two, filiform.

39. Coniferous Plants. CONIFERÆ.

* *Fruit a dry cone, with two winged seeds under each scale.*

(89) **Pinus**—male catkins cylindrical, with two anther-cells to each scale.

** *Fruit small, succulent, containing two or three hard seeds.*

(90) **Juniperus**—male catkins small, with four anther-cells to each scale.

40. Taxaceous Plants. TAXACEÆ.

(91) **Taxus**—male catkins small, with three to eight anther-cells to each of the upper scales.

41. Araceous Plants. ARACEÆ.

(92) Arum—flowers densely packed on the lower part of a columnar spike (spadix), which is enclosed in a spathe or leafy sheath.

42. Iridaceous Plants. IRIDACEÆ.

* *Stigmas 3-parted slender, deeply 2-cleft.*

(93) Trichonema—perianth with a very short tube, and a limb of six deep equal segments.

** *Stigmas 3-parted; segments widening upwards, convolute.*

(94) Crocus—perianth with a slender tube, which is twice as long as the six-parted equal limb.

43. Amaryllidaceous Plants. AMARYLLIDACEÆ.

* *Perianth with a cup-shaped coronet or crown in the mouth of its tube.*

(95) Narcissus—perianth tubular at the base, the six-parted limb spreading.

** *Perianth without a coronet.*

(96) Galanthus—perianth divided to the base, the three outer segments larger than the three inner.

(97) Leucojum—perianth divided to the base, the six segments equal.

44. Orchidaceous Plants. ORCHIDACEÆ.

* *Plants scaly, leafless, brownish; (pollen simple or in slightly cohering granules.)*

(98) Neottia—flowers small; lip spurless, two-cleft, longer than the sepals.

** Plants having green leaves; (pollen cohering in grains or masses, which are indefinite in number and waxy.)
† *Central anther perfect, the two lateral ones abortive.*
‡ *Lip spurred at the base.*

(99) **Orchis**—lip lobed; pollen-masses with two glands enclosed in a common pouch.

‡‡ *Lip not spurred.*

(100) **Ophrys**—lip lobed, convex, usually hairy and figured resembling the body of an insect; pollen-masses with two glands each, enclosed in a separate pouch.

†† *Central anther abortive, the two lateral ones perfect.*

(101) **Cypripedium**—lip large, inflated, pouch-like.

45. Liliaceous Plants. LILIACEÆ.

* *Stem leafy, branching, shrub-like.*

(102) **Ruscus**—leaves stiff, prickly, bearing the flowers on their upper reversed surface.

** *Stem leafy, simple.*
† *Fruit a berry (succulent).*

(103) **Polygonatum**—flowers axillary, jointed with their stalk; perianth cylindrical, shortly six-cleft.

†† *Fruit a capsule (dry).*
‡ *Flowers solitary, terminal.*

(104) **Fritillaria**—perianth-segments converging into a bell-shape, drooping; segments chequered, the inner three with an oval honey-pore near their base; anthers attached above their base.

(105) **Tulipa**—perianth-segments converging into a bell-shape, erect; segments not chequered, nor having honey-pores at the base; anthers erect, attached by their base.

†† *Flowers umbellate.*

(106) **Gagea**—perianth-segments spreading; anthers erect; scape more or less leafy.

*** *Stem leafless, the leaves all radical.*

† *Fruit a berry (succulent.)*

(107) **Convallaria**—perianth bell-shaped, with short recurved segments, deciduous; flowers racemose on a naked scape.

†† *Fruit a capsule (dry).*

‡ *Perianth-segments combined throughout.*

(108) **Muscari**—perianth globose or subcylindrical, contracted at the mouth, shortly six-toothed.

‡‡ *Perianth-segments combined below.*

(109) **Hyacinthus**—perianth tubular-bell-shaped, the segments reflexed at the extremity.

‡‡‡ *Perianth-segments distinct, spreading.*

(110) **Scilla**—perianth deciduous; filaments filiform.

(111) **Ornithogalum**—perianth persistent, withering; filaments (of three alternate stamens) dilated.

46. Juncaceous Plants. JUNCACEÆ.

(112) **Luzula**—leaves flat, grass-like; capsule one-celled, with three seeds.

47. Cyperaceous Plants. CYPERACEÆ.

(113) **Carex**—flowers unisexual, the stamens and pistils under separate glumes; ovary enclosed in a little boat-shaped utricle, through the top of which the style protrudes.

48. Graminaceous Plants. GRAMINACEÆ.

* *Inflorescence dense, spike-like, ovoid or cylindrical.*

(114) **Sesleria**—spikelets two-flowered; pales (flowering glumes) without awns; stamens three.

(115) **Anthoxanthum**—spikelets **one-flowered** (*i. e.* one perfect flower, and two neuter ones consisting of empty pales) pales of the neuter flowers awned; stamens two.

** *Inflorescence lax, spike-like, one-sided.*

(116) **Chamægrostis**—glumes two; pales **membranous, awnless.**

*** *Inflorescence lax, panicled.*

(117) **Aira**—spikelets two-flowered; pales awned, the awn scarcely protruding beyond the glumes.

(118) **Poa**—spikelets three- or more flowered; glumes shorter than the spikelets; pales awnless, keeled at the back.

[III.—SPECIES AND VARIETIES.]

₊ *The numbers in parentheses correspond with those in the foregoing list.*

(1) **Anemone.**

* *Sepals covered outside with silky hairs.*

A. Pulsatilla: flowers large, solitary, violet-purple, with the involucre deeply cut into linear segments; carpels with feathery tails.—Pasque-flower.—Limestone pastures. Fl. April, May.

** *Sepals smooth on both sides.*

A. nemorosa: flowers small, solitary, white tinged with red outside; the involucre of stalked leaves, with the leaflets lobed and cut; carpels downy with a point not feathery.—Wood Anemone.—Woods and thickets. Fl. April.

(2) **Myosurus.** MOUSETAIL.

M. minimus: annual, dwarf; radical leaves linear; flowers solitary, small, greenish; receptacle becoming much elongated.—Damp sandy or gravelly fields. Fl. May, June.

(3) Ranunculus. CROWFOOT.

* *Flowers white; (aquatics.)*

R. aquatilis: plant floating; leaves capillary below three-parted above water, the lobes wedge-shaped and toothed; carpels transversely wrinkled.—Ponds, ditches, and streams. Fl. May, June.

There are several plants more or less closely resembling this, which are by some regarded as mere varieties, induced by varied conditions of growth.

** *Flowers yellow; (terrestrial plants.)*
† *Leaves undivided, angulate.*

R. Ficaria: rootstock with many oblong tubers; leaves mostly radical, heart-shaped, angulate, smooth, often mottled; flowers yellow, glossy; petals variable, usually 8 or 9.—Pilewort or Lesser Celandine.—Damp shady places. Fl. April.

†† *Leaves deeply cut.*

R. auricomus: stem erect, many-flowered; leaves smooth, three-cleft, the leaflets notched; flowers large, bright yellow; carpels downy.—Goldilocks.—Woods and coppices. Fl. April, May.

R. bulbosus: stem bulb-like at the base, erect, many-flowered; leaves hairy, three-cleft, with trifid cut segments; flowers large bright-yellow, the calyx-lobes reflexed; carpels smooth.—Buttercups.—Meadows and pastures. Fl. May.

R. repens: stems producing creeping shoots; leaves hairy, three-cleft, with lobed or toothed segments; flowers large, bright yellow, the calyx-lobes spreading; carpels smooth.—Buttercups.—Meadows and pastures. Fl. May to August.

R. parviflorus: annual; stem prostrate or ascending; leaves hairy, the lower 5 the upper 3–5-lobed, the segments lobed;

flowers small, yellow; carpels tuberculate.—Gravelly fields.—
Fl. May, June.

(4) Caltha. MARSH MARIGOLD.

C. palustris: leaves heart-shaped, rounded; flowers large,
golden-yellow.—Marshy places. Fl. March to May.

(5) Helleborus. HELLEBORE.

H. viridis: flowers 3–4, large, with spreading yellowish-
green sepals.—Woods and thickets in calcareous soil. Fl.
March, April.

H. fœtidus: flowers numerous, panicled, with large ovate
bracts, and converging green sepals often tipped with purple.
—Bear's-foot.—Thickets in calcareous soil. Fl. March, April.

(6) Actæa. BANEBERRY.

A. spicata: radical leaves large, twice ternate, with ovate
three-lobed coarsely toothed leaflets, flowers nearly white, in
a loose raceme.—Herb Christopher.—Mountain woods and
thickets in the north. Fl. May.

(7) Berberis. BERBERRY.

B. vulgaris: shrub; branches with 3-lobed thorns at the
base of leaf-tufts; leaves obovate, sharply toothed; flower
small, yellow, in drooping racemes, disagreeably scented.—
Hedges and thickets. Fl. May.

(8) Cheiranthus. WALLFLOWER.

C. Cheiri: subshrubby; leaves lance-shaped, acute, slightly
hoary; flowers large, racemose, yellow or reddish-brown, very
fragrant.—Walls and rocky places. Fl. April, May.

(9) **Barbarea.** WINTER-CRESS.

B. vulgaris : stem erect; lower leaves lyrate, with a broad rounded terminal lobe; flowers, small, yellow, racemed.—Belleisle Cress or Yellow Rocket.—Waste places. Fl. May to July.

(10) **Arabis.** ROCK-CRESS.

A. Thaliana : annual; stem slender, erect; lower leaves stalked, oblong, with a few coarse teeth, upper few, small, sessile; flowers small, white.—Thale-cress.—Banks and waste places. Fl. April.

A. Turrita : biennial; stem stiff, erect; lower leaves stalked, upper sessile, stem-clasping, oblong-lanceolate, slightly toothed; flower small, whitish.—Tower-cress.—Walls, rare. Fl. May.

(11) **Cardamine.** BITTER-CRESS.

C. amara : stems weak, ascending; leaves pinnate with 5–7 large ovate or orbicular segments, the upper ones angular toothed; flowers largish, white, in few-flowered racemes.—Sides of streams and watercourses. Fl. April, May.

C. pratensis : stems stiff, erect; leaves pinnate the segments of the lower roundish, those of the upper ones narrow and entire; flowers largish, pale pink or blush-coloured. —Ladies'-smock or Cuckoo-flower.—Moist meadows. Fl. April, May.

C. hirsuta : stems ascending or erect, more or less hairy; leaves pinnate, with small roundish-oblong angularly toothed segments; flowers small, white.—Moist shady places. Fl. March to June.

(12) **Alliaria.** GARLIC MUSTARD.

A. officinalis : annual or biennial; stem erect; lower leaves roundish, crenate, upper heart-shaped, ovate, or triangular, with an odour of garlic; flowers small, white.—Sauce-Alone or Jack-by-the-Hedge.—Hedges and waste places. Fl. May.

(13) Dentaria. Tooth-cress.

D. bulbifera: stem tallish, weak; lower leaves pinnate, upper simple; flowers few, large, purplish.—Coral-root.— Woods and shady places, rare. Fl. April, May.

(14) Draba. Whitlow-grass.

* *Flowers yellow.*

D. aizoides: dwarf, tufted; leaves sessile, linear, ciliated; flowers few, rather large.—Rocks, rare. Fl. March, April.

** *Flowers white.*

D. muralis: annual; stem weak, ascending, leafy; lower leaves ovate or oblong, toothed; flowers white with minute entire petals.—Rocks and walls, rare. Fl. April, May.

D. verna: annual; stem dwarf, erect, leafless; leaves radical, ovate or oblong; flowers small, white, with deeply cloven petals.—Walls and dry banks. Fl. April, May.

(15) Alyssum. Madwort.

A. calycinum: annual, dwarf, branching; leaves oblong-linear narrowed to the base; flowers yellow, inconspicuous.— —Cornfields. Fl. May and June.

(16) Hutchinsia.

H. petræa: annual, dwarf; lower leaves pinnate; stem-leaves few and smaller; flowers minute, white.—Walls, rocks, and stony places. Fl. March, April.

(17) Teesdalia.

T. nudicaulis : annual, dwarf; leaves radical, pinnate, with a larger terminal lobe; flower-stems leafless, bearing very small white flowers.—Dry gravelly places. Fl. May.

(18) Thlaspi. PENNY-CRESS.

T. arvense : annual; stem erect; leaves oblong, toothed; flowers small, white.—Mithridate Mustard.—Waste and cultivated ground. Fl. May to July.

T. perfoliatum : annual; stem erect; root-leaves ovate or orbicular, upper ones oblong, stem-clasping; flowers small, white.—Limestone pastures, rare. Fl. April, May.

19. Viola. VIOLET.

* *Stem very short, the flowers and leaves appearing to grow from the root.*

V. palustris : leaves kidney-shaped, smooth; flowers pale blue, with purple streaks, scentless.—Marshy places. Fl. April.

V. odorata : leaves heart-shaped, downy or shortly hairy; flowers bluish-purple or white, rarely purple reddish or lilac, sweet-scented.—Hedge-banks and thickets. Fl. March, April.

V. hirta : leaves heart-shaped, rough with hairs; flowers pale blue, scentless.—Limestone woods and pastures. Fl. April, May.

** *Stem elongated, evident.*

V. canina : primary stem short, flowering branches elongated ascending or erect; leaves ovate-cordate; flowers large, pale blue, scentless. There are several varieties.—Hedges, groves, and thickets. Fl. April to July.

V. tricolor : biennial or perennial; stem branching; leaves

oblong, deeply crenate, with large lyrately pinnatifid stipules; flowers purple or yellow, or with mixed colours.—Pansy or Heart's-ease, and the origin of the garden Pansies.—Cultivated fields. Fl. May to September.

Var. *lutea :* flowers large, rich yellow; habit more decidedly perennial.—Mountain pastures.

Var. *arvensis :* flowers much smaller, with very short whitish petals; annual.—Cornfields.

(20) Lychnis.

L. **Flos-cuculi:** leaves narrow, lance-shaped, the lower ones stalked; flowers pale red, the petals cut into four linear segments.—Ragged Robin.—Moist meadows. Fl. May, June.

L. **diurna :** leaves oval-oblong or ovate-lanceolate; flowers deep red, subdiœcious, the petals half-cloven, with narrow divergent lobes.—Hedges. Fl. May, June.

(21) Sagina. Pearlwort.

S. **procumbens :** minute, forming branching decumbent tufts; leaves small, subulate; flowers small, whitish, often without petals.—Common in sandy wastes. Fl. May to August.

(22) Arenaria. Sandwort.

A. **verna :** leaves awl-shaped, bluntish; flowers small, white, —Mountain pastures. Fl. May to August.

A. **trinervis :** annual; leaves ovate acute, ribbed; flowers small, white.—Moist shady places. Fl. May, June.

(23) Mœnchia.

M. **erecta :** annual; stem short, erect; leaves linear; flowers few, rather large, white.—Sandy wastes. Fl. May.

(24) **Holosteum.**

H. umbellatum : annual; leaves ovate, acute; flowers white. umbellate.—Walls and roofs.   Fl. April.

(25) **Cerastium.** MOUSE-EAR CHICKWEED.

C. **vulgatum**: annual, branched, usually viscid; leaves roundish-ovate, stalked; flowers minute, in dichotomous panicles.—A very variable plant.—Waste places, very common. Fl. April to September.

C. **arvense** : leaves crowded below, lanceolate-linear; flowers large, white, in loose cymes.—Dry sandy fields.   Fl. May to August.

(26) **Stellaria.** STARWORT.

S. **nemorum** : lower leaves heart-shaped, upper ovate sessile; flowers large, white, in elegant spreading cymes, the petals narrow and deeply cleft.—Moist woods.   Fl. May, June.

S. **media**: annual, smooth; stems weak, branched; leaves small, ovate, flowers small, white.—Chickweed.—Cultivated and waste ground.   Fl. all the year.

S. **Holostea** : leaves lanceolate, finely saw-toothed; flowers large, white, with inversely heart-shaped petals.—Stitchwort.— Groves, thickets, and hedge-sides.   Fl. May.

(27) **Geranium.** CRANE'S-BILL.

G. **molle** : annual; leaves orbicular, cut into 7–11 wedge-shaped lobes; peduncles 2-flowered; flowers small, purple, with deeply notched petals.—Waste places.   Fl. April to August.

(28) **Oxalis.** WOOD-SORREL.

O. **Acetosella** : peduncles radical, 1-flowered; flowers white. —Groves and shady places.   Fl. April, May.

E

O. corniculata: annual, procumbent, branched; peduncles axillary, supporting an umbel of yellow flowers.—Waste ground. Fl. May to October.

(29) Acer. MAPLE.

A. campestre: tree; leaves 5-lobed, with obtuse, entire or sinuate lobes; flowers in loose erect corymbs; wings of the carpels diverging horizontally.—Hedges and thickets. Fl. May, June.

A. Pseudo-platanus: tree; leaves larger, 5-lobed, the lobes toothed; flowers in pendent racemes; wings of the carpels parallel or but slightly divergent.—Sycamore.—Hedges and plantations. Fl. May.

(30) Polygala. MILKWORT.

P. vulgaris: stems diffuse or ascending; leaves obovate below, oblong-lanceolate above; flowers racemed, blue, pink, or white.—Gravelly and heathy pastures. Fl. May to September.

(31) Ulex. FURZE.

U. europæus: shrub; branches erect, the numerous short branchlets all ending in a thorn, the leaves generally reduced to thorns; flowers large, yellow, in showy racemes, intermixed with thorns, the calyx very hairy.—Whin or Gorse.—Sandy heaths and commons. Fl. March to July.

(32) Genista. GREEN WEED.

G. pilosa: shrub, without thorns; leaves obovate or lanceolate, hairy; flowers yellow, succeeded by narrow flattened hairy pods.—Sandy and gravelly heaths. Fl. May and September.

G. anglica: shrub, thorny; leaves ovate or lanceolate; flowers yellow, succeeded by short inflated pods.—Petty Whin. —Moist boggy heaths. Fl. May, June.

(33) Sarothamnus. BROOM.

S. scoparius: shrub; branches erect, strait, wiry, angled; leaves ternate, the upper ones simple; flowers large, bright yellow, axillary.—Dry gravelly thickets. Fl. May, June.

(34) Trifolium. CLOVER.

T. subterraneum: annual; stem short, decumbent; leaflets obovate; heads lateral, stalked, few-flowered, the peduncles becoming lengthened after flowering, and the heads buried in the earth; flowers long and slender, almost white.—Dry pastures and heaths. Fl. May.

T. suffocatum: annual; stems short, decumbent; leaflets obovate; heads dense, lateral, sessile; flowers minute, closely sessile, inconspicuous.—Sandy sea-coasts. Fl. May to July.

(35) Vicia. VETCH.

V. sativa: annual or biennial; leaves pinnate, tendrilled; flowers axillary, sessile, solitary or in pairs, large reddish-purple; calyx gibbous at the base on one side; seeds smooth. —There are several forms.—Cornfields and cultivated ground. Fl. May, June.

V. lathyroides: annual or biennial; leaves pinnate, tendrilled; flowers axillary, sessile, solitary, smaller than in *V. sativa*, purple; calyx equal at the base; seeds granulated.— Dry pastures and roadsides. Fl. April, May.

(36) Arthrolobium.

A. ebracteatum: annual; stems slender, with axillary peduncles bearing umbels of minute yellow flowers.—Sandy ground in the Channel and Scilly Isles. Fl. May to July.

(37) Hippocrepis. HORSESHOE VETCH.

H. comosa : stems branched, procumbent ; leaflets 9–15, small, obovate or oblong, smooth ; flowers 5–8 in the umbel, pale yellow.—Dry chalky banks. Fl. May to August.

(38) Ornithopus. BIRD'S-FOOT.

O. perpusillus : annual ; stems procumbent ; leaflets 5–10 or more pairs with an odd one, small oval or oblong ; flowers 2–3, white, with red lines.—Sandy pastures. Fl. May.

(39) Prunus.

P. communis : shrub ; leaves ovate or oblong ; flowers small white, nearly sessile, solitary or in pairs, appearing before the leaves ; drupes covered with a bluish bloom.—Blackthorn.—Woods and hedges. Fl. April, May.

Different forms of this are called : *P. spinosa*, the Sloe, with spiny branches and small austere globular fruits ; *P. insititia*, the Bullace, less spiny, and with larger less austere fruits ; *P. domestica*, the Plum, without spines, and with still larger oblong fruit.

P. Cerasus : shrub or small tree ; leaves ovate or ovate-lanceolate ; flowers large, white, on long pedicels, 2–3 or more together, appearing before the leaves ; drupes without bloom. —Cherry or Gean.—Woods and hedges. Fl. April, May.

There are two forms of this : *P. Cerasus*, representing the austere Morello cherries, and *P. Avium*, the sweet cherries of our gardens.

P. Padus : shrub ; leaves oval or ovate-lanceolate ; flowers small, white, in long drooping racemes ; drupes small, without bloom.—Bird Cherry.—Woods and hedges. Fl. May.

(40) **Fragaria.** STRAWBERRY.

F. vesca : leaflets 3, ovate, toothed, sessile at the end of a long leaf-stalk; flower-stems erect, leafless, bearing a few pedicellate white flowers.—Wood **Strawberry**.—Groves and thickets. Fl. May, June.

(41) **Potentilla.** CINQUEFOIL.

P. Fragariastrum : stems tufted; leaves digitately divided; flowers white.—Dry gravelly pastures. Fl. March, April.

P. verna : stem short, tufted, sometimes procumbent; leaves digitately divided; flowers yellow.—Mountain pastures. Fl. April, May.

(42) **Alchemilla.**—LADY'S MANTLE.

A. vulgaris : radical leaves 7–9-lobed, plaited; flowers in terminal panicles, on the decumbent or ascending stems.— Mountain pastures. Fl. April to August.

A. arvensis : annual; leaves 3-lobed, cut; flowers in sessile axillary heads.—Parsley Piert.—Sandy fields. Fl. May to August.

(43) **Mespilus.** MEDLAR.

M. germanica : small tree; leaves nearly sessile, lanceolate or oblong; flowers large, white, solitary, sessile on short leafy branches.—Hedges, rare. Fl. May.

(44) **Cratægus.** HAWTHORN.

C. Oxyacantha : shrub or small tree, thorny; leaves obovate, wedge-shaped, stalked, cut into 3–5 toothed lobes; flowers white or pink, scented, in sessile corymbs on short leafy branches.—Whitethorn or May.—Hedges and woods. Fl. May, June.

(45) Pyrus.

* Leaves simple, rarely pinnatifid.

P. communis: tree; leaves ovate or obovate, finely saw-toothed, smooth; flowers large, white, in short racemes or bunches of 6–10; fruit large, top-shaped.—Pear-tree.—Woods and hedges. Fl. April, May.

P. Malus: tree; leaves ovate or obovate, saw-toothed, downy; flowers large, pinkish, several together, in a sessile umbel; fruit large, roundish.—Apple-tree or Crab.—Woods and hedges. Fl. May.

P. Aria: shrub or tree; leaves ovate or obovate, sharply serrated, white and cottony beneath; flowers white, in dense flat corymbs; fruit small, globular, red. There are varieties: *P. intermedia,* and *P. pinnatifida* with the leaves variously cut.—Beam-tree.—Mountainous woods. Fl. May.

P. torminalis: tree: leaves broad, cut into about 7 broad pointed toothed lobes, white beneath while young; flowers white, in corymbs; fruit small, globular, greenish-brown.—Wild Service-tree.—Woods and hedges. Fl. April, May.

** Leaves pinnately divided.

P. Aucuparia: tree; leaflets numerous, oblong, toothed, downy beneath; flowers small, very numerous, in large corymbs; fruit small, globular, bright red.—Rowan-tree, Quicken-tree, or Mountain Ash.—Mountainous woods. Fl. May.

(46) Cotoneaster.

C. vulgaris: shrub; leaves ovate or orbicular; flowers small, greenish-white, solitary, or in short drooping racemes on short leafy branches; fruit small, reddish.—Limestone cliffs, Carnarvonshire. Fl. May, June.

(47) **Saxifraga.** SAXIFRAGE.

* *Leaves opposite, densely crowded.*

S. oppositifolia : stems creeping ; leaves small, opposite, obovate, overlapping, fringed ; flowers large, purple, solitary, on short erect branches, but so numerous as to cover the stems. —Alpine rocks. Fl. April.

** *Leaves alternate, the upper ones distant.*

S. granulata : stem erect, hairy ; lower leaves reniform, bluntly crenate ; flowers largish, white, in terminal cymes.— Meadows and pastures. Fl. May.

S. tridactylites : annual ; stem short, erect, usually red ; leaves small, entire or three-lobed ; flowers small, white.— Walls and dry barren places. Fl. April, May.

(48) **Chrysosplenium.** GOLDEN SAXIFRAGE.

C. alternifolium : leaves alternate ; flowers yellow, inconspicuous.—Moist shady places. Fl. May.

C. oppositifolium : leaves opposite ; flowers yellow, inconspicuous.—Moist shady places. Fl. May.

(49) **Rhamnus.** BUCKTHORN.

R. catharticus : shrub ; leaves ovate, minutely toothed ; branches often thorny ; flowers small, green, dioecious, with four petals and four stamens ; fruit black.—Thickets. Fl. May.

R. Frangula : shrub ; leaves more obtuse, entire ; branches without thorns ; flowers small, hermaphrodite, with five petals and five stamens ; fruit dark purple.—Thickets and woods.— Fl. May.

(50) **Euonymus.** SPINDLE-TREE.

E. europæus : shrub ; leaves ovate-oblong, minutely toothed ;

flowers small, yellowish-green; fruit red, angular, opening at
the angles and exposing a brilliant orange-coloured arillus,
which envelopes the seeds.—Hedges and thickets. Fl. May.

(51) Montia. Blinks.

M. fontana : subaquatic, annual; leaves opposite, obovate
or spathulate; flowers white, inconspicuous.—Water Chickweed.—By rills and in springy places. Fl. April, May.

(52) Ribes. Currant.

* Branches without prickles.

R. rubrum : shrub; leaves palmately 3–5-lobed, inodorous;
flowers complete, greenish, in pendulous racemes, the stalks
all short; berries smooth, red or white.—Common Currant.—
Mountainous woods. Fl. April, May.

R. nigrum : shrub; leaves palmately 3-lobed, strongly
scented; flowers complete, greenish, in pendulous racemes, the
lower stalks longer than the upper ones; berries smooth, black.
—Black Currant.—Cool shady thickets. Fl. April, May.

** Branches armed with prickles.

R. Grossularia : shrub; leaves palmately 3–5-lobed, inodorous; flowers greenish, hanging, the peduncles 1–2-flow-
ered; berries yellowish, oblong, much varied by cultivation,
smooth or hairy.—Gooseberry.—Thickets. Fl. April, May.

(53) Arctostaphylos. Bearberry.

A. Uva-ursi : shrub; leaves evergreen, obovate, entire,
shining; flowers ovate, rosy, in short drooping terminal
racemes; berries bright red.—Mountain heaths. Fl. May.

A. alpina : shrub; leaves withering, rugged, serrated; flowers ovate, white, in terminal drooping clusters or racemes,
berries black.—Mountain heaths. Fl. May.

(54) Vaccinium. WHORTLEBERRY.

V. Myrtillus: shrub; stems acutely angular; leaves ovate, serrated; flowers solitary in the leaf-axils, nearly globular, pale greenish-white, tinged with red, nodding; fruit a roundish bluish-black berry, covered with bloom.—Bilberry.—Heathy wastes. Fl. May.

V. uliginosum: shrub; stems terete; leaves obovate, entire; flowers axillary, somewhat clustered, flesh-coloured, nodding; fruit a black berry.—Great Bilberry.—Mountain bogs. Fl. May.

(55) Adoxa. MOSCHATEL.

A. Moschatellina: root-leaves two or three times ternate, with 3-lobed segments; flowers small, pale-green, in little globular heads, terminating the short flowering stems.— Shady hedge-banks and groves. Fl. April, May.

(56) Asperula. WOODRUFF.

A. odorata: stems angular; leaves lance-shaped or obovate, 8 in a whorl; flowers small, white, 4-lobed, in cymes; fruit globular, hispid; plants smelling like new hay in drying.— Sweet Woodruff.—Dry mountainous woods. Fl. May.

(57) Valerianella. CORNSALAD.

V. olitoria: annual; leaves oblong, entire, or with a few coarse teeth, those of the stem narrower and clasping; flowers inconspicuous, in little terminal cymes.—Lamb's Lettuce.— Cornfields. Fl. April to June.

(58) Bellis. DAISY.

B. perennis: leaves all from the root, obovate or oblong;

flower-heads single, on radical peduncles; the ray-florets white, tinged with pink, those of the disk small, yellow, tubular.— Pastures and meadows. Fl. March to November.

(59) Senecio. GROUNDSEL.

S. vulgaris : annual; leaves pinnatifid, toothed, smoothish; flower-heads without ray-florets, consisting wholly of yellow tubular disk-florets.—Simson.—A weed everywhere. Fl. at all seasons.

(60) Tussilago. COLTSFOOT.

T. Farfara : leaves heart-shaped, angular or toothed, cottony beneath; flower-heads solitary, appearing before the leaves, with bright yellow ligulate ray-florets.—Moist waste places. Fl. March, April.

T. Petasites : leaves heart-shaped, unequally toothed, cottony; flower-heads in panicles, nearly diœcious, pinkish; the florets all tubular (males), or nearly all small and filiform (females), not ligulate.—Butterbur.—Moist meadows. Fl. April.

(61) Doronicum. LEOPARD'S-BANE.

D. Pardalianches : leaves heart-shaped, toothed, the radical ones stalked, the rest stem-clasping; flower-heads usually 3 to 5, with numerous long narrow ligulate ray-florets.— Mountain pastures. Fl. May.

(62) Leontodon. DANDELION.

L. Taraxacum : leaves all radical, milky, oblong-lanceolate, runcinately pinnatifid; flower-heads large, solitary, on hollow peduncles, with numerous yellow florets, all ligulate.—Cultivated and waste ground everywhere. Fl. March to July.

(63) Primula. PRIMROSE.

P. vulgaris : leaves oblong-ovate, wrinkled, crenate, narrowing gradually into the footstalks; scapes single-flowered; flowers erect, pale yellow, with a flat limb.—Woods and thickets. Fl. April, May.

Var. *umbellata*, flowers umbellate, oxlip-like.—Pastures.

P. veris : leaves ovate, contracted below, wrinkled, crenate; scapes umbellate, many-flowered ; flowers small, nodding, deep yellow, with a concave limb, the segments cordate.—Cowslip or Paigle.—Meadows and pastures. Fl. April, May.

These two plants are sometimes, perhaps correctly, regarded as varieties of one species. It is probable that many hybrids are formed between them, many of which are mistaken for the true *P. elatior*

P. elatior : leaves ovate, contracted below, wrinkled, denticulate ; scapes umbellate, many-flowered ; flowers large, nodding, yellow, with a somewhat concave limb, the segments cordate-oblong.—Oxlip.—Pastures, rare. Fl. April, May.

(64) Vinca. PERIWINKLE.

V. major : stems ascending ; leaves broadly ovate, fringed ; flowers large, purplish-blue, the calyx-segments ciliate at the margin.—Woods and shady banks. Fl. May.

V. minor : stems procumbent ; leaves narrow-ovate, smooth-edged ; flowers smaller, blue, the calyx-segments quite smooth —Thickets. Fl. May.

(65) Gentiana. GENTIAN.

G. verna : dwarf, tufted ; lower leaves ovate or oblong, crowded ; flowers large, vivid blue, salver-shaped, solitary on the short simple flower-stems, 5-lobed, with small intermediate bifid segments.—Mountain pastures. Fl. April.

(66) Menyanthes. BUCKBEAN.

M. trifoliata : stems creeping or floating ; leaves trifoliate, the leaflets obovate or oblong ; flowers in upright racemes, bell-shaped, with five lobes, white, tinged externally with red, and fringed within.—Marsh Trefoil.—Boggy meadows and pools. Fl. May to July.

(67) Pulmonaria. LUNGWORT.

P. officinalis : root-leaves ovate-oblong, usually spotted ; flowers in terminal forked cymes, the corolla tubular-bell-shaped, purple ; plant hairy.—Woods and thickets. Fl. May.

(68) Symphytum. COMFREY.

S. officinale : stems 2–3 feet high ; leaves large, broad-lanceolate, tapering to a long point, roughly hairy, those of the stems decurrent ; flowers in forked cymes on the branching stems, the corollas ventricosely tubular, purple or yellowish-white.—Damp meadows.—Fl. May, June.

(69) Scrophularia. FIGWORT.

S. vernalis : biennial ; stems 2 feet high ; leaves orbicular-cordate, pale green, coarsely toothed ; peduncles mostly axillary, bearing a small cyme of yellow flowers.—Hedge-banks and thickets. Fl. April, May.

(70) Lathræa. TOOTHWORT.

L. Squamaria : leafless ; stems erect, scaly ; flowers in a dense spike, nodding, flesh-coloured or slightly bluish, streaked with purple.—Dry shady places, on the roots of hazel and other trees. Fl. April.

(71) **Nepeta.** CAT MINT.

N. Glechoma: stems creeping and rooting; leaves orbicular crenate; flowers blue, in axillary whorls of about 6, the corolla-tube twice as long as the calyx.—Ground Ivy or Alehoof.—Hedge-banks. Fl. April, May.

(72) **Lamium.** DEAD NETTLE.

* *Annuals, with small flowers in terminal leafy whorls.*

L. amplexicaule: lower leaves long-stalked orbicular, upper floral ones closely sessile, deeply crenate; flowers in 1–2–3 whorls, purplish-red, with a slender tube.—Henbit.—Sandy fields. Fl. March to June.

L. purpureum: lower leaves long-stalked orbicular, upper floral ones crowded, shortly stalked, ovate or triangular, often pointed; flowers dull purplish-red, the corolla-tube broader and more open than in the last.—Red Dead-Nettle.—Waste and cultivated ground, common. Fl. April to October.

Var. *incisum*: upper leaves deeply cut.—Waste ground.

** *Perennials, with large flowers in axillary whorls.*

L. album: leaves heart-shaped, strongly serrated, hairy; flowers numerous, elongated, white.—White Dead-Nettle.—Waste ground. Fl. May, June.

L. Galeobdolon: leaves stalked, ovate, toothed; flowers yellow.—Archangel.—Groves and hedges. Fl. May.

(73) **Ajuga.** BUGLE.

* *Flowers axillary, growing in pseudo-whorls.*

A. reptans: glabrous or slightly hairy; stem forming creeping runners; leaves obovate, entire or coarsely toothed; flowers blue, the upper whorls crowded, forming a cylindrical leafy spike.—Woods and moist pastures. Fl. May.

** *Flowers axillary, solitary or in pairs.*

A. Chamæpitys: annual; leaves crowded, deeply cut into three linear lobes, the lateral of which are sometimes again divided; flowers yellow, shorter than the leaves.—Ground Pine or Yellow Bugle.—Sandy fields. Fl. April, May.

(74) Daphne.

D. Mezereum: shrub; leaves deciduous, narrow-oblong; flowers purple or white, sweet-scented, in clusters of 2–3 along the preceding years' wood before the leaves appear; berries red or yellow.—Spurge Olive or Mezereon.—Woods. Fl. March.

D. Laureola: shrub; leaves obovate-lanceolate, evergreen; flowers green, scentless, in clusters of 3–5 in the leaf-axils; berries bluish-black.—Spurge Laurel.—Woods and thickets. —Fl. March.

(75) Ulmus. Elm.

U. montana: tree; leaves obovate cuspidate, doubly coarse-toothed, scabrous; flowers in dense clusters before the leaves, reddish, succeeded by the flat winged fruits, slightly notched at the top, the seed-bearing cavity considerably below the notch.—Wych or Witch Elm.—Woods and hedges. Fl. March, April.

U. campestris: tree; leaves smaller, rhomboid-ovate, acuminate, oblique at the base, doubly coarse-toothed, scabrous; flowers as in the former; fruits deeply notched almost to the cavity of the seeds.—English Elm.—Hedges. Fl. March, April.

(76) Asarum. Asarabacca.

A. europæum: stems short; leaves about two, roundish,

heart-shaped or kidney-shaped, smooth; flowers terminal, solitary, on recurved stalks, dull greenish-brown, bell-shaped, three-cleft.—Mountainous woods. Fl. May.

(77) **Hippophae.** SALLOW THORN.

H. rhamnoides : shrub, clothed with leprous scales; leaves alternate, linear-lanceolate, entire; flowers diœcious, very small and inconspicuous, the females succeeded by small yellowish berries.—Sea Buckthorn.—Sandy cliffs. Fl. May.

(78) **Buxus.** Box.

B. sempervirens : shrub; leaves evergreen, ovate, convex, shining; flowers small, green, sessile.— Chalky hills. Fl. April.

(79) **Empetrum.** CROWBERRY.

E. nigrum : dwarf shrub; leaves small, crowded, evergreen, linear-oblong; flowers quite inconspicuous; fruits black.— Crakeberry.—Mountain heaths. Fl. May.

(80) **Myrica.** GALE.

M. Gale : shrub; leaves deciduous, lanceolate, serrated, tapering and entire at the base, sweet-scented; catkins short, sessile along the ends of the branches, appearing before the leaves.—Sweet Gale, Bog Myrtle, or Dutch Myrtle.—Bogs. Fl. May.

(81) **Alnus.** ALDER.

A. glutinosa : tree; leaves roundish, wedge-shaped, obtuse, wavy, serrated, glutinous, deciduous; catkins 2–3 together in loose terminal clusters; the males long, pendent, the females ovate or oblong, and in fruit somewhat resembling miniature fir-cones.—River-sides and other wet places. Fl. March.

(82) Betula. BIRCH.

B. alba: tree; leaves deciduous, ovate-deltoid, varying to broadly cordate, and with a rhomboidal tendency, acute, unequally serrated, glabrous, often with resinous spots; male catkins cylindrical, drooping, 1–2 inches long, females shorter and more compact.—Woods. Fl. March, April.

The Common Birch, B. glutinosa, and the White Birch, B. alba, are two slightly different forms: the first having the fruits broadly obovate, and the lateral lobes of the 3-lobed scales of the female catkins ascending; the second having the fruits obovate-elliptical, and the lobes falcate reflexed.

B. nana: shrub; leaves deciduous, small, nearly orbicular, crenate, not pointed; catkins small, oblong, erect.—Spongy Bogs. Fl. May.

(83) Salix. WILLOW.

* Male and female catkins on short leafy shoots.

S. pentandra: small tree; leaves thick, smooth, shining, deciduous, broadly lanceolate or ovate, pointed, finely toothed; catkins cylindrical, loose; stamens about five, sometimes more. —Bay Willow.—River-sides. Fl. May, June.

S. fragilis: tree; leaves deciduous, ovate-lanceolate, pointed, smooth, serrated throughout; catkins long, loose; flowers large; stamens two; capsules stalked.—Crack Willow.— Marshy ground. Fl. April, May.

S. alba: tree; leaves deciduous, ashy-grey, silky on both sides, elliptic-lanceolate, pointed, serrated; catkins loose, cylindrical; stamens two; capsules nearly sessile.—Moist woods. Fl. May.

S. triandra: small tree; leaves deciduous, linear-oblong,

serrated, smooth, paler beneath; catkins cylindrical, loose; stamens three.—Wet woods and hedges. Fl. May, and often again in August.

** *Male catkins sessile, females sessile or on very short peduncles, with or without leafy bracts.*

S. purpurea: shrub; leaves deciduous, lanceolate or oblong-lanceolate, broader upwards, serrated, smooth, whitish beneath, sometimes opposite; catkins narrow cylindrical, closely packed; stamens one under each scale, consisting of an entire filament with double anther, or a forked filament with an anther on each branch. There are several allied forms.—Marshes and river-banks. Fl. March, April.

S. viminalis: shrub or small tree; leaves deciduous, linear, inclining to lanceolate, elongated, taper-pointed, entire, wavy, white and silky beneath; catkins cylindrical; stamens two, distinct.—Common Osier.—Marshy places. Fl. April, May.

S. Caprea: shrub or small tree; leaves deciduous, roundish-ovate, serrated, wrinkled, finely toothed, glaucous and downy beneath; catkins cylindrical; stamens two, distinct.—Common Sallow.—Woods and hedges. Fl. April.

S. repens: dwarf creeping shrub; leaves deciduous, small, ovate-oblong or lanceolate, nearly entire, glaucous and silky beneath; catkins short, cylindrical; stamens two, distinct.—There are several varieties.—Heaths, moors, and sandy wastes. Fl. May.

(84) Populus. POPLAR.

P. alba: tree; leaves deciduous, cordate-ovate, angularly toothed, snow-white and densely downy beneath; catkins sessile, about two inches long.—White Poplar or Abele. The Grey Poplar is a variety with smaller leaves.—Damp woods. Fl. March.

P. tremula: tree; leaves deciduous, orbicular or rhomboidal, smooth on both sides, irregularly and rather coarsely toothed, with slender stalks which twist readily, so as to keep the leaves in motion; catkins small.—Aspen.—Woods. Fl. March, April.

P. nigra: tree; leaves deciduous, triangular-ovate, tapering at the point, serrated, smooth on both sides; catkins two inches long, lax, cylindrical.—Damp places. Fl. March.

(85) **Fagus.** BEECH.

F. sylvatica: tree; leaves deciduous, ovate, obscurely toothed, silky when young, afterwards glabrous; male catkins globular on pendulous stalks, females erect globular softly hairy.—Woods.—Fl. April, May.

(86) **Quercus.** OAK.

Q. Robur: tree; leaves deciduous, cuneately oblong, irregularly sinuated or almost pinnatifid, sessile or shortly stalked; fruits clustered above the middle of a long peduncle (2–6 inches long).—Woods and hedges. Fl. April.

Q. sessiliflora: tree; leaves deciduous, cuneately-oblong, irregularly sinuated, stalked, the stalks ½–1 inch long; fruits solitary or clustered, sessile on the branch, or seated on a short peduncle (rarely 1 inch long).—Woods. Fl. April.

(87) **Corylus.** HAZEL.

C. Avellana: shrub or small tree; leaves deciduous, roundish heart-shaped, pointed, obscurely lobed, doubly coarse toothed; male catkins long drooping cylindrical; females resembling ovate leaf-buds, with a few crimson threads.—Nut, or Hazel-nut.—Coppices and thickets. Fl. April.

(88) Carpinus. HORNBEAM.

C. Betulus : tree ; leaves deciduous, ovate acute, parallel-veined, plaited when young ; catkins drooping, about 1½ inch long, the female ones slender, often several inches long when in fruit ; fruit-bracts 3-lobed, leafy, the middle lobe elongated.—Woods and hedges. Fl. May.

(89) Pinus. PINE.

P. sylvestris : tree ; leaves evergreen, rigid, subulate, growing in pairs, each pair surrounded by a membranous sheath ; male catkins small, oblong, clustered ; young cones small ovate-conical stalked recurved, in the mature state enlarged formed of woody close-set scales.—Scotch Fir or Common Pine.—Highlands. Fl. May.

(90) Juniperus. JUNIPER.

J. communis : shrub ; leaves evergreen, linear, spreading, with a prickly point, growing in whorls of three ; catkins minute, axillary ; fruits globose, dark purple.—Calcareous hills and downs. Fl. May.

(91) Taxus. YEW.

T. baccata : tree ; leaves evergreen, crowded, linear, disposed in two ranks so as to form flattened spray ; catkins small, axillary ; fruit roundish, bright red, forming a fleshy cup around the hard seed.—Mountain woods and limestone cliffs. Fl. March.

(92) Arum. CUCKOO-PINT.

A. maculatum : tuberous ; leaves ovately halberd-shaped or almost arrow-shaped, entire, often purple-spotted ; spathe obliquely campanulate, pale-green, contracted above the base, the limb tapering to a point, the upper club-shaped naked part of the purplish or yellowish spadix only visible; berries bright red.—Cuckow-pint, Wake-robin, or Lords-and-Ladies. —Hedge-banks, and groves. Fl. May.

(93) Trichonema.

T. Bulbocodium : bulb-tuberous, dwarf; leaves narrow-linear or grass-like, longer than the flower-stalks; flowers solitary, erect, pale purplish-blue, with a yellow centre.—Sandy grassy hillocks near the sea, rare.—Fl. March, April.

(94) Crocus.

C. vernus : bulb-tuberous, dwarf; leaves narrow-linear, grassy; flowers solitary, bluish-purple, with rich orange wedge-shaped jagged stigmas.—Meadows, rare. Fl. March.

(95) Narcissus.

N. Pseudo-Narcissus : bulbous; leaves broadly linear, blunt, not keeled; flowers solitary, large, scentless, with a broad deep yellow tubular coronet as long as the segments, which are paler.—Daffodil or Daffy-down-dilly.—Woods and thickets. Fl. March.

N. biflorus : bulbous; leaves broadly linear, bluntly keeled, their edges reflexed; flowers usually two, pale straw-colour or nearly white, sweet-scented, the coronet short, broadly cup-shaped, yellow.—Primrose Peerless.—Sandy fields. Fl. April, May.

(96) Galanthus. SNOWDROP.

G. nivalis : bulbous, dwarf; leaves narrow-linear; flowers

solitary, drooping, white, the inner segments much the shorter and tipped with green.—Meadows and thickets. Fl. February.

(97) **Leucojum.** SNOWFLAKE.

L. æstivum: bulbous; leaves broadly linear; flowers several, elevated on a scape, drooping, the segments nearly equal, white, with slight greenish tips.—Moist meadows. Fl. May

(98) **Neottia.**

N. Nidus-avis: leafless, the stems clothed with pale brown scales; flowers in a dense terminal spike, dingy-brown.— Shady woods.—Fl. May, June.

(99) **Orchis.**

* *Spur very short, reduced to a small pouch.*

O. ustulata: tuberous, the tubers entire; leaves lanceolate; spike dense, 1–2 inches long; flowers small, the lip not longer than the sepals, white with purple spots, three-lobed, the middle lobe deeply bifid.—Chalky downs. Fl. May, June.

** *Spur lengthened, varying from the half to the whole length of the ovary.*
† *Tubers not divided.*

O. Morio: leaves lanceolate; flowers few in a loose spike, the sepals purple, all converging; lip longer, broadly and shortly three-lobed, pinkish-purple, paler in the centre, with dark spots; spur nearly as long as the ovary.—Meadows and pastures. Fl. May, June.

O. militaris: leaves broadly oval to oblong; flowers in a dense oblong spike; the sepals purple, all converging; lip longer, pale-coloured, spotted with purple, three-lobed, the lateral lobes small, the middle one larger, two-cleft; spur scarcely half the length of the ovary.—Pastures and borders of woods. Fl. May.—Mr. Bentham includes as varieties the following forms, often regarded as distinct :—

Var. *fusca :* larger and stouter; sepals dark brownish-purple, variegated; lip paler, its middle lobe broad and short. —Chalky bushy hills.

Var. *tephrosanthos :* smaller and more slender; sepals dark purplish; lip purple, its middle lobe long and narrow.— Chalky hills.

O. mascula: leaves elliptic-lanceolate, usually spotted with purple; flowers in a loose spike, 3–6 inches long, pinkish-purple varying to flesh-colour; upper sepal arching, the lateral ones spreading; lip three-lobed with the middle lobe emarginate; spur as long as the ovary.—Pastures. Fl. April, May.

†† *Tubers palmately divided.*

O. maculata: leaves varying from nearly ovate to narrow-lanceolate, often spotted; flowers in a dense spike, 2–3 inches long, pale pink; lip variously spotted with purple, broadly orbicular, irregularly 3-lobed; spur slender, shorter than the ovary.—Meadows and woods. Fl. May, June.

(100) Ophrys. INSECT ORCHIS.

O. muscifera: tuberous; slender; flowers few, small; sepals greenish; lip oblong, considerably lengthened beyond the sepals, purplish-brown, with pale blue or white central marks, convex, with the lateral lobes turned down, and the central one deeply notched.—Fly Orchis.—Chalky pastures. Fl. May, June.

O. aranifera: tuberous; flowers few, distant; sepals greenish; lip scarcely longer than the sepals, broad, convex, hairy, dull brown, inscribed with pale-yellowish markings in the

centre, the edges obscurely lobed and scarcely turned under.—
Spider Orchis.—Chalky pastures. Fl. April.

Var. *fucifera :* lip usually undivided, obovate, longer than
the sepals.—Chalky pastures.

(101) Cypripedium. LADY'S SLIPPER.

C. Calceolus : leaves large, ovate, pointed, ribbed ; flowers
1–2 ; sepals broadly lanceolate, and as well as the linear
petals, brown-purple ; lip large inflated yellow, netted with
dark veins.—Mountain woods in the north. Fl. May, June.

(102) Ruscus. BUTCHER'S BROOM.

R. aculeatus : shrub-like ; leaves (or flattened shoots) evergreen, ovate with a prickly point; flowers small, white, the
axillary pedicel adnate halfway along the leaf, so that the
flower appears to grow from the centre of the leaf, on the
upper side, which is turned downwards by a twist in the stalk ;
fruit red.—Woods and bushy heaths. Fl. March, April.

(103) Polygonatum. SOLOMON'S-SEAL.

P. multiflorum : stem round, 2 feet high, arching ; leaves
alternate, ovate-oblong ; flowers axillary, on short branched
one- or many-flowered peduncles, drooping, white with greenish tips; filaments hairy.—Woods and thickets. Fl. May,
June.

P. officinale : stem angular, 1 foot high ; leaves alternate,
ovate-oblong ; flowers axillary, on short one- or two-flowered
peduncles; filaments smooth.—Mountain woods. Fl. May,
June.

(104) Fritillaria. FRITILLARY.

F. Meleagris : bulbous; stem with 3–4 linear-lanceolate

leaves, and a terminal drooping flower of a dull red brighter inside, marked with chequered lines and spots.—Snake's-head or Chequered Daffodil.—Moist meadows. Fl. April.

(105) Tulipa. TULIP.

T. sylvestris: bulbous; leaves 1-3, lanceolate, glaucous; flower terminal, yellow, slightly fragrant, drooping in the bud, nearly erect when mature.—Chalk-pits and pastures. Fl. April.

(106) Gagea.

G. lutea : bulbous, dwarf; leaves one, rarely two, linear-lanceolate; flowers 3-4, corymbosely racemed, yellow, star-like; the leafy bracts as long as the pedicels or longer.— Groves and pastures. Fl. April.

(107) Convallaria.

C. majalis : leaves oblong, stalked, usually two ; peduncle leafless, radical, shorter than the leaves, terminating in a loose raceme of smallish white drooping globosely bell-shaped flowers, which are pure white and very fragrant.—Lily of the Valley.—Groves and thickets. Fl. May.

(108) Muscari. GRAPE HYACINTH.

M. racemosa : bulbous, dwarf; leaves narrow-linear, thickish, channeled, recurved, longer than the scape; flowers in a close terminal raceme or head, nodding, small, dark blue, ovate, the uppermost paler and erect.—Starch Hyacinth.— Sandy fields. Fl. May.

(109) Hyacinthus. HYACINTH.

H. non-scriptus : bulbous; leaves linear, shorter than the

scapes; flowers in terminal nodding or one-sided racemes, drooping, each with a pair of bracts at its base.—Bluebell.—Woods and thickets. Fl. May.

(110) Scilla. SQUILL.

S. verna : bulbous, dwarf; leaves narrow-linear, channeled ; flowers small, erect, blue, in short terminal corymbose racemes, the pedicels furnished with a linear bract.—Maritime cliffs.—Fl. April.

(111) Ornithogalum.

O. umbellatum : bulbous, dwarf; leaves long, linear, flaccid; flowers corymbose, on a short scape, shorter than the leaves, white with green streaks outside.—Star of Bethlehem. —Meadows and pastures. Fl. April, May.

O. nutans : bulbous ; leaves linear-lanceolate ; flowers large, nodding, racemose, white, greenish outside.—Fields and orchards. Fl. April, May.

(112) Luzula. WOOD-RUSH.

L. campestris : leaves flat, grass-like, fringed with long white hairs; flowers collected 6–8 together in close ovoid heads or clusters, 3–6 of which are arranged to form a panicle, the upper head sessile, the rest stalked; flowers scarious, shining brown.—Barren pastures. Fl. April, May.

L. pilosa : leaves flat and fringed as in the last; flowers panicled, each on a separate stalk, scarious, shining brown.—Shady groves. Fl. March, April.

(113) Carex. SEDGE.

* *Spikes or flower-clusters simple, solitary.*

C. dioica : creeping, dioecious, grass-like, 6–8 inches high ; female spikes shorter, ovoid.—Spongy bogs. Fl. May, June.

** *Spikelets arranged in a compound continuous or interrupted spike; stigmas 2.*

　† *Rootstock creeping.*

C. **intermedia**: stems 1–2 feet high; spikelets large ovoid, simple, crowded into a terminal spike.—Wet meadows. Fl. May.

　†† *Rootstock tufted.*

C. **vulpina**: stems 2–4 feet high; spikelets numerous, densely crowded into a terminal spike, interrupted at the base. —Wet places. Fl. May, June.

C. **remota**: stems 1–2 feet high; spikelets small, pale-coloured, widely separated; bracts leaf-like.—Moist shady places. Fl. May, June.

C. **stellulata**: stems 6–12 inches high; spikelets three or four, oval-oblong or roundish, rather distant; bracts not leaf-like.—Boggy places. Fl. May, June.

*** *Spikes separate, one or more terminal ones wholly (rarely partially) barren, the rest fertile.*

　† *Beak of the fruit entire, emarginate or shortly 2-toothed.*

　‡ *Nut plano-convex; stigmas 2; barren spikes one or more.*

C. **acuta**; stems 2–3 feet high; spikes cylindrical, dark-coloured, the fertile 3 inches long or more; glumes narrow and pointed.—Wet places. Fl. May, June.

C. **vulgaris**: stems 1 foot high; spikes oblong, ½–2 inches long, dark-brown; glumes blunt.—Marshes. Fl. May, June.

　‡‡ *Nut three-cornered; stigmas 3.*

　(*a*) *Fruits smooth; barren spikes terminal, solitary.*

C. **panicea**: stem 1–2 feet high; fertile spikes about two, erect, cylindrical, ½–1 inch long, loosely imbricated.—Carnation Grass.—Marshy places. Fl. May, June.

(b) *Fruits smooth; barren spikes terminal, several.*

C. pendula : stems 3–6 feet high; fertile spikes drooping, 4–6 inches long, cylindrical, densely flowered.—Damp woods. Fl. May.

C. glauca : stems ½–1½ feet high; fertile spikes two or three, ½–1 inch long, more or less drooping; fruit slightly scabrous.—Damp places. Fl. May, June.

(c) *Fruits downy.*

C. præcox : stem ¼–1 foot high; barren spike solitary, fertile spikes one to three, oblong-ovate, near together, sessile, or the lowermost sometimes slightly stalked; lowest bract shortly sheathing, with a leafy point.—Dry heaths. Fl. April, May.

C. pilulifera : stems ½–1 foot high; barren spike solitary; fertile about three, roundish, near together, sessile; lowest bract with a leafy point, sheathless.—Wet heaths. Fl. May.

†† *Beak of the fruit long, 2-toothed or bifid; (nut three-cornered: stigmas 3).*

‡ *Bracts sheathing, leaf-like; beak of the fruit plano-convex.*

C. flava : stems 6–12 inches high; barren spike solitary, cylindrical; fertile ones about three, roundish-oval, subsessile, near the barren, and often one lower down on a longish stalk. —Wet places. Fl. May, June.

C. sylvatica : stem two feet high; barren spike solitary; fertile about four, distant, drooping, linear.—Damp woods. Fl. May.

C. hirta : stem 1½–2 feet high; barren spikes two or three; fertile ones two or three, remote, erect, oblong-cylindrical, stalked.—Hammer-sedge.—Damp places. Fl. April.

‡‡ *Bracts not sheathing ; beak of the fruit terete.*

C. ampullacea : stem 1–2 feet high ; barren spikes several ; fertile ones 2–4, remote, cylindrical, erect, stalked ; fruit inflated, abruptly contracted into a long beak.—Wet bogs. Fl. May, June.

C. vesicaria : stem 2 feet high ; barren and fertile spikes nearly as in the last, but rather shorter; fruit inflated, gradually tapering into a short beak.—Wet bogs. Fl. May.

C. riparia : stem 3 feet high ; barren spikes several ; fertile two or three, rather distant, cylindrical, 2–3 inches long ; fruit much flattened or convex on both sides narrowed into a short broad beak.—Watery places. Fl. April, May.

(114) Sesleria. Moor-grass.

S. cœrulea : dwarf, densely tufted ; stem 6–12 inches high, supporting a solitary ovate-oblong spike-like panicle, ½–¾ inch long, of a bluish-grey.—Alpine limestone rocks. Fl. April to June.

(115) Anthoxanthum. Vernal-grass.

A. odoratum : stems 1–2 feet high, tufted ; spike-like panicle, 1½–2 inches long ; herbage fragrant when withering having the well-known smell of new hay, which is attributable to this common pasture grass.—Sweet Vernal Grass.—Meadows and pastures. Fl. May, June.

(116) Chamagrostis.

C. minima : annual, tufted, 2–3 inches high ; spikelets small, purplish, almost sessile, in a simple slender spike, ½ inch long.—Sandy pastures on the coast. Fl. March, April.

(117) **Aira.** HAIR-GRASS.

A. præcox : annual, tufted, 3-6 inches high ; panicle con-tracted, ½-1 inch long.—Dry gravelly ground. Fl. April, May.

(118) **Poa.** MEADOW-GRASS.

* Rootstock tufted.

P. annua : annual, stems spreading below, 3-12 inches high ; panicle loosely spreading, 1½-3 inches long ; florets not webbed.—Meadows and cultivated ground everywhere. Fl. all the year.

P. bulbosa : stems swollen and fleshy or bulbous at the base, 6 inches high ; panicle ovate or oblong, close, about an inch long ; florets webbed at the base.—Sandy sea shores. Fl. April, May.

P. trivialis : stems 1-2 feet high, and as well as the leaf-sheaths roughish to the touch ; panicle spreading 3-6 inches long ; spikelets ovate of 2-3 acute webbed florets.—Meadows and pastures, common. Fl. May and June.

** Rootstock creeping.

P. pratensis : stems 1-2 feet high, and as well as the leaf-sheaths smooth ; panicle spreading 2-3 inches long ; spikelets ovate or oblong, of 3-4 webbed florets.—Meadows and pas-tures, common. Fl. May and June.

SUMMER FLOWERS.

—◆—

"Behold those brightly-tinted Roses,
How fresh the blush upon their silken leaves,
With the clear dewdrop glancing in the sun
As bright as diamond, with its ray intense,
Shining the most when most 'tis shone upon!
Does it not glad thy heart to look on them?
Are they not glorious ministers of Heaven,
Shedding their sweetness on the summer earth
To tell us of His love who sent them here?"

Countess of Blessington.

ILLUSTRATIONS.

THE Rose is *par excellence* the flower of summer. Summer indeed " brings the Roses back to us, and their rich fragrance loads the golden air," as many a wanderer through our rural lanes and bye-ways can testify. This is true of the Wild Roses —to say nothing of our garden beauties.

The Dog Rose* is one of the commonest of our wild Roses, being found in almost every hedge and thicket: " vaulting o'er banks of flowers." It forms a somewhat straggling bush, armed with strong curved prickles; and the branches, which are furnished with elegant pinnated leaves, bearing stipules or

* *Rosa canina*—Plate 11 D.

little wing-leaflets on either side at the base, are profusely decorated with spreading five-petaled flowers.

The flowers—those of the single Dog Rose—consist of an egg-shaped smooth-surfaced calyx-tube contracted towards the tip, and dividing into a spreading limb of five, often unequal, sometimes lobate and almost leafy, segments. The five petals are obcordate and generally pink, and within them the numerous stamens are inserted around the mouth of the calyx-tube, which latter encloses the numerous one-seeded carpels. This part, that is, the calyx-tube, enlarges and acquires as it ripens a certain degree of succulency, becoming converted into the bright scarlet hips so commonly seen on wild Rose bushes, and with which the branches are often rendered gay after leaves and blossoms have passed away.

There is considerable resemblance in the general aspect of many of the plants of the Rosaceous family, which this Dog Rose may be taken to illustrate, and certain Ranunculaceous plants, of which the Buttercup is typical; but they are really very different both in structure and properties, and may always be known by the position of the stamens, which in the Ranunculaceous plants are set on quite distinct from the floral envelopes—that is, calyx and corolla, but in the Rosaceous family are set on to the calyx itself.

> " 'There is no flower that blows '
> —Such are the words of song—
> 'So lovely as the Rose ;'
> Nor thus, perchance, we wrong
> The fairest blossoms that around may throng.

> " O'er hedgerow green, in spring,
> Where the mild breezes play,
> The pale Wild Roses fling
> Their lightly wreathing spray,
> And show their petals fair by rude or lonely way.

" When shineth summer light—
　In every garden glade,
　Flush forth the blossoms bright :
　And sweetest is the shade
Where clustering roses twined a bower of rest have made.

" And oft some lovely Rose
　Doth linger last of all,
　When wind of autumn blows,
　When frosts of autumn fall ;
Like memory sad and sweet, past summer to recall."

The Rose belongs to the perigynous division of Calycifloral Exogens, the peculiarities of which have been already explained.

Reverting to something like systematic order while giving a brief explanation of our Illustrations of Summer Flowers, we commence again with the Thalamifloral Exogens, a group of plants which, it will be remembered, have the petals all distinct, and the stamens hypogynous or inserted beneath the young seed-vessels, and distinct from the surrounding parts of the flower.

During the summer months, an abundance of common Ranunculaceous plants will be found in blossom in meadows and waste places; but as the peculiarities of this family have been already pointed out, we must pass on to another allied group of Thalamiflores, the Berberidaceous plants, which are represented by the common Barberry.* This is a deciduous shrub, having the branches armed with long three-lobed thorns at the base of the tufts of leaves, which are alternate or clustered, oblong-ovate, and sharply toothed. The flowers grow from the leaf-axils in short drooping racemes, and are yellow, with a disagreeable smell. They consist of a small calyx of six sepals; a corolla of six concave petals, each having

* *Berberis vulgaris*—Plate 7 A.

two glands at the base; six distinct stamens, which have the peculiarity of opening by valves; and a peltate stigma. The flowers are succeeded by bright orange-red, oblong, succulent berries, which are sometimes used for making preserves. The recurved anther-valves are very peculiar.

In lakes and slowly moving waters, will be seen floating the broad peltate roundish heart-shaped glossy leaves of the White Water Lily,* and amongst them, just lying above the water surface, the beautiful rosette-like flowers, which are among the most lovely of our wildings: it is indeed

> "A water-weed, too fair
> Either to be divided from the place
> On which it grew, or to be left alone
> To its own beauty."

This "water-weed" represents the Nymphæaceous family, a small assemblage of Thalamiflores, all having aquatic predilections. The plant has a thick rootstock, which is submerged, rooting into the mud, and producing annually the broad floating leaves which are attached by a slender petiole or stalk fixed near their centre and which elongates sufficiently to bring the leaves to the surface. The flowers also have separate stalks, which elongate to a sufficient extent to elevate the chaste and noble blossoms above the water. Thus, in all her classic purity, "upon her throne of green, sits the large Lily as the Water's Queen," and there "she seems, all lovely as she is, the fairy of the stream." The blossoms are rather peculiar, having numerous sepals and petals and stamens all present; but the transition from one series of organs to the other is so gradual, that it is difficult to determine where the one ends and the other begins. The calyx is usually set down as consisting of about four sepals, like the outer petals in form, but green

* *Nymphæa alba*—Plate 7 B.

exteriorly; then come the numerous petals, set in several rows, and passing gradually into the also numerous stamens, the anthers of which are adnate or fixed by their whole length. The ovaries too, are numerous, imbedded in the thick receptacle, forming separate cells radiating from a common centre, while the petals and stamens are attached to the outside of the receptacle nearly as high as the top of the cells.

One interesting peculiarity has been recorded of the White Water Lily, namely, that its flowers expand only in bright weather, and close towards evening, when they either recline on the surface of the water, or sink beneath it; so that it has been called "day's own flower," and described as "drooping its head beneath the waves," there "watching, weeping, through the live-long night, impatient for the dawn."

From this water-fairy let us pass to the flaunting Poppy* of the cornfields—type of the Papaveraceous family. Here we have an erect branched annual plant, with hairy leaves and stems, the leaves deeply cut in what is called a pinnatifid manner, and the flowers large and specious, but fugacious. The calyx consists of a pair of hairy sepals, quickly deciduous, being pushed off by the expansion of the corolla, which latter consists of four broad, spreading, nearly equal, overlapping bright scarlet petals, that are crumpled not folded in the bud. The ovary, which is somewhat top-shaped, with the stigma radiating on its flat sessile disk, is imperfectly many-celled and filled with small seeds, which escape by means of a series of apertures beneath the rim of the stigmatic disk.

In waste places, especially in limestone districts, will be found the spike-like inflorescence of the Wild Mignonette,† a member of the Resedaceous family, forming a dwarfish herb,

* *Papaver Rhœas*—Plate 7 C.
† *Reseda lutea*—Plate 7 D.

with many branches clothed with variously divided leaves, and terminating in long erect racemes of greenish-yellow dull-looking flowers, which have no pretension to beauty, and are entirely without the delicious fragrance which in the Mignonette of the gardens atones for the absence of graceful form or attractive colouring. Here there are from four to six sepals; as many petals, of which the lowest is entire or two-cleft, and the others irregularly divided; an indefinite crowd of stamens; and an oblong ovary, having three short apical teeth.

The common Rockcist* represents another Order of regular-flowered Thalamiflores,—the Cistaceous family. It is a low, diffuse-growing, slender, shrubby plant, having its branches furnished with small opposite oblong leaves, which have each a pair of minute stipules at their base; and bearing loose terminal racemes of pretty yellow flowers, which are however of very short duration. These flowers consist of five sepals, the two outer of which are smaller than the rest; five broad obovate spreading petals; numerous stamens; and a one-celled capsule, which opens in three valves. The species of *Helianthemum* are very pretty summer-flowering plants, of which many forms are cultivated on sunny rockwork in gardens.

On dry hilly pastures and peaty fens, during the summer months, the Milkwort† may very frequently be gathered. This little herb represents a genus of exotic shrubs, much more showy than itself, and also the type of the irregular-flowered Polygalaceous family. Its stems are slender, diffuse, generally a few inches long, with small leaves, obovate below and lanceolate above. The flowers, which form terminal racemes, are usually of a bright blue, and very pretty. The calyx consists of five sepals, of which the two innermost,

---

* *Helianthemum vulgare*—Plate 8 A.

† *Polygala vulgaris*—Plate 8 C.

larger than the others, petal-like and elegantly veined, are
commonly called wings. The corolla is very irregular; the
petals much smaller than the sepals, the two lateral ones oblong-linear, the lowest keel-shaped tipped with a little crest,
and all more or less united with the stamens, which form two
parcels, each with four anthers. The ovary and capsule are
both flat, and contain a single seed in each of the two cells.

Another small family, that of the Frankeniaceous plants, is
represented by the Sea Heath or Common Frankenia * a diffuse-growing much-branched perennial, with the small leaves
crowded in little opposite clusters along the branches, the
flowers very few, pink, sessile among the upper leaves, and
regular in structure. The sepals are combined into a tubular
calyx with four or five teeth; the petals, four or five in number, have each a long claw, and a spreading limb; there are
four or five stamens alternating with the petals and usually
two or three opposite to them; the stigma is three-cleft; and
the capsule, which opens in two or three or four valves, contains very small seeds. It is found on maritime sands, and
salt marshes on the south-eastern coasts.

A comprehensive family of Thalamiflores, that of the Caryophyllaceous plants, is illustrated by the Pink,† a plant found
in a half-wild state in a few localities in England, but probably
an escape from cultivation. This handsome plant, the original
doubtless from which some at least of the favourite double
florists' Pinks have descended, is a perennial, with peculiar
rigid grassy glaucous leaves, and a flower-stem 6–12 inches
high, branched, and bearing from two to five pinkish fragrant
aromatic flowers, the calyx of which is tubular, with five teeth,
and clasped at the base by two opposite pairs of green scales,

* *Frankenia lævis*—Plate 8 D.
† *Dianthus plumarius*—Plate 9 A.

which are very much shorter than the tube. The corolla, which is regular, consists of five long-clawed spreading petals which are deeply multifid at the margin ; and there are ten stamens, and two long recurving styles.

Another plant of the same family, a common cornfield weed, and at the same time a very pretty flower, is that known as the Corn Cockle.* This is a tall erect-growing annual, slightly branched, hairy, with long narrow leaves, and large showy regular flowers, on long leafless peduncles, borne on the upper part of the stems. These flowers have a tubular calyx with five long linear lobes projecting much beyond the petals, which latter are broad undivided and spreading, tapering into a long claw at the base, and forming together a large inodorous flower, of circular outline, and of a pale reddish-purple colour. There are ten stamens and five or rarely four styles. The capsule, which opens in five teeth, contains numerous seeds.

The Malvaceous family, another group of the same great subdivision as the foregoing, is represented by the Common Mallow † found abundantly in waste places. This plant is a biennial, that is, of two seasons' duration, and has tall, erect or ascending branching stems, clothed with roundish slightly-lobed leaves, and bearing axillary clusters of reddish purple regular flowers of the true mauve colour, these being the plants whence the French name of that fashionable hue is derived. These flowers will be found to have, external to the five-lobed calyx, an involucel or little involucre of three small bracts inserted on the lower part of the calyx ; within comes the whorl of five wedge-shaped petals notched at the end, and the staminal column, which is unlike anything previously described. It consists of the filaments of the stamens united into a tube around

---

* *Lychnis Githago*—Plate 9 B.
† *Malva sylvestris*—Plate 9 C.

the pistil, the lower part of the tube being bare, like a shaft, and
the upper decorated with the numerous one-celled anthers,
arranged around it in several series, above which the cluster of
ten styles projects. The fruit consists of about ten carpels united
into a flattish disk or ring, seated within the persistent calyx.

The Linden, or Lime-tree,* found in woods over the greater
part of Europe, and wild in some parts of England, represents
a large tropical Order of plants, the Tiliaceous family, which is
related to that of Malvaceous plants, but differs in having the
two-celled anthers free, or, at least, not consolidated into a
column. The Lime forms a large and handsome deciduous
tree, furnished with stalked broadly heart-shaped leaves, extended into a point. The inflorescence consists of small cymes
produced on the current year's shoots and hanging down
on axillary peduncles among the leaves, the peduncle being
winged halfway up by a long leaf-like bract, with which it is
so far confluent. The flowers themselves are small, greenish,
very sweet-scented, consisting of five small sepals and petals,
and numerous stamens, which cohere at the very base into several clusters. The globular ovary becomes a small nut, containing one or two seeds.

The Hypericaceous family is confined among British plants
to the single genus *Hypericum*, called St. John's Wort, of which
the Small Upright St. John's Wort † is a very good representative. This little plant grows with stiff erect slender stems,
one to two feet high, bearing a few short lateral branches, and
furnished with opposite cordate leaves, clasping the stem at the
base; those of the branches are smaller and narrower, but all
are marked with pellucid dots, which become evident on hold-
ing up the leaf to the light. The flowers form an oblong

* *Tilia europæa*—Plate 9 D.
† *Hypericum pulchrum*—Plate 10 A.

panicle at the top of the stem, and are yellow, with five broad obtuse sepals united nearly to the middle, and fringed at top with black glandular teeth; five oblique but regular petals; an indefinite number of stamens, clustered and united at the base into from three to five parcels, which is one of the most particular marks of the Order. This little plant is frequently found in dry woods, and on open heaths and wastes.

The Geraniaceous family finds a representative in the Meadow Crane's-bill,* a handsome species, often cultivated in gardens, but also met with in the wild state in meadows, woods, and thickets. It is a perennial herb, of vigorous habit, with five-parted leaves, the lobes of which are multipartite, with numerous acute segments. The flowers are large, circular, bluish-purple, and loosely panicled. They consist of a five-leaved calyx; a five-petaled regular corolla, the petals of which are broad and obovate, with ciliated claws; ten stamens of unequal length, having the filaments flattened out in the lower part; and a five-lobed ovary, with elongated styles which are joined to a central axis, from which they partially separate when the ripe fruit breaks up. Before this takes place, the fruit has a long tapering beak, which has suggested for the genus the name of Crane's-bill. The *Pelargonium* is often falsely called *Geranium.*

Extensively cultivated for its fibre and its seeds, but sowing itself readily as a weed of cultivation, the Flax or Linseed † may be occasionally met with in a semi-wild condition. This belongs to the Linaceous family, a regular-flowered polypetalous group, consisting of herbs and undershrubs, having entire or simple leaves. The Common Flax is a tall erect annual, with smooth slender stems, slightly branched towards the top,

* *Geranium pratense*—Plate 10 B.
† *Linum usitatissimum*—Plate 10 C.

occasionally producing a few spindly branches from the base, and everywhere furnished with narrow-lanceolate leaves. At the top of the stem is produced a loose leafy corymb of bright blue flowers, which have five ovate-lanceolate three-nerved acute sepals, five obovate spreading regular petals, and five stamens united below into a hypogynous ring surrounding the roundish ovary, which is crowned by five styles. The capsule is globular or slightly depressed, really five-celled, but the cells being divided into two by a nearly complete partition, it is apparently ten-celled. The stems of this plant furnish the valuable flax fibre, and its seeds yield linseed oil.

The common Tamarisk * is now found in several parts of the southern coast of England, apparently established, though it is probably only an introduced plant. It is a shrub of maritime habitat, and being one of the few which thrive in the vicinity of the sea, is very often planted in such situations. It forms a very elegant shrub of five or six feet in height, with slender erect twiggy branches, covered, in place of leaves, with what have the appearance of scales—little green imbricating bodies lying close over each other, with a loose spur at their base. The flowers are pinkish, and grow in little spikes or racemes. They are small, and have a calyx of four or five sepals, a corolla with an equal number of petals, as many or twice as many stamens, and a free ovary with three styles. The seeds are each crowned with a tuft of cottony hairs.

One more illustration of the Thalamiflores, a flower of very irregular structure, must be noted. In moist shady places, chiefly in the north of England and in Wales, may be found in the height of summer, a tall annual plant, with succulent branching stems, swollen at the joints, and bearing stalked ovate pale-green flaccid leaves. This is the Touch-me-

* _Tamarix anglica_—Plate 11 C.

not, or Yellow Balsam,* which will serve as a representative of the Balsaminaceous family, a group of plants of which the species are chiefly tropical. The flowers grow on slender peduncles from the axils of the leaves; they are large and showy, yellow, spotted with orange-red, and curiously spurred behind; one or two of them are perfected on each peduncle, which besides bears a few others which are minute and imperfect, but producing the essential organs are those which usually mature seed. The perfect flowers are very curious in structure. The sepals and petals are all coloured, and usually consist of six pieces, of which three represent the calyx, and three the corolla. The three outer or sepals consist of two small flat opposite pieces, and a third much larger, which is really the upper sepal, but, from a twist in the stalk, hangs lowest; this sepal is hood-shaped, and prolonged behind into a curved conical spur. The petals are equally without symmetry; the lower one, or that which from the twisting before mentioned actually stands uppermost, is much smaller than the other two, but still broadish and concave, while the other two, which stand innermost of the whole six, are large broad and irregular in shape, oblique, and more or less divided into two unequal lobes. These flowers have five stamens whose anthers cohere round the pistil, and five minute sessile stigmas. The fruit is a long pointed capsule, which when ripe bursts open into five valves with great elasticity, the valves suddenly curling up, and the seeds being hurled to some distance. Hence the plant has received its generic name of *Impatiens,* and also its specific name, *Noli-me-tangere,* or Touch-me-not.

We next come to the Calycifloral division, in one portion of which, now to be noticed, it will be remembered the stamens are generally perigynous, and the petals distinct.

* *Impatiens Noli-me-tangere*—Plate 10 D.

Of these we find flowering during the early summer, a shrub growing in hedges and thickets, known as the Spindle Tree,* and belonging to the Celastraceous family. It is rather an insignificant plant, except when in fruit, but the curious form and bright colour of this fruit render it later in the summer a rather conspicuous object. It is a smooth shrub, of about five feet high, with ovate-oblong or lanceolate pointed deciduous leaves, and axillary cymes of small green flowers ; these have a flat calyx of four or five short lobes, and an equal number of larger petals, an equal number also of stamens alternating with the petals and united with them on a slightly thickened disk which covers the base of the calyx. The ovary is immersed in this disk with a short protruding style, and becomes a four-angled (sometimes three- or five-angled) red capsule, which opens, when ripe, at the angles, and exposes the seeds enveloped in a bright orange-coloured arillus—the arillus being a part of the fruit corresponding with what is known as mace in the fruit of the nutmeg.

Another unattractive Calyciflore blossoming early in the summer is the Common Buckthorn,† found occasionally in hedges and bushy places. This belongs to the Rhamnaceous family, and is a shrub or small tree, with spreading branches, which sometimes become spiny. It has ovate toothed leaves marked by a few prominent veins, mostly originating below the middle. The flowers are small, green, staminate or pistillate, clustered in the axils of the leaves ; they have each four or five small calyx-teeth, and within these as many still smaller petals. The staminate flowers have an abortive ovary, broader petals, and four or five stamens alternating with the calyx-teeth, and inserted on a disk which lines the base of the calyx.

* *Euonymus europæus*—Plate 11 A.
† *Rhamnus catharticus*—Plate 11 B.

The pistillate flowers have narrow petals, the rudiments of stamens, a deeply four-cleft style, and each a free three- or four-celled ovary, which becomes one of the roundish nearly black berries. Though itself an unattractive plant—not however without its uses—the Buckthorn represents the handsome *Ceanothus* shrubs grown in our gardens.

The great family of the Leguminous plants, or Pod-bearers (also called Papilionaceous, from the flowers, by a stretch of imagination, being taken to represent a butterfly), is illustrated by the Meadow Vetchling,* a weak branching perennial herb, found abundantly in moist meadows and pastures. It has smooth, angular, straggling or half-climbing stems, from one to two feet long, furnished with branched tendrils, each bearing a pair of narrow lance-shaped leaflets, and furnished at the base with a pair of large broadly lanceolate sagittate stipules. From the axils of these tendril-bearing leaves grow the elongated flower-stalks, supporting a short raceme of from six to ten yellow flowers; they consist of a small five-toothed calyx and a five-petaled papilionaceous corolla, which is made up of a large upper petal, called the standard, two narrow lateral ones called the wings, and two other narrow ones, more or less united along the lower edge, called the keel. These flowers are succeeded by small smooth pods. The plant represents the Sweet Pea and Everlasting Pea cultivated for ornamental purposes in almost every flower-garden.

Of the Lythraceous family we have an example in the beautiful Purple Loosestrife,† a willow herb, as it might well be called, whose tall upright stems, supporting long spikes of rich purple flowers, may be seen adorning the sides of wet ditches, and the banks of streams. This very showy plant is

---

* *Lathyrus pratensis*—Plate 12 D.

† *Lythrum Salicaria*—Plate 12 A.

a perennial with annual stems two to four feet or more in
height, and furnished with lance-shaped entire leaves growing
in opposite pairs, or sometimes in threes. These leaves become
smaller in the upper part of the stem, where, in their axils,
the flowers appear, in masses of crowded whorls, the whole
forming a whorled spike, more or less leafy below. The individual flowers consist of a small tubular cylindrical calyx,
having from eight to twelve teeth, arranged in two series, the
inner ones being broader than the outer; a corolla of from
four to six large showy petals, inserted in the upper part of
the calyx-tube; twelve stamens, inserted near the base of the
calyx-tube; a two-celled free or superior ovary, surmounted
by a filiform style, and growing into a many-seeded capsule.

The Water Chickweed,* an insignificant succulent annual,
found in springy places and on the edges of rills, represents
the Portulacaceous or Purslane family, another group of the
Calyciflores, of which the genus *Calandrinia* of our flower-
gardens affords much handsomer illustrations. This little plant
grows in tufts a few inches high, and has small obovate or
spathulate leaves, and solitary flowers in the axils of the upper
leaves, the flowers being minute, with a calyx or cup of two
sepals, and a corolla of five white petals united into one but
split open in front. There are three stamens, three stigmas, and
a capsule opening in three valves and containing three seeds.

In the Wall Pepper † we have an illustration of the Crassulaceous family, a Calyciflorous group, containing many very
beautiful species of flowering plants familiar in gardens under
the names of *Sedum, Sempervivum, Crassula*, etc. The present native species is a common perennial British plant, found
on walls and on rocks, and in stony and sandy places in a

* *Montia fontana*—Plate 13 E.
† *Sedum acre*—Plate 13 C.

wild state, and not unfrequently in gardens upon artificial rockwork. It forms a close spreading flat tuft or patch of the brightest green when not in flower, and during the flower-ing season is literally covered with its bright yellow star-like blossoms. The leaves are small and thick, ovoid, spurred at the base, those of the barren shoots usually imbricated in about six rows. The flowers are wholly bright yellow, in short terminal cymes; they have five short sepals, and five longer distinct petals which spread out in the form of a star, ten stamens, and five carpels. The whole family is distinguished for the succulent or fleshy character of its leaves or stems.

In boggy places on open heaths, growing amongst sphagnum moss, may generally be found a profusion of little rosulate plants furnished with curious glandular hairs. They are the Sundews, and represent the family of Droseraceous plants, which some learned botanists place among the Thalamiflores, and others along with the Calyciflores, according as the inser-tion of the stamens is regarded as hypogynous or perigynous. Its position is thus quite an unsettled point. We here follow Mr. Bentham in regarding it as associated with *Saxifraga* and *Parnassia*. The Common Sundew * has a short slender rootstock, encircled by round or orbicular leaves attached by long stalks, and covered on the upper surface with long red viscid hairs, each hair bearing a gland at the top. In the centre rise two or three slender flower-stems supporting an undivided or once-forked one-sided raceme of small whitish flowers, which have five small sepals, five somewhat larger petals, and five stamens,—the latter being considered almost or by some bota-nists quite hypogynous. They are very interesting little plants, and are called Sundews from the little glands secreting a pel-lucid fluid which sparkles like dew-drops in the sunshine.

* *Drosera rotundifolia*—Plate 8 B.

We come next to a group of Calyciflores having this diversity of structure from the preceding—the petals and stamens, instead of being perigynous, are what is called epigynous, or, in other words, they appear to grow from the top of the ovary. We must describe a few illustrations of the group.

The Onagraceous family, known among the British Calyciflores with an inferior ovary, by the parts of the flower being all in twos or fours, is very well illustrated by the Great Willow Herb,* a handsome perennial, found by the side of ditches and watercourses: a common plant, having stout branched hairy stems, three to five feet high, furnished with lanceolate leaves clasping the stem at the base, and toothed along the margin. The flowers grow from the axils of the upper leaves, and are stalked, the calyx and corolla being elevated on a long slender quadrangular ovary which looks like a thickened stalk. The calyx is divided into four small teeth, and the corolla consists of four broad deeply-notched petals, forming a large handsome flower of a pinkish rose-colour, within which are eight stamens and a deeply four-lobed stigma, all these growing at the apex of the long narrow four-angled ovary, which becomes a hairy four-celled four-valved capsule, containing numerous small seeds, each crowned by a tuft of hair.

The Mare's-tail † represents a small group of insignificant plants, sometimes considered as a separate group, called Haloragaceous plants, and sometimes regarded as a subdivision of the Onagraceous or Evening Primrose family. It is a water plant, found in shallow ponds and ditches, and has erect annual simple stems springing from a perennial rootstock. These stems are furnished with whorls of from eight to twelve linear entire leaves, in the axils of which grow the minute flowers,

* *Epilobium hirsutum*—Plate 12 C.
† *Hippuris vulgaris*—Plate 12 B.

which are entirely without petals, and have a **scarcely per-ceivable** calyx, a single stamen, one subulate style, and a single ovule. The plant has considerable superficial resemblance to the Equisetums or Horsetails, but has no affinity with them.

In the Red or Common Bryony,* we have a British example of the Cucurbitaceous family, to which the Melon and Cucumber belong. It is a scrambling plant, as are most of the family, trailing over the surface of the ground if no support is at hand, or clambering by means of its tendrils over any adjacent herbage. In the southern parts of England, it is common in hedges and thickets. It has a thick tuberous perennial root-stock, from the crown of which the annual stems are produced. These are hairy, branched, very much elongated, and furnished with broadish leaves, divided into five or seven angular lobes, of which the middle one is the longest. The flowers are produced separately by young plants, the staminiferous on one and the pistilliferous on another; but as they become older, both forms, though still separate, are borne by the same plant. The staminiferous flowers grow several together in long-stalked racemes from the leaf-axils, while the fertile or pistilliferous ones are on short stalks; both have a five-toothed calyx, and five petals just united at the base into a single five-lobed corolla, which is inserted in the margin of the calyx. The stamens are combined into three sets, of which two are double, comprising two stamens each, and one single. The pistilliferous flowers are succeeded by small scarlet berries, which are fetid when bruised. In the younger stages of growth, the fruits, in that state called ovaries, have the flowers growing from their apices, the same as may be seen in a young Cucumber.

In the family of the Umbellifers, or Umbel-bearing plants, we have further illustrations of this epigynous group of Calyci-

---

* *Bryonia dioica*—Plate 12 E.

flores. One of them, called the Hemlock Dropwort,* a tall perennial herb, growing commonly in wet ditches and by the sides
of streams, is a virulent poison. Its root-fibres form thickish
elongated tubers close to the stock ; its stems are branched and
grow four or five feet high ; its leaves are twice or thrice pinnated with lozenge-shaped or broadly wedge-shaped leaflets,
deeply cut into three or five lobes. The flowers grow on long
terminal peduncles, and form umbels as in the rest of this
family : that is to say, a number of divisions start off from a
common point, similar to the rays of an umbrella, and then
bear each of them a smaller tuft which grows out on the same
plan, all the little divisions being here terminated by a flower.
This double branching constitutes a compound umbel, whereas
if the flowers terminate the first series of ramifications, the
umbel is simple. The branches of the umbels are called rays,
the secondary heads partial umbels, and either the general
umbel or the partial umbels or both, may have at the point
whence they start a whorl of usually narrow leaflets, called an
involucre. In the present instance, which is that of a coarse-
growing plant, the general umbel produces from fifteen to
twenty rays, two inches long or more, and surrounded by
an involucre (rarely wanting) of few linear bracts, whilst the
partial umbels have an involucre of several bracts. The
flowers are small, whitish, those at the circumference stalked,
and mostly but not always barren, while the central ones are
fertile and almost sessile. They have a stiff leafy calyx of five
short teeth ; and a corolla of five notched petals with an in-
flected point; five stamens alternating with the petals, and
with them inserted round a little fleshy disk which crowns the
ovary, and from the centre of which arise two styles. The
fruit is cylindrical or oblong, crowned by the stiffened styles,

* *Œnanthe crocata*—Plate 13 A.

and collected into close hard heads, each consisting of two carpels—each having the appearance of a seed and being called
a mericarp—and marked outside with five prominent ridges,
with a vitta or oil-cyst in each furrow. When separate, the
two carpels of each fruit have a general semicircular section ;
a section of the whole showing an oval figure. Such fruits
are said to be laterally compressed, and they separate across
their narrow diameter. The examination of the fruit in the
ripe state is essential to a thorough knowledge of this Umbelliferous family. A thin cleanly cut horizontal slice examined
by a magnifying glass, will show both the outline, and the
ridges and oil-cysts. If very hard, the fruits may be soaked
in hot water before cutting them.

In another example of this large and important family,
the common Parsnip,* we have, instead of a poison, a bland
and nutritious esculent—not indeed in the wild form of the
species, found abundantly in chalky fields and thickets, but
in that form of it which has been produced by cultivation.
Here too we have a coarse-growing herb, but of annual or
biennial duration only, furnished with a long tap-root, which
has been improved into the edible Parsnip of our gardens.
The stem is two or three feet high, furnished with pinnate
leaves, having from five to nine large, sharply toothed, more
or less deeply lobed segments. The umbels of flowers are
compound, of from eight to twelve rays, and usually without
involucres. The flowers themselves are yellow, but otherwise
very much like those just described. The fruits however are
very different, being flattened from front to back, and broadly
winged, hence they appear flat and oval ; they have three fine
scarcely prominent ribs, and a vitta in each of the interstices,
and they separate along their greatest diameter, forming two

* *Pastinaca sativa*—Plate 13 B.

H

very thin scale-like bodies. Besides the Parsnip, this family comprises some of our most useful esculents, as the Carrot, Celery, Parsley, Fennel; many valuable medicines, as Asafœtida, Opopanax, Ammoniacum ; and some of the most virulent of poisons, as Hemlock and Cowbane.

The Dogwood, or Common Cornel,* another of the epigynous Calyciflores is a deciduous shrub of five or six feet high, found in hedges and thickets in the southern parts of England. It has opposite, broadly ovate, stalked leaves, which are silky with closely appressed hairs while young, and the numerous flowers form terminal cymes of a couple of inches across, and consist of a four-toothed calyx on the summit of the ovary, a four-petaled corolla of a dull white, and four stamens. The flowers are succeeded by globular, almost black, very bitter drupes. This plant affords a capital illustration of a cymose inflorescence.

From these we pass to the Monopetalous series, commencing with a subdivisional group, in which the one-leaved corolla is also epigynous, bearing the stamens.

Of this series we have an example in the Common Honeysuckle, or Woodbine,† which also illustrates the Caprifoliaceous family. The Woodbine is common throughout Britain, in woods, thickets, and hedgerows, and forms a woody climber, scrambling over the bushes and trees to a considerable height :

"Wound on the hedgerow's oaken boughs,
    The Woodbine's tassels float in air ;
And, blushing, the uncultured Rose
    Hangs high her beauteous blossoms there."

The leaves of the plant are opposite, smooth above, and generally slightly hairy beneath, ovate or oblong, the lower ones

---

* *Cornus sanguinea*—Plate 13 D.
† *Lonicera Periclymenum*—Plate 14 A.

stalked, the upper ones sessile, but not united at the base as in our other native Honeysuckles. The flowers are stalk-less, forming close terminal stalked heads, yellowish, tinged externally with red, and, as every one knows, deliciously scented, the pale wan flowers, well compensating their sickly looks " with never-cloying odours." They consist of a calyx with five small teeth; a monopetalous corolla with a narrow elongated tube, and a two lipped limb, of which the upper lip has four lobes and the lower one is entire; five stamens; a filiform style with a capitate stigma; and a two- or three-celled ovary growing into a small one- or few-seeded berry.

The Ladies' Bedstraw* is one of a genus of weak straggling herbs, representing the Galiaceous family; which itself is sometimes, under the name of Stellates, referred as a section to the Rubiaceous or Cinchonaceous family. The Galiums have spreading or straggling stems, and narrow leaves placed in whorls around the stem. The species here represented is a smooth branching herb, with decumbent or ascending stems, which have the small linear leaves in whorls of six or eight, and which terminate in oblong panicles of bright yellow flowers. These have the calyx completely consolidated with the ovary, without any visible border, and a rotate four-lobed corolla, of which the tube is hardly perceptible; four stamens alternating with the lobes of the corolla; and a style cleft in two, with a capitate stigma on each branch. The fruit is small, dry, smooth, two-lobed, with one seed to each lobe.

The Red Valerian,† another of the epigynous Monopetal﹦, represents the Valerianaceous family. This plant, a perennial herb, native of the southern parts of Europe, is now natu-ralized in many localities, on old walls and on chalky cliffs,

---

* *Galium verum*—Plate 14 B.

† *Centranthus ruber*—Plate 14 C.

as well as cultivated in flower-gardens. It forms a much branched, almost sub-shrubby, mass, one to two feet high, quite smooth, the stems furnished with broadish, ovate-lanceolate, slightly-toothed leaves, and terminating in close cymose panicles of red, rarely white, flowers. In these, the calyx at the time of flowering consists of a border rolled inwards and entire; the corolla has a slender tube, projected in the form of a spur at the base, and divided at top into five segments, in a somewhat two-lipped manner; and there is one stamen, and a slender style. The fruit is seed-like, and the incurved border of the calyx becomes, in the mature state, unrolled into an elegant feathery pappus. An approach towards the structure of Composite plants is very evident in these pappus-crowned fruits.

The Common Teasel,* which illustrates the Dipsacaceous family, has in outward aspect a kind of intermediate position between the Composites and Umbellifers, as seen in certain capitate-flowered plants of those families, such as *Echinops* and *Eryngium*. The free condition of the anthers, however, separates them from the Composites, and the opposite leaves and monopetalous corollas from the Umbellifers. The Teasel is a vigorous growing, erect, branched biennial, of some four or five feet high, armed on the stems, midribs, flower-stalks, and involucres with numerous short prickles. The leaves are sessile, elongate lanceolate, toothed, opposite, the upper ones broadly connate, that is, joined together by their base. The flower-heads, which come on long stalks, have at their base a spreading involucre of from eight to twelve stiff narrow linear prickly leaves or bracts, and are at first ovoid, gradually acquiring a cylindrical form; the flowers are crowded over their surface, each standing in the axil of a scale which is rather longer than

* *Dipsacus sylvestris*—Plate 14 D.

the flowers, broad and hairy at the base, narrowed into a thin prickly point. The flowers are small, pale lilac, each inserted in a small angular involucel having the appearance of an outer calyx with a small thickened border; the calyx has a small cup-shaped border appearing above the involucel, and the monopetalous corolla is four-lobed and oblique. The flowers have four free stamens inserted in the tube of the corolla, and the ovary becomes a dry single-seeded fruit, crowned by the border of the calyx. The united bases of the opposite leaves of this plant form a hollow around the stem, in which water collects, and hence the plant was called *dipsakos*, or thirsty; hence also it obtained the name of Venus's Bath. Superstitious persons have fancied that the water thus collected from the rains and dews was good for bleared eyes.

In the Musk Thistle * we have a further example of the Composite family, not however belonging to the same divisions as the Daisy and Dandelion already noticed. These respectively belong to the Corymbiferous and Ligulate groups, but the Thistles belong to a very distinct subdivision of the family, which has been appropriately called Thistleheads. This latter is distinguished by the florets being all tubular, and by the style being swollen below its two arms. The Musk Thistle is a biennial plant, producing in the first year a spreading tuft of very pretty oblong-lanceolate sinuately pinnatifid leaves, the edges of which are prickly-toothed—and very sharply prickled too, like other Thistles. In the second year, the branching stem grows up two or three feet high, furnished with smaller pinnatifid prickly leaves, whose edges are decurrent, that is, running down the stem, forming narrow prickly wings. The flower-heads are terminal, large, drooping, and handsome. The involucres are globular, formed of numerous

* *Carduus nutans*—Plate 15 A.

closely imbricated bracts, and surrounding a thick receptacle bearing bristles between the florets; these bracts have a stiff narrow-lanceolate appendage, ending in a spreading or reflexed prickle. The florets are crimson, all equal, with a long slender tube, and five erect narrow divisions. The fruits, called achenes, are glabrous, with a pappus of simple hairs longer than the achene itself. The plant is found commonly in waste places in many parts of Britain, most frequently in the south.

In another subdivision of epigynous Monopetals, the stamens are inserted on the calyx, free from the corolla. This is the case in the Harebell Campanula,* a beautiful little wildflower found in hilly pastures and heathy wastes, and extending in its range from the Mediterranean to the Arctic Circle. The plant is a dwarf perennial herb, with a slender creeping rootstock. At the base of its stems, which vary from six to eighteen inches in height, the leaves are long-stalked, roundish or heart-shaped, and sparingly toothed; those higher up the stem are lance-shaped or linear and entire. The stems are variously branched, according to their luxuriance, forming usually a loose raceme or panicle of elegant drooping blue flowers: "little bells of faint and tender blue, which gracefully, bend their small heads in every breeze." Sir Walter Scott describes the elastic tread of his fair "lady of the lake" as not even disturbing the position of the slender Harebell:—

> " A foot more light, a step more true,
>   Ne'er from the heath-flower dashed the dew;
>   E'en the slight Harebell raised its head
>   Elastic from her airy tread."

These flowers consist of a calyx adherent to the ovary, and having five narrow spreading lobes; a regular bell-shaped five-lobed corolla, inserted within the lobes of the calyx; five

* *Campanula rotundifolia*—Plate 15 B.

stamens with distinct anthers, inserted within the base of the corolla but free from it; and a style cleft at the top into two or three stigmatic lobes. The capsule, which is of course inferior, is pendulous, and opens by short clefts near its base. It is a really elegant little plant.

The family of Ericaceous plants, represented by what Mr. Bentham calls the Scotch Heath,* is another Monopetalous group in which the stamens are free from the corolla, but in this case they are hypogynous. This very common and very beautiful plant, though it has been distinguished as the Scotch Heath, is by no means confined to that country, but ranges over the whole of Britain, and is common in Western Europe, covering immense tracts of moorland, which in the flowering season are sheeted with the rich purple of the heather-bells. It is a dwarf bushy shrub, of about a foot in height, clothed with fine linear leaves which are usually set three in a whorl, with clusters of smaller leaves in their axils. The flowers are numerous, in dense terminal elongated whorled racemes, and are furnished with a calyx of four small sepals; an ovoid corolla, with a contracted mouth, and four very small lobes or teeth; eight stamens enclosed in the corolla, and remarkable for their opening by two pores at the top, and also for having a small toothed appendage at the point where the anthers are joined to the filament; and a long style thickened at the stigmatic end. The fruit is a free four-celled capsule. It will be observed that the Heaths are plants in which the parts of the flower are made up in fours or in multiples of four; hence they belong to what are called tetramerous plants, those in which the parts are governed by the more usual number, five, being called pentamerous. It has already been pointed out that in the great group of Monocotyledons the number three is that which governs the part of the flowers.

* *Erica cinerea*—Plate 15 C.

We now come to a group of perigynous Monopetals in which the corolla is nearly or quite regular, and is moreover attached beneath the ovary. It is the perigynous condition of the flowers, the corolla bearing the stamens, which constitutes the chief technical distinction between these and the hypogynous-flowered Heaths previously described.

Of these, the Common Privet,* a member of the Oleaceous or Olive family, though by some referred to the Jasmines, is an example. This well-known shrub, employed in gardens in making hedges, and as an undergrowth in shrubberies, is found wild in hedgerows and thickets in the southern parts of England. It is a subevergreen shrub, of six or eight feet high, with long slender branches, which bear opposite lance-shaped leaves, and are terminated by short compact panicles of flowers, consisting of a small four-toothed calyx, a four-lobed short-tubed corolla, and a pair of short stamens. The flowers are succeeded by blackish globular berries, which are two-celled, with one or two seeds in each cell. The Ash is a somewhat anomalous member of the same family, wanting both calyx and corolla in the flowers of our native species, but yielding in some exotic kinds flowers which have a four-lobed calyx and corolla. The Lilac is another well-known cultivated plant of the same group.

The Gentianaceous family, already referred to, is another of these perigynous Orders, with the stamens growing directly on the corolla. Of this family we have another illustration in an aquatic plant found in ponds and still waters in many habitats, known under the name of Nymphæa-like Villarsia,† or sometimes under that of *Limnanthemum nymphæoides*, the Common Limnanth. This is a perennial plant, with long slender stems, which creep and root at the base, and becoming

---

* *Ligustrum vulgare*—Plate 15 D.
† *Villarsia nymphæoides*—Plate 16 A.

branched rise to the surface of the water. They bear a single leaf at each upper branch, and a terminal floating tuft of leaves and flowers. The leaves are long-stalked, and are smooth, roundish, deeply cordate, lying on the water's surface like those of the Water Lily, of which they are miniatures. The flowers grow several from the tuft, and are just elevated above the water. They are rather large, yellow, consisting of a five-cleft calyx; a nearly rotate (that is, short-tubed and spreading, so as to become wheel-like) plaited corolla, slightly fringed at the base within, and finely toothed round the margin; five stamens; and a five-cleft stigma. The capsule breaks open unequally, not having any regular valves.

The Greek Valerian, or Jacob's Ladder,* belongs to the same series, and represents the Polemoniaceous family, better known in gardens by the fine hardy herbaceous genus *Phlox*, and the annual *Gilia*. The Greek Valerian, common in cottage gardens, and found in some places apparently wild, is a perennial herb, producing at the base tufts of leaves, which are pinnate or divided into separate leaflets or little leaves, from eleven to twenty-one in number, of a lance-shaped figure, and quite entire. These leaves are what are called impari-pinnate, that is to say, pinnate with the leaflets in opposite pairs along the sides, and having an odd terminal one. The flower-stems, in vigorous plants reaching two feet in height, bear a few small pinnate leaves, and terminate in a kind of corymbose panicle of showy blue flowers, which have a five-lobed calyx, a regular rotate five-lobed purplish-blue corolla, five stamens, the filaments of which are dilated into hairy scales, and a simple style with three stigmatic lobes or stigmas. The fruit is a three-celled capsule, containing many seeds. This is a very widely diffused plant, being scattered "over the higher

* *Polemonium cœruleum*—Plate 16 B.

northern latitudes of Europe, Asia, and America, extending
also into the mountain regions of central Europe and Asia."

The Convolvulaceous family must be known to every one, being rendered familiar in our gardens by the fine exotic genus
Pharbitis, which contains the annual Convolvulus major of
the seed-shops; in our hedges by the Common or Larger
Bindweed (*Calystegia sepium*), whose large white bells are so
beautiful as almost to plead an excuse for the intrusion of
so really troublesome a weed as this is generally held to be;
and in our cornfields and waysides by the Lesser Bindweed,*
which is the subject of our illustration. This little prostrate
plant has a slender perennial rootstock, which creeps extensively underground; from this grow out numerous trailing
slender stems, which either spread on the surface or reach a
couple of feet or so in height by twining up the stems of the
corn plants and other herbage about them : "although the
field is bare, fringing the path, or scattered near, a few neg-
lected ears we find, round which Convolvulus hath twined."
They have alternate ovately arrow-shaped or sometimes hastate leaves, from the axils of which grow the usually two-
flowered peduncles. These flowers, which are fragrant, and
close at night and in dull weather, have five small blunt sepals;
a bell-shaped corolla an inch or more in diameter, beautifully
variegated with pink and white, or sometimes cream-coloured
nearly white; five stamens attached near the base of the corolla; and a simple style with two linear stigmatic lobes. The
ovary is two-celled,—the cells two-seeded,—and is surrounded
by an annular hypogynous disk having the appearance of a
fleshy ring around its base. This plant has the property of
expanding its gaily-coloured blossoms in the sunshine, and
closing them at the approach of night :—

* *Convolvulus arvensis*—Plate 16 C.

"As the sun retires in seas of gold,
Though yet thy twining stem, where'er it grows,
Hanging in rich festoons, no langour shows,
Thy fragile cup its beauties doth enfold,
To shun the damp and coldness of the night,
Until awakened by the orb of light."

Of the Boraginaceous family we have a familiar and lovely example in "that blue and bright-eyed flow'ret of the brook, Hope's gentle gem, the sweet Forget-me-not," or the Water Scorpion-grass,* so commonly met with " by rivulet or spring or wet roadside." The name ' Forget-me-not ' is said to have originated in this manner :—Two betrothed lovers were strolling by the banks of the Danube, on a pleasant summer evening in the flowery month of June, occupied in agreeable and affectionate converse, when they observed the pretty flower of the Water Scorpion-grass apparently floating on the water. The bride elect looked upon the flower with admiration, and supposing it to be detached, regarded it as being carried to destruction. Her lover, regretting its fate and wishing to preserve it, jumped into the river with this object; but as he seized the flower, he sank beneath the stream. Making a final effort, he threw the flower upon the bank, repeating, as he was sinking for the last time, the words Vergiss mich nicht. Hence the Germans have called the flower by a name which we translate ' Forget-me-not.'

"That name! it speaks in accents dear
Of love and hope and joy and fear ;
It softly tells an absent friend
That links of love should never rend ;
Its whispers waft on swelling breeze,
O'er hill and dale, by land and seas,
                    Forget-me-not !"

This pretty Myosotis is a perennial herb, with stems more or

* *Myosotis palustris*—Plate 18 A.

less creeping, and rooting at the base, and then ascending, from half a foot to a foot and a half in height, angular owing to the prominent decurrent lines which pass down from the margins of the leaves, and generally more or less pubescent or hairy, but sometimes nearly smooth. The leaves are oblong, bluntish, smooth, or with hairs appressed to the surface, and borne alternately along the stems. The flowers grow in incurved, one-sided, or, as they are technically termed, scorpioid racemes, which are very frequently forked, and gradually straighten as the flowers are developed, the lowest flower opening first, and the rest in succession towards the point. They consist of a small five-toothed or five-cleft calyx, and a salver-shaped monopetalous regular corolla, of which the tube is short as well as straight and narrow, and half-closed at its mouth by five short scaly appendages, while the limb is spreading and somewhat concave; inserted on the corolla-tube are five short stamens, and enclosed within its base is a deeply four-lobed ovary, having a simple style inserted between the lobes, which ultimately become hard shelly seed-like fruits, called nuts, surrounded by the persistent calyx. The flowers are of a pretty clear light or azure blue, with a golden-yellow centre.

It has been remarked of this plant, which constantly grows in wet places, that affectionate remembrance will always moisten the eye of sensibility, and hence no dry habitat can be allowed to the Forget-me-not. Mr. Lees relates of a nearly allied plant, the *M. repens,* which grows in swamps and quaking bogs, that it was once forcibly impressed upon his recollection, thus: perceiving it blooming in the midst of a bog, on the bleak deceptive sides of Plinlimmon, he dashed after it, but received only a *cool* reception from the beauty, though his knees bent before her dripping shrine, and after all he retired with but a very inadequate specimen of the favours she had, at first, appeared so disposed to offer.

Another regular-flowered group of perigynous Monopetals is the Solanaceous family, our illustration of which is the Bitter-sweet,* a deleterious plant, for which it is to be regretted that Mr. Bentham has used the name Deadly Nightshade, thus diverting from the much more virulent *Atropa Belladonna*, to which this premonitory title properly belongs, the caution which is necessary to prevent poisoning, by means of its really tempting-looking, but deadly cherry-like black berries. The Bitter-sweet is a shrubby plant, with straggling branches, often growing to a considerable height, but being killed back for some distance by the frosts of winter. These bear alternate leaves, which are of variable shape, sometimes ovate or ovate-lanceolate, broadly cordate at the base and entire, sometimes with the base angular on one side or unequally on both sides, or sometimes with a small lobe or segment on one or both sides at the base; they are also sometimes quite smooth and sometimes downy. The flowers are produced in loose cymes, terminating short lateral peduncles, and they consist individually of a five-toothed calyx, a rotate five-lobed star-shaped purple corolla, and five stamens which are united into an erect cone around the simple prolonged style, each anther opening by a small pore at the top. The fruit is a small roundish red berry, containing several seeds.

The Plumbaginaceous family, represented by the Common Thrift,† belongs to this same regular-flowered series. This well-known plant, a sea-side resident, and frequently used as an edging plant in gardens, is a perennial of tufted habit, producing numerous narrow-linear, almost grass-like leaves,

* *Solanum Dulcamara*—Plate 16 D.
† *Armeria maritima*—Plate 18 C.

from among which the flowering stems, simple and leafless, grow
up to the height of from three or four to six or eight inches,
and terminate in a round head of numerous flowers, these
flowers being intermixed with scarious or dry membrane-like
scales, of which the outer series form themselves into a kind
of involucre, and the two outermost of all are lengthened below their insertion, so as to form a sheath around the upper
part of the stalk. The flowers are pink, sometimes varying
to white or deep rose-pink. They have a tubular funnel-
shaped calyx, of a dryish scarious texture, with a petal-like
border crowned by five short slender teeth, and a five-lobed
corolla, of which the lobes are scarcely united in the lower
parts, so that the plants are barely monopetalous; there are
besides, five stamens, and a one-celled ovary surmounted by
five simple styles, which are hairy in the lower part.

In dry limestone pastures will be seen, numerous in many
localities, rosulate tufts of broad-ovate leaves, spreading close
to the ground, and producing from among them upright spikes
of insignificant flowers. These are the Plantains, represen-
tatives of the Plantaginaceous family, another group of regular-
flowered perigynous Monopetals. The Hoary Plantain* has
the leaves ovate sessile, their surface hoary from the presence of numerous whitish downy hairs, and marked with five
or seven longitudinal ribs; they spread in a compact tuft close
to the ground. The flowers are in cylindrical spikes, one to
two inches long, closely packed, the spikes terminating simple
leafless scapes or flower-stalks, which issue from among the
rosette of root-leaves, and rise six or eight inches high. These
flowers consist of a calyx of four sepals, a small whitish scarious corolla with a short tube and four spreading lobes, four
much-protruded stamens with purplish anthers, a long simple

* *Plantago media*—Plate 18 D.

style, and a two-celled ovary with two ovules in each cell. The longitudinal ribs in the leaves of these plants are peculiar, and have procured the name of Ribwort for one of the species.

Of the irregular-flowered perigynous Monopetals we find an illustration in the Scrophulariaceous family, here represented by the Purple Foxglove,* a flower commonly met with on dry hilly wastes, by the forest side, and along the banks of sandy lanes. In situations such as these, "the Foxglove rears its pyramid of bells, gloriously freckled." The plant is a biennial, that is to say, it springs up and forms a tuft of leaves one season, and shoots up its flowering stem the following year, and then perishes. The leaves which are produced in the first year are rather large long-stalked coarsely-veined and downy, of an ovate or ovate-lanceolate figure, and from their midst springs the erect flowering-stem of from two to four feet high, having a few shortly-stalked leaves on the lower part, and terminating in a long, stately, pyramidal, one-sided raceme of purple flowers, which are hairy and beautifully spotted inside. These have a calyx of five unequal segments, and an oblique tubular corolla an inch and a half long, contracted above the base and then much inflated, the mouth oblique and having five lobes, of which four are short and the remaining lower one is about twice the length of the others. The stamens are four in number, didynamous, that is, ranging in two pairs, one pair being longer than the other. There is a simple style with a two-cleft stigma; and a two-celled ovary becoming a capsule, containing numerous seeds. The Purple Foxglove is one of the most beautiful of our wild-flowers.

Near to the Scrophulariaceous plants rank those of the Orobanchaceous family, represented by the Lesser Broom-

* *Digitalis purpurea*—Plate 17 A.

rape.* These are very singular plants, forming dwarfish herbs
of a brownish or purplish colour, never green, the place of the
leaves being occupied by dull-coloured scales. One of their
leading peculiarities is that they grow parasitically on the
roots of other plants; that is to say, instead of forming roots
as most other plants do, to obtain their nutriment from the
earth, the Broomrapes form a junction with the roots of certain selected plants growing near them, and derive their nourishment directly from those plants on which they fix themselves. The species here selected grows from six to nine
inches or even a foot in height, the stem furnished with
brownish scales below, and terminating in an oblong spike of
dull bluish-purple flowers. The calyx is divided to the base
on the upper side, and often also on the lower, so as to form
two lateral sepals, which are usually two-cleft, the segments
ending in long slender points. The corolla is tubular and
curved, hairy outside, with a two-lipped limb of five rounded
lobes, and having the four stamens, which form two pairs,
fixed to its inner surface near the base; the anthers of these
stamens have the cells pointed at the lower end, and the style
is simple with a two-lobed stigma. These curious plants, in
consequence of the total absence of green, and the dull brownish hue, might when growing be readily mistaken for dead
flowers.

The Labiate or Lamiaceous plants form another family, and
a prominent one too, of irregular-flowered perigynous Monopetals. They are generally remarkable for aromatic properties, as in the Mint, Lavender, Rosemary, Sage, etc., and
include the splendid *Salvia* of our greenhouses and many
handsome border flowers. These *Salvias* or Sages, which are a
very numerous group, are represented among our field plants

* *Orobanche minor*—Plate 17 B.

by the Meadow Sage,* a perennial with a root-tuft of stalked ovate-oblong leaves, which are coarsely toothed and much wrinkled, and a flower-stem from a foot to a foot and a half high, furnished with a few smaller leaves near its base. The flowers grow in a handsome elongated terminal spike, in which they are arranged in whorls at short intervals. They have a two-lipped calyx, the upper lip of which is split into three small teeth, and the lower one cleft in two divisions. The corolla is remarkably irregular in form; it is much longer than the calyx, and of a rich deep purple-blue, two-lipped (whence the name labiate or lipped), the upper lip long arched and convex, the lower spreading three-lobed, with the side lobes minute, and the middle one large notched at the point. The stamens are two in number and of peculiar form. Usually a stamen consists of a slender thread called the filament, and a small oblong case (the anther), consisting of two cells held together by a central part called the connective, to which the top of the filament is attached. This part, the connective, is usually small and unnoticeable, but in the *Salvia* it is very much enlarged, the real filament is short, while the connective is long and slender, having a filament-like appearance, forming two unequal arms, and bearing one of the anther-cells, a perfect one, at the end of the longer arm, and a smaller cell, usually deformed, on the shorter arm. The ovary is four-lobed, with an erect ovule in each lobe, and from between these lobes grows the slender style, shortly cleft at top into two stigmatic branches. The four lobes become separated into four small seed-like nuts, which are enclosed in the permanent calyx. The plant is rare in England. A commoner species, *Salvia Verbenaca*, also bears purple flowers, but they are smaller in proportion to the other parts.

* *Salvia pratensis*—Plate 17 D.

I

Closely allied to the Labiates is the Verbenaceous family, represented by the Common Vervain,* a weed plentiful by roadsides and in waste places in the southern parts of England. The blossoms bear no comparison with the handsome *Verbenas* of our gardens, which belong to the same genus. The plant is an erect-growing perennial, having stems one to two feet high, with long spreading wiry branches. The leaves are opposite, those on the lower part of the stem, where they are most numerous, obovate or oblong, stalked, and coarsely toothed or cut, the upper ones being few sessile and lance-shaped. The flowers are small, in long slender spikes terminating the stem and branches, crowded at first, but becoming distant below by the elongation of the spike; they consist of a five-toothed calyx, a tubular corolla with an unequal five-cleft spreading limb, four stamens included in the tube, and a two- or four-celled ovary bearing the style at top, and dividing into four one-seeded nuts.

Another of these irregular-flowered perigynous Monopetals is the Common Butterwort,† which belongs to the Lentibulariaceous family. It is a little herb, found not uncommonly on wet rocks by mountain rills and in boggy places. It forms a rosulate tuft of spreading flat ovate or broadly oblong light green somewhat succulent leaves, which are involute at the margin and covered with soft glandular points which give them a clammy feel. The flower-stalks grow up among these to the height of four or five inches, and each terminate in a solitary bluish-purple flower. This consists of a two-lipped calyx, three-toothed in the upper lip and bifid in the lower; a two-lipped corolla spurred at the base, with a broad open mouth, a short broad two-lobed upper lip, and a longer three-

---

* *Verbena officinalis*—Plate 17 C.
† *Pinguicula vulgaris*—Plate 18 B.

lobed lower lip; two stamens; and a one-celled ovary opening in two valves and containing many seeds. The common name of Butterwort appears to have arisen from the property which the leaves are said to possess, of coagulating milk.

The peculiar features of the Monochlamydeous group have been already pointed out. During the summer period many flowers having this peculiar structure will be met with, but the limited number of our figures will, as before, only afford a few selected illustrative examples.

We have here, first, the Bistort or Snakeweed,* a sample of the Polygonaceous family, which is distinguished among Monochlamyds by having sheathing stipules. The Bistort is a perennial herb, found growing in moist pastures in various parts of Britain. It has a thick rootstock, from which spring up the patches of long-stalked ovate or cordate leaves, which are remarkable for their sudden contraction at the base into a narrow wing which borders the stalk. These are the leaves springing from the base of the plant, and are called the radical or root-leaves. The flower-stem grows quite upright, one to two feet high, with a few leaves on the lower part similar to the others, but smaller and with little or no stalk; at their base however is a kind of sheath surrounding the stem, so that the latter appears to grow through it. The sheath is formed of the united stipules—stipules being a pair of appendages assuming a variety of forms, produced on either side of the petiole or leaf-stalk at its base in certain families of plants. Here they coalesce and form a sheath or tube. The flowers form a close oblong spike at the top of the stem, and are of a pretty pink or light rose colour. The flowers have but one floral envelope, which is called the perianth, and this consists of five nearly equal segments, eight stamens considerably

* *Polygonum Bistorta*—Plate 19 A.

I 2

longer than the perianth, and three styles which are united at
their base, surmounting a free ovary with a single ovule. The
fruit is a small triangular seed-like nut, enclosed in the persistent perianth.

The family of the Aristolochiaceous plants is another of
these Monochlamydeous groups. It is represented by the
Common Birthwort,* a South European plant, naturalized
among ruins and in stony rubbishy places in the east and
south of England. This is a perennial, with a root creeping
so extensively underground as to become a rather troublesome
weed in situations which are congenial to it. The stems are
erect, two to three feet high, several often springing up in a
tuft, simple, striated, clothed with stalked, broadly heart-
shaped leaves, reniform at the base. The flowers are yellow,
collected in little groups in the axils of these leaves, clustered,
or aggregated, as it is called; they are tubular, erect, or a
little arching, globosely tumid at the base, and with a long
slender slightly widening tube above, the upper side of which
is prolonged into an oblique ovate concave emarginate limb.
Within the globular portion at the base of perianth, are placed
the anthers and stigmas: the latter being six-lobed, ray-like,
terminating the short thick style, around which are fixed the
six sessile anthers. The fresh plant when bruised has a strong
disagreeable smell, resembling that of Elder. The tumid
part of the corolla is covered inside with stiff hairs pointing
downwards. When expanded, the flowers are frequently
visited by a little insect, called *Tipula pennicornis*, which enters
them, and is prevented by the hairs from making its egress
until it has brushed off the pollen from the anthers on to
the stigma: "the perianth then withers, the hairs become
flaccid, and the insect makes its escape." This genus is one

* *Aristolochia Clematitis*—Plate 19 B.

of very remarkable appearance, and some very striking exotic species are occasionally seen in hothouses. One of them, *Aristolochia Gigas*, has flowers of enormous size.

A very peculiar group referred to this Monochlamydeous series, is that of the Pinaceous or Coniferous plants, in which there is no perianth, and what is more remarkable, the ovules have no covering whatever, as they have in the case of other Orders of plants. The Scotch Fir* is a familiar example. This, as is well known, forms a large evergreen tree, which is renowned for its valuable timber. The branches are clothed with persistent leaves, which grow two together within little membranaceous or scarious sheaths, and in this manner they are thickly distributed over the branches; they are stiff, dark-green, awl-shaped or linear bodies, with a sharp point, and are straight and directed forwards or towards the point of the branch. The flowers grow in catkins, the two sexes separate. Those containing the male flowers consist of closely imbricated scales, on the inner face of which are two adnate anther-cells; the scales in this case are the connectives of the anthers, so that the catkins are in reality formed simply of closely imbricated anthers. In like manner the female catkins consist of closely packed scales, having two ovules on the inner face of each, these ovules having the open pore at their upper end, technically called the foramen, turned downwards. The male catkins fall away, but the female ones grow into the cone-like fruit, from which the family has acquired the name of Conifers, or Cone-bearers. This fruit, which is sessile, ovoid, conical, and recurved, consists, when mature, of hard woody scales, thickened upwards, and having a short thick point, which is often turned backward in the lower scales. Each scale encloses two seeds, which have an oblique membranaceous wing.

* *Pinus sylvestris*—Plate 19 C.

Separated by some botanists from the great group of Monocotyledons, with which they are associated by others, is a small intermediate group, called Dictyogens, the peculiar features of which are, that they combine with the floral structure of Endogens (Monocotyledons) the netted venation and woody structure of Exogens (Dicotyledons). Of this group we have two illustrations.

First, of the family of Dioscoreaceous plants, which contains the Yam, we have a native species, the Black Bryony,* which bears very much the general aspect of the plants of this family. This Black Bryony, like the cultivated Yams, has thick tuberous root-stocks, and rather slender stems, which twine over hedges and bushes to a considerable extent, in many parts of the country. It is rather an elegant plant in its general aspect, though wanting in beauty of inflorescence. The stems are quite smooth, and furnished with alternate glossy bright green leaves of a heart-shaped figure, tapering to a slender point, and attached by longish stalks. These leaves are traversed by a few longitudinal strongly-marked ribs, between which they are occupied by a network of smaller veins. The plant is diœcious, that is, it bears only staminiferous flowers on one plant, and pistilliferous or fertile ones on another. The former grow in long slender racemes, which are often branched and longer than the leaves; they are individually small, consisting of a perianth of six small green segments, and six stamens—quite inconspicuous. The latter, which are equally wanting in attractiveness, and are even less striking on account of their growing in much shorter and less elegant racemes, consist of six segments and a three-branched style, the ovary being inferior; these however are succeeded by bright scarlet berries of globular form.

* *Tamus communis*—Plate 19 D.

Another Dictyogen is found in the Herb-Paris,* which belongs to the Trilliaceous family,—though this is by some botanists regarded as a section only of the Liliaceous Order. It is a dwarf herb with a creeping rootstock, producing a simple erect stem, six to nine inches or rarely somewhat more in height, furnished with a few scales at the base, but otherwise naked to the top, where grows a whorl of broadly ovate or obovate leaves, two to three inches long, strongly marked with a few longitudinal ribs, and netted between them with finely reticulated veins. In the centre of this guard of leaves stands a single erect flower on a stalk of moderate length, and consisting of a perianth of eight segments, of a yellowish-green colour, the outer series narrow-lanceolate and much broader than the inner ones, which are quite linear; within this eight erect stamens, which are awl-shaped, with the anther-cells affixed one on each side near the middle. The ovary is superior, four-celled, with four styles, and becomes a succulent bluish-black berry. The name Paris is said to come from *par*, *paris*, equal, in allusion to the regularity of numbers occurring in the parts—four leaves, four sepaline and four petaline divisions, twice four stamens, four styles, and a four-celled ovary, —which latter becomes a lurid-purple berry, whence rustics give the plant the name of One-berry, or True-Love.

We must now select a few illustrations from the different families of Monocotyledons, commencing with those in which the sexes are separated.

Of this group, the Common Frog-bit,† itself the type of the Hydrocharidaceous family, furnishes an illustration. This is a pretty water plant, with rather slender stems, producing here and there tufts of floating leaves and submerged roots.

* *Paris quadrifolia*—Plate 20 A.
† *Hydrocharis Morsus-ranæ*—Plate 20 B.

The leaves are roundish, entire, somewhat fleshy, smooth, and
lie flat on the water. The flowers rise up amongst and above
the leaves, and consist of three small outer green segments,
representing a calyx, and three larger inner white ones representing the petals of a corolla, and large enough to give the plant
a rather attractive character, being produced freely on the surface of ponds and ditches. In the staminate flowers, a variable
number of stamens, from three to twelve, is produced ; while in
the female flowers, which have an inferior ovary, there are six
styles with two-cleft stigmas. The flowers issue from spathes
formed of two thin bracts, those of the male plants being
shortly stalked, and those of the females sessile among the
leaves ; in the latter the pedicel is enlarged at the top into a
sort of tube to the perianth, enclosing the ovary, which becomes a dry six-celled fruit, containing several seeds. Mr.
Lees has observed of this common water-plant · "The economy
of this almost unregarded tenant of the water is not unworthy
of notice, nor when closely examined is it devoid of beauty.
Its floating reniform leaves are purple beneath, and it in-
creases almost entirely by floating runners, so that small retired pools are sometimes entirely covered with the thick-set
foliage, affording an impervious retreat to thousands of *Lymneæ*
and aquatic insects. The stainless flowers are of so delicate a
structure that they are injured by contact with the water, and
instead therefore of floating on its surface, they are providently
provided with elevating stalks, around whose basis is a pellucid
protecting bract. About wild commons and shady untrodden
lanes, little shield-like pools often appear, whose waters are
entirely hidden, roofed over, with a verdant covering of the
Hydrocharis, and scattered about this emerald table appear the
numerous white and delicate tripetaled blossoms, as if Titania
and her fairy court had there prepared a picnic banquet in the

shadowy retreat. On such a picture I have gazed in the silence of a summer's evening, when, as these silvery flowers are long conspicuous in the twilight, the splendour of the broad rising moon has increased and harmonized the illusion of the scene."

Another of the same series, is the Reed-mace or Cat's-tail, often called Bulrush,* a type of the Typhaceous family, in which the flowers, which are collected into dense spikes, have no perianth, but are monœcious, that is, separate, though growing on the same individuals. This plant is an aquatic perennial herb, with a thickish creeping rootstock, and erect reed-like stems, four to six feet high, having very long erect linear leaves, sheathing at the base, but flat and of a glaucous-green upwards. The flower-spike is terminal on the stem, and is often a foot long or more. When in flower, the upper portion, which is continuous with the lower, not separated by a short interval as appears in the figure, and consists of male flowers, is rather the thickest, and is yellow from the numerous, closely packed, linear anthers; the minute ovaries of the lower part, which is of a deep brown, a colour given to it by the protruded stigmas, are also closely packed, and enveloped in tufts of soft hairs. When in fruit, the upper part becomes bare, or appears smaller from the shrinking of the dead stamens, and the lower part much thicker by the enlargement of the nuts, which are still enveloped in the thick, dark-coloured felt, formed by the hairs and stigmas, and at length become stalked. The fruit is a small seed-like nut, which continues small, and enveloped in the downy hairs. The plants are very stately objects in damp situations—these *Typhas* "marshalled in battalions, like grenadiers with hairy caps of the olden day." There is but little difference be-

* *Typha latifolia*—Plate 21 B.

tween the two species found in this country, the chief points
of dissimilarity being the slightly smaller size, the narrower
leaves, and the slightly interrupted flower-spike.

The Ivy-leaved Duckweed* is a curious little plant, coming
in the same category as the last. The group of Duckweeds
consists of singular aquatic plants, resembling little green
scales floating on the surface of stagnant waters, and forming
part of the Pistiaceous, or, as Mr. Bentham calls it, Lemna-
ceous family. They are but rarely met with in flower, and
the flowers are so small and simple, that even when present
they are not readily detected. The Ivy-leaved species represented in our plate, has, it will be seen, no distinct stem or
leaves, but consists of small, leaf-like fronds, of a lance-shaped
figure, minutely toothed at one end, and tapering to a stalk-
like base at the other. Usually two young fronds grow from
opposite sides of the older one near its base, each one producing
eventually a single root from beneath. In this way growth goes
on, the fronds becoming detached, and themselves producing
others from their sides. The roots, which strike down perpendicularly, are in all the species capped by a small calyptra,
or sheath. They increase " not only by seeds, but more abundantly by buds concealed in the lateral clefts of the parent
frond, which growing out on two opposite sides into new
plants, and these again producing offspring in the same way,
while still attached to their parents, present a most curious
appearance." The minute flowers grow from a fissure in the
edge of the frond, two together, the inflorescence consisting
of a membranaceous bract or spathe enclosing two stamens,
and a single one-celled ovary, both without trace of a perianth.
Dr. Lindley describes this structure as follows :—the flowers
are two in number, one male and the other female, lying

* *Lemna trisulca*—Plate 21 E.

concealed in a slit of the frond; they have neither calyx nor corolla, but are enclosed in a delicate, membranous bag. The other species have rounder and thicker fronds than our example, and branch out in a similar manner, but less regularly, or at least, from their crowded condition, more confusedly than that we have described. Usually there is but one root to each frond, but in the Greater Duckweed, *Lemna* (or *Spirodela*) *polyrhiza*, a cluster of roots is produced under each.

Still bearing imperfect, but in some cases hermaphrodite flowers, there is the family of the Naiadaceous plants, represented by the Broad Pondweed.* This is a perennial herb, with long submerged branching stems, furnished with alternate stalked leaves, of which the uppermost float on the surface of the water, and the lower are sometimes reduced to a mere stalk. These floating leaves are largish, of an ovate form, thick in texture, marked by longitudinal nerves, and having a sheathing scarious stipule in the axil of their stalks. The flowers are small, sessile, arranged in spikes, which terminate the axillary flower-stalks, and which stand up above the water; they consist of four small green scales, representing a perianth; four stamens, with sessile anthers opposite these scales; and four distinct carpels, each with a sessile stigma. It is a common plant in stagnant waters and slow streams. Some botanists refer this genus to the Juncaginaceous family.

Passing on to the series in which the flowers are furnished with a calyx and corolla, and both stamens and pistil, the ovary being free, we come to one or two groups in which the perianth nevertheless bears a very inconspicuous character. The first of these is represented by the Sweet Flag,† by some referred to the Araceous family, but more fittingly associated

* *Potamogeton natans*—Plate 21 D.
† *Acorus Calamus*—Plate 21 C.

with that of the Orontiaceous plants. It is a reed-like plant
with thick, shortly creeping rootstocks, and everywhere highly
aromatic. The leaves are linear, sword-shaped, erect, two to
three feet long. The flowering stem is also erect, simple, and
very much resembling the leaves, its long linear, leaf-like
spathe forming a flattened continuation of the stalk portion
which supports the dense cylindrical green spike or spadix
of flowers. The spike, which pushes out sideways, while the
spathe grows erect, appears, indeed, as if it grew out of the
side of a leaf; it is sessile, stoutish, two to three inches long,
consisting of numerous hermaphrodite flowers closely packed,
the flowers consisting of six short green scales, six stamens,
and a two- or three-celled ovary. The aromatic herbage of this
plant is sometimes used for flavouring beer and spirits, and in
Norfolk, where it abounds, it is said to be strewed on the
floors of the churches on festival days.

Somewhat similar to this in the minutiæ of its structure,
though differing considerably in aspect, is the Common Rush
(*Juncus communis*), a plant of the Juncaceous family. This
well-known plant, one form of which is frequently called
Juncus effusus, has a short, creeping, matted rootstock, which
produces dense tufts of cylindrical leafless stems, two to three
feet high, sheathed at the base by a few brown scales, and
tapered above into a fine point. Some of these stems are
barren and seem to resemble leaves, whilst others bear on one
side, towards the top, a loosely clustered, irregular, compound,
bracteated panicle of small brownish flowers. These consist
of a regular, dry, calyx-like perianth of six pointed segments,
usually three, but sometimes six, stamens, a single style, with
three stigmas, and a many-seeded very obtuse capsule, opening in three valves. The plant is very abundant in wet situations, almost all over the northern hemisphere, and in some

parts of the southern. Another form of the same species, called *Juncus conglomeratus*, has the clusters dense and compact, so as to form roundish heads. Both forms are used for platting into mats and chair-bottoms, and the pith of their stems is formed into wicks for candles.

The Flowering Rush,* belonging to the Butomaceous family, which it typifies, is of a more ornamental character than the preceding; indeed, it has truly been said to be a greater adornment to the banks of our rivers than any other British wild-flower. It is a perennial aquatic herb, with a thick creeping rootstock, from which spring up the clusters of long, erect, triangular, sedge-like leaves, which are broad and sheathing at the base. The flower-stem is stout, leafless, rush-like, three or four feet high, bearing a large, simple umbel of numerous showy pale rose-coloured or pinkish flowers, and having an involucre of three lance-shaped bracts at the base of the umbel. The flowers are nearly an inch in diameter, the perianth formed of six ovate nearly equal spreading segments; within this are nine stamens, and, in the centre, six erect carpels, connected below, tapered above into short styles, and each containing numerous small seeds. It is a very handsome aquatic plant, quite deserving of cultivation. By some authorities, the family to which it is referred is included in that of the Alismaceous plants.

The beautiful family of Iridaceous plants, another of the regular-flowered groups of Monocotyledons to which allusion has already been made, is further illustrated by the Yellow Flag,† found abundantly in marshy places and by the sides of watercourses. There, as Shelley writes, " where the embowering trees recede, and leave a little space of green expanse, the cove is closed by meeting banks, whose yellow flowers for ever

---

* *Butomus umbellatus*—Plate 21 A.
† *Iris Pseud-acorus*—Plate 20 D.

gaze on their own drooping eyes reflected in the crystal calm."
This plant has a good deal the aspect of the Sweet Flag (*Acorus Calamus*), excluding the flowers, however, which are remarkably showy and petaloid, instead of being reduced to the appearance of green closely packed scales. The plant has a thick,
horizontal root-stock or rhizome, as it is called, from which grow
up the long sword-shaped leaves, three or four feet long, stiff
erect and of a glaucous green, equitant or alternately bestri-
ding each other at the base, those on the flower-stem (which
does not grow so high as the leaves) being shorter. The
flowers, which are large and showy, are produced two or three
in succession, from the axils of sheathing bracts near the upper part of the stem; they are erect, of a bright yellow colour,
the three outer or sepaline segments of the perianth being
large and reflexed, broadly ovate, contracted into a claw at the
base, and the three inner or petaline divisions, small, oblong,
and erect. There are three stamens opposite the sepaline segments, and over these the petal-like appendages of the three
stigmas are arched; these appendages are longer and larger
than the petaline segments, yellow, two-cleft at top, and
toothed on the edge. The ovary is inferior, becoming a three-
cornered, oblong capsule. Many beautiful garden plants belong
to this family, *Iris* itself being an extensive and ornamental
genus, besides which there are *Gladiolus*, *Ixia*, *Crocus*, and
numerous others of equal beauty.

The Orchidaceous family is a group of plants having, like
the foregoing, an inferior ovary, but singularly irregular
flowers. It is a very extensive and, including the exotic species, a remarkably varied group of plants. Our illustration
among summer flowers, the Bee Ophrys,* or Bee Orchis, as it
is more commonly called, gives but a faint idea of the gro-
tesque beauty of many tropical species. The peculiar charac-

* *Ophrys apifera*—Plate 20 C.

teristics of the group have been adverted to in describing the Spotted Palmate Orchis and the Ladies' Slipper, so that we may now confine our remarks to the Bee Orchis before us. This plant is furnished with a pair of tubers, of which one is growing and the other waning: the former being in process of formation as a store of nutriment for the stem of the succeeding year, while the latter has been exhausted by the growth of the stem of the present year, which has, as it were, sucked out its vitals. The stem rises from nearly a foot to a foot and a half in height, erect, leafy near the base, and terminating in a loose spike of curiously coloured flowers. The leaves are oblong or lanceolate, the upper ones being the smaller. The flowers have at the base a bract as long as the ovary, which latter simulates a flower-stalk, the flowers being really sessile; they consist of three ovate spreading or reflexed sepals, which are always more or less tinged with pink, two petals, which are smaller than the sepals and nearly erect, and a part, very unlike all the rest, called the lip or labellum, which is broad and convex, of a rich velvety-brown, downy at the sides, smooth in the middle, and variously marked by paler lines or spots; while on each side is a small downy lobe turned under, and at the point three terminal ones, which are turned under so as to be quite concealed. It is this lip which is supposed to resemble a bee. The column is erect, with a distinct curved beak above the anther. The plant is found plentiful in some of the southern or eastern parts of England, growing in dry pastures in limestone districts.

Of the Liliaceous family, which differs from many of the foregoing Monocotyledons in having a superior ovary, the well-known Lily-of-the-Valley* affords a good example. This sweet little plant—

* *Convallaria majalis*—Plate 20 E.

> "Fair flower, that lapt in lowly glade
> Dost hide beneath the greenwood shade,
>   Than whom the vernal gale
> None fairer wakes on bank or spray,
> Our England's Lily of the May,
>   Our Lily of the Vale!"

—this little gem among flowers, has a creeping rootstock, forming buds and tufts of roots at intervals. From these grow the radical leaves, two or three in a scaly sheath, and having the long stalks enclosed one within the other, so as to resemble a stem ; they are four to six inches long, oblong, tapering to both ends, and somewhat striated. The flower-stalks issue from one of the lower scales, and are slender, shorter than the leaves, supporting the loose racemes of droop-ing, bell-shaped, pure white, deliciously scented blossoms. "No flower amid the garden fairer grows, than the sweet Lily of the lowly vale, the queen of flowers." Its blossoms consist of a bell-shaped shortly six-cleft perianth, six erect stamens, in-serted near the base, and a simple style, with a blunt three-cornered stigma. The three-celled ovary becomes a scarlet berry, with one-seeded cells.

Another extensive and important division of the Monocoty-ledons has what are called glumaceous flowers, that is to say, their parts are not at all petal-like, as in those we have been considering, but dry and husky-looking, as seen in the husks of the grasses and the corn plants. This, however, is not their chief peculiarity, or at least not that by which they may be most readily known, for in some of the groups to which we have already referred, something of the same texture in the parts of the flower has been spoken of. The most obvious and characteristic difference consists in the position of the parts of the flower, which in those we have already considered are arranged in whorls, whilst in the glumaceous series they are imbricated or arranged so as to overlap each other like the tiles of a roof, the endogenous structure common to all Monocotyledons being therewith combined.

Forming a considerable group in this glumaceous series is the family of the Cyperaceous plants, represented by the Great Common Sedge,* a tall grassy-looking plant found commonly by the sides of rivers and watercourses. It is a stout-growing erect perennial, with a creeping rootstock, triangular acute-angled leafy stems three to four feet high, and long broadish grassy leaves tapering to a narrow point. The flowers grow in longish spikelets, arranged in a racemose manner at the top of the stem, which from their weight they incline gracefully to one side. Of these spikelets, which are several in number, the two or three upper ones are composed of staminiferous or male flowers only, while the others consist entirely of pistilliferous or female flowers. The former are cylindrical, upwards of an inch long, formed of acute glumes (chaff-like scales) lying closely over each other, and each producing in its axil three stamens, the anthers of which have a long point, and while fresh impart to the spikelet a yellowish colour. The latter are cylindrical tapered at the point, purplish, the lower ones stalked, all having leafy bracts, and consisting of imbricated pointed glumes, each enclosing in its axil an oblong-ovate ovary; this is narrowed into a short, broad, cloven beak, terminated by a three-cleft style, and becomes hardened into a somewhat three-cornered nut. The Sedge family is a very extensive one, always having grass-like foliage and something of the aspect of the species here described, but differing considerably in stature, and in the details of growth and structure.

The other principal family of the glumaceous Monocoty-

* *Carex riparia*—Plate 21 G.

ledons is that of the Graminaceous plants or Grasses, one of the most important to man in the whole range of the vegetable world, as affording the staple food both of himself and the animals subservient to his use. It is a family presenting great variation of structure as well as of aspect, yet combined with certain constant and obvious characteristics which render the plants easily distinguishable from all others though they are by no means so readily distinguished among themselves. Some of these characteristic features are the hollow roundish stems or culms, separated into lengths by somewhat thickened joints or nodes, and the narrow alternate parallel-veined leaves or leaf-blades which sheath the stem by their base, the sheath being split open on the side opposite the blade, and usually terminated just within the base of the blade by a small scarious appendage called a ligule. The flowers are arranged in spikelets consisting of chaffy scales imbricating over each other, the outer of which are called glumes, and the inner pales. The fruit is a seed-like grain, or caryopsis.

Of this important race of plants we have an example in the Soft Brome-Grass,* which is one of our commonest species in open waste places. It is an annual or biennial plant, with a culm one to two feet high, everywhere clothed with soft short hairs, and producing an ovate slighly compound flowering panicle two to three inches long. The spikelets are ovate, somewhat compressed, pubescent, standing nearly erect; they are made up of a pair of glumes at the base, the outer of which is considerably the larger of the two, and within these from five to ten florets ranged alternately on either side of the axis of the spikelet, each floret consisting of two pales or paleæ, the outer of which is larger, rounded on the back, and having a straight awn or bristle as long as the floret, growing from just

* *Bromus mollis*—Plate 22 B.

below its bifid extremity, the inner one smaller and narrower and conspicuously ciliated or fringed on its ribs or nerves; these pales enclose three stamens, and a roundish ovary crowned by a pair of feathery styles.

Another illustration is afforded by the Common Reed,* which belongs to a different subdivision of the family. This is a stout perennial grass, with a culm varying from five to ten feet high surmounted by a plume of flowers often a foot in length, more or less drooping, and of a purplish-brown colour. The plant has a stout creeping rootstock, and its stems or culms are clothed all the way up with broad grassy leaves often an inch in width. The inflorescence is a large compound panicle, with very numerous small narrow spikelets. These spikelets are formed of two very unequal lance-shaped sharp-pointed glumes, within which are developed about five florets, the pales of which are narrower, ending in an almost awl-like point, and surrounded by long silky hairs developed from the rachis, which lengthen as the seed ripens, and give to the panicle at that stage a beautiful silky appearance. The lower floret is triandrous (bearing three stamens) and barren, but the rest of the florets are perfect, with three stamens, and an ovary with two feathery styles. The Reed, which is generally a common plant in wet places, forms patches of very great extent, called Reed-ronds in some parts of England, and the culms are much used for thatching and for making garden screens, as well as for the walls of sheds and huts. When applied to the latter use, they are generally plastered with well tempered clay. Even without this plastering they last for a considerable time.

Thus we complete our slight descriptive sketch of the illus-

* *Arundo Phragmites*—Plate 22 A.

K 2

trations of Summer Flowers which have been selected as examples of the much more comprehensive galaxy of beauties with which field and wood is adorned at this flowery season. Who would not catch up the strain of Campbell's pleasant song with its love of Nature's wildings?

> " Ye field-flowers! the gardens eclipse you 'tis true,
>   Yet, wildings of nature, I dote upon you,
>   For ye waft me to summers of old,
>   When the earth teemed around me with fairy delight,
>   And daisies and buttercups gladdened my sight,
>   Like treasures of silver and gold.

> " Even now what affections the violet awakes—
>   What loved little island, twice seen in the lakes,
>   Can the wild water lily restore!
>   What landscapes I read in the primrose's looks!
>   What pictures of pebbles and minnowy brooks
>   In the vetches that tangle the shore!"

We may, however, add in the words of another poet—

> " Who loves not Summer's splendid reign,
>   The bridal of the earth and main?
>   Yet who would choose, however bright,
>   A dog-day noon without a night?"— *Montgomery.*

So pass we on to the blossoms and the fruits of Autumn, pausing by the way to make record in a summary form, of those of the Summer season.

SUMMARY OF SUMMER FLOWERS.

[I.—GROUPS AND ORDERS.]

EXOGENOUS PLANTS or DICOTYLEDONS.

Leaves with netted veins. *Flowers* usually quinary—the parts in fives, or quaternary—the parts in fours. *Embryos* with two (rarely more) cotyledons; hence dicotyledonous. This group includes the Thalamiflores, Calyciflores, Monopetals, and Monochlamyds.

Thalamiflores: Polypetalous dichlamydeous plants, with petals distinct (*i. e.* separable) from the calyx, and the stamens hypogynous; Orders numbered 1 to 18.

> * *Carpels more or less distinct (*i. e. *apocarpous), sometimes solitary with one lateral placenta.*

1. **Ranunculaceous plants**—herbs or climbing shrubs; stamens indefinite, usually numerous, inserted on the receptacle.

> ** *Carpels combined into an undivided (*i.e. *syncarpous) ovary.*
> † *Seeds attached to the spongy dissepiments (*i. e. *placentas dissepimental).*

2. **Nymphæaceous plants**—aquatic herbs, with the carpels imbedded into the receptacle, or combined into a single many-celled ovary.

> †† *Seeds attached to the sides of the carpels (*i. e. *placentas parietal).*
> ‡ *Stamens indefinite.*

3. **Papaveraceous plants**—herbs; flowers regular, with two sepals and four petals; stamens numerous.

4. **Cistaceous plants**—shrubs or herbs; flowers with three (or five) sepals and five regular petals; **stamens numerous.**

5. **Resedaceous plants**—herbs; flowers irregular, with four to six sepals, and several small unequal petals, some of which are divided; stamens few.

‡‡ *Stamens definite.*
§ *Stamens tetradynamous (4 long, 2 short), distinct.*

6. **Cruciferous plants**—herbs; flowers regular; sepals four; petals four, arranged **crosswise.**

§§ *Stamens 6, united in two sets.*

7. **Fumariaceous plants**—herbs; flowers very irregular; sepals two; petals four.

§§§ *Stamens 4 or 5, alternating with the petals, sometimes with accessory or additional ones opposite the petals.*

8. **Frankeniaceous plants**—herbs; flowers regular, with four or five sepals and petals, and usually with accessory stamens.

9. **Tamaricaceous plants**—shrubs, with scale-like imbricating leaves; flowers regular, with four or five sepals and petals.

††† *Seeds attached in the axis or centre of the carpels (i. e. placentas axile).*
‡ *Flowers regular (i. e. the parts equal as to size and form).*
§ *Sepals overlapping at the edge (i. e. imbricate).*
‖ *Ovary one-celled.*

10. **Caryophyllaceous plants**—herbs; leaves opposite, undivided, without stipules; flowers symmetrical, with four or five sepals and petals, and definite stamens.

11. **Illecebraceous plants**—herbs; leaves opposite or alternate, furnished with stipules; flowers inconspicuous; calyx 3–5-lobed; petals 3–5, often rudimentary or wanting; stamens as many as the sepals, rarely fewer.

||| *Ovary many-celled.*

(a) *Stamens distinct.*

12. **Elatinaceous plants**—minute aquatic herbs; the symmetrical flowers with three to five sepals and petals, and as many or twice as many stamens.

(b) *Stamens monadelphous.*

13. **Linaceous plants**—herbs; leaves entire; flowers symmetrical, with five sepals and petals, and as many stamens united at the base into one parcel; carpels separating without leaving a central axis.

14. **Geraniaceous plants**—herbs; leaves divided; flowers symmetrical, with five sepals and petals, and five or ten stamens united into one parcel; carpels fixed around a persistent central axis.

(c) *Stamens polyadelphous.*

15. **Hypericaceous plants**—shrubs; leaves opposite, often dotted; flowers regular, with five sepals and petals; stamens indefinite, united below into 3–5 parcels.

§§ *Sepals parallel at the edge* (i. e. *valvate*).

16. **Malvaceous plants**—herbs or shrubs; flowers regular, of five sepals and petals, surrounded by an involucre of three or more bracts; stamens numerous, their filaments united into a tube around the pistil.

17. **Tiliaceous plants**—trees; flowers attached to a leaf-like bract; sepals and petals five; stamens numerous, shortly cohering in several clusters.

‡‡ *Flowers irregular.*

18. **Balsaminaceous plants**—herbs; flowers of six pieces, very irregular, the sepals and petals all coloured, one of the sepals spurred; stamens five, the anthers cohering round the pistil.

Calyciflores: Polypetalous dichlamydeous plants, with the petals usually distinct, and the stamens perigynous or epigynous; Orders 19 to 27.

    * *Stamens perigynous.*
      † *Carpels more or less distinct, or single.*
      ‡ *Ovary superior, the calyx distinct from the carpels.*

19. **Leguminous plants**—herbs or shrubs; flowers very irregular, papilionaceous; stamens 10, all or 9 of them united; ovary single, becoming a legume.

20. **Rosaceous plants**—herbs or shrubs; flowers regular, rarely without petals; stamens indefinite; fruit 1–2-seeded nuts or drupes, sometimes enclosed within the fleshy tube of the calyx, or follicles containing several seeds.

21. **Crassulaceous plants**—herbs, with succulent leaves; flowers regular; the sepals and petals usually five (sometimes 3–4 or 6–20), isomerous; carpels as many as the petals, free.

    ‡‡ *Ovary superior or half-inferior, the calyx adhering more or less to the carpels.*

22. **Saxifragaceous plants**—herbs; flowers regular; stamens definite; ovary syncarpous at the base, with a separate style for each carpel.

    †† *Carpels combined into an undivided ovary, with more than one placenta; (ovary superior).*

23. **Lythraceous plants**—herbs; leaves opposite; sepals four, five, or more, united below into a tube; petals as many, inserted at the top of the calyx-tube; stamens as many or twice as many as the petals.

    ** *Stamens epigynous; (ovary inferior).*
      † *Flowers not umbellate.*

24. **Onagraceous plants**—herbs; leaves simple, usually opposite; flowers complete; sepals and petals four each, with four

or eight stamens, or two each, with two stamens; seeds numerous.

25. **Haloragaceous plants**—herbs; leaves whorled; flowers minute, the calyx four-lobed or wanting; stamens 8-6-4, or sometimes reduced to one; fruit dry, containing a solitary seed.

26. **Cornaceous plants**—shrubs; leaves opposite; calyx-teeth petals and stamens four each; fruit 1-2-celled, the cells containing a solitary seed.

†† *Flowers in umbels.*

27. **Umbelliferous plants**—herbs; leaves alternate; fruit dry, of two carpels adhering by their face (commissure) to a common central axis; seed solitary.

Monopetals: Dichlamydeous plants, with the petals united (from the base more or less upwards) into a single piece; Orders 28 to 52.

* *Stamens epigynous; (ovary inferior).*

† *Stamens attached to the corolla, the filaments cohering in three parcels.*

28. **Cucurbitaceous plants**—herbs; leaves lobed; flowers unisexual, the calyx and corolla each five-parted; fruit succulent.

†† *Stamens attached to the corolla, separate, alternating with its lobes; (ovules mostly solitary).*

‡ *Fruit three- to five-celled.*

29. **Caprifoliaceous plants**—shrubs or herbs; leaves opposite; corolla five- rarely four-lobed, regular or irregular; fruit-cells one- or few-seeded.

‡‡ *Fruit double, formed of two united one-seeded carpels.*

30. **Galiaceous plants**—herbs; leaves and leaf-like stipules whorled around the four-angled stems; corolla four-lobed.

‡‡‡ *Fruit single or one-seeded.*

31. **Valerianaceous plants**—herbs; flowers not in close heads; corolla five-lobed; stamens fewer than the corolla-lobes.

32. **Dipsacaceous plants**—herbs; flowers in compact heads or spikes; stamens equalling in number the divisions of the four- or five-lobed corolla.

††† *Stamens attached to the corolla, cohering by their anthers.*

33. **Composite plants**—herbs; florets collected in heads; stamens as many as the corolla lobes; fruit dry, one-seeded.

†††† *Stamens free from the corolla, the anthers cohering.*

34. **Lobeliaceous plants**—herbs; flowers not in heads; stamens as many as the lobes of the corolla; fruit many-seeded.

††††† *Stamens free from the corolla, the anthers distinct.*

35. **Campanulaceous plants**—herbs; stamens as many as the lobes of the corolla; anthers opening longitudinally.

36. **Vacciniaceous plants**—shrubs; stamens twice as many as the lobes of the corolla; anthers opening by two terminal pores.

** *Stamens hypogynous.*

37. **Ericaceous plants**—shrubs; stamens equal to or twice as many as the lobes of the corolla; anthers with an appendage, opening by two pores.

*** *Stamens perigynous; (ovary superior).*
† *Seeds attached to a free central placenta.*
‡ *Stamens opposite the corolla-lobes.*

38. **Primulaceous plants**—herbs; stamens equalling in number the lobes of the corolla; style one; ovary many-seeded.

39. **Plumbaginaceous plants**—herbs; stamens equalling in number the lobes of the corolla; styles five; ovary containing

one ovule, which is pendulous from the point of a thread (umbilical cord) rising from the base of the cavity.

   ‡‡ *Stamens alternating with the corolla-lobes.*

40. **Plantaginaceous plants**—herbs; stamens equalling in number the lobes of the four-cleft regular corolla; style one; ovary 1–2-celled, with one or several ovules.

41. **Lentibulariaceous plants**—herbs; corolla irregular, two lipped; stamens two, fewer than the corolla-lobes; ovary 1-celled, many-seeded.

   †† *Seeds attached to the sides of the cells (placentas parietal), or in their axial angle.*

   ‡ *Corolla regular or nearly so.*

   § *Ovary entire (not lobed), its cells with one or two ovules.*

42. **Aquifoliaceous plants**—trees or shrubs; corolla 4–6-parted; stamens as many, alternating with its lobes; fruit fleshy, 4-celled, with one ovule in each cell.

43. **Oleaceous plants**—trees or shrubs; corolla 4-parted or wanting, with two stamens alternating with its lobes; ovary two-celled, the cells two-seeded.

44. **Convolvulaceous plants**—herbs; corolla plaited, 4–5-lobed, with as many alternating stamens; ovary 2–3-celled, the cells two- (rarely one-) seeded.

   §§ *Ovary entire (not lobed), its cells containing many seeds.*

45. **Polemoniaceous plants**—herbs; corolla five-lobed, with as many alternating stamens; ovary three-celled.

46. **Gentianaceous plants**—herbs; corolla usually five-lobed (sometimes 4, 6. 8 or 10) with as many alternating stamens, twisted; ovary 1–2-celled.

47. **Solanaceous plants**—herbs; corolla five, rarely four-lobed, with as many alternating stamens, plaited; ovary two-celled.

   §§§ *Ovary four-lobed.*

48. **Boraginaceous plants**—herbs; corolla five, rarely four-lobed,

with as many alternating stamens; ovary four-celled, with a single ovule in each cell affixed at the base of the cavity.

‡‡ *Corollas irregular.*
§ *Ovary four-lobed.*

49. **Labiate plants**—herbs; corolla two-lipped; stamens usually four didynamous (2 long, 2 short), or sometimes only two; lobes of the ovary one-seeded.

§§ *Ovary undivided* (i. e. *not lobed*).

50. **Verbenaceous plants**—herbs; stamens four didynamous, rarely two; ovary four-celled, dividing when mature into four one-seeded nuts.

51. **Scrophulariaceous plants**—herbs; stamens four, didynamous; ovary two-celled, the cells many-seeded; placentas axile.

52. **Orobanchaceous plants**—leafless parasitical herbs; stamens four, didynamous; ovary one-celled, with two or more many-seeded parietal placentas.

Monochlamyds: Perianth single (*i. e.* consisting of a calyx only), or altogether wanting, and replaced by scaly bracts; Orders 53 to 59.

* *Carpels solitary, simple.*
† *Calyx inferior.*

53. **Chenopodiaceous plants**—herbs; leaves without stipules; perianth small, the stamens (4–5, rarely 1–2), alternating with its lobes; ovule 1-celled, with one ovule.

54. **Ceratophyllaceous plants**—aquatic floating herbs; leaves without stipules; perianth none; ovary and fruit 1-seeded, or 4-lobed, with one seed in each lobe.

55. **Urticaceous plants**—herbs; leaves with distinct stipules; flowers unisexual; perianth small, green, the stamens, usually 4, opposite its segments; styles two, rarely one.

56. **Polygonaceous plants**—herbs; leaves with ochreate or sheathing stipules; perianth small; styles two or more.

†† *Calyx superior.*

57. **Santalaceous plants**—herbs; perianth with four or five small valvate lobes, and with four or five stamens opposite its lobes; styles simple; ovules 2–3; seed solitary.

** *Carpels combined, two- or more-celled.*

† *Calyx inferior.*

58. **Euphorbiaceous plants**—herbs or shrubs; flowers unisexual; fruit consisting of three carpels, each containing one or two pendulous seeds.

†† *Calyx superior.*

59. **Aristolochiaceous plants**—herbs; perianth irregular or three-lobed; stamens 12; ovary 3–6-celled, the cells many-seeded.

DICTYOGENOUS PLANTS.

Leaves with netted veins. *Flowers* usually ternary—the parts in threes. *Embryos* as in *Endogens*. This group consists of the Orders numbered 60 and 61.

* *Perianth inferior.*

60. **Trilliaceous plants**—herbs; leaves whorled; flowers hermaphrodite; fruit 3–5-celled, succulent; seeds indefinite.

** *Perianth superior.*

61. **Dioscoreaceous plants**—climbing herbs; leaves alternate; flowers unisexual; fruit 3-celled, the cells 1–2-seeded.

ENDOGENOUS PLANTS or MONOCOTYLEDONS.

Leaves with parallel veins. *Flowers* usually ternary—the parts in threes. *Embryos* with one cotyledon; hence monocotyledonous. This group includes the Orders numbered from 62 to 73.

> * *Flowers imperfect or naked* (i. e. *without perianth or glumes), or consisting of scales.*
>
> † *Inflorescence forming dense spikes or heads.*

62. **Orontiaceous plants**—herbs; flowers hermaphrodite, spadiceous, the spadix issuing from a two-edged leaf; perianth scales six, herbaceous; ovary free, one- or more-celled, with erect ovules.

63. **Typhaceous plants**—aquatic herbs; flowers unisexual, arranged on a naked spadix; scales thin.

> †† *Flowers distinct, or in loose spikes.*

64. **Pistiaceous plants**—floating, stemless, aquatics; flowers borne on the edge of the small scale-like fronds, quite naked.

65. **Naiadaceous plants**—floating aquatics, with distinct stem and leaves; perianth of four scales, or wanting; ovaries 1, 2, or 4.

> ** *Flowers perfect, with a petal-like whorled perianth.*
>
> † *Ovary inferior.*

66. **Hydrocharidaceous plants**—floating or submerged herbs; flowers regular; dioecious; stamens nine or more, free.

67. **Iridaceous plants**—herbs; flowers six-leaved, hermaphrodite; stamens three, free.

68. **Orchidaceous plants**—herbs; flowers six-leaved, irregular; stamens gynandrous (combined with the style).

> †† *Ovary superior.*
>
> ‡ *Ovary syncarpous.*

69. **Liliaceous plants**—herbs; perianth six-leaved, the sepals and petals alike, or nearly so (sometimes combined), regular.

‡‡ *Ovary apocarpous.*

70. **Alismaceous plants**—herbs; flowers three-petaled, the sepals and petals usually unlike or distinct, regular.

*** *Flowers perfect, with a dry calyx-like whorled perianth.*

71. **Juncaceous plants**—perianth regular, six-leaved, **brown.**

**** *Flowers glumaceous, i. e. formed of imbricated chaffy scales or bracts (glumes).*

72. **Cyperaceous plants**—herbs; leaves grassy, with entire sheaths; bracts one to each flower or floret.

73. **Graminaceous plants**—herbs; leaves grassy, with their sheaths split on the side opposite the blade; bracts two to each **flower** or floret.

[II.—GENERA OR FAMILIES.]

1. **Ranunculaceous Plants.** RANUNCULACEÆ.

* *Anthers bursting outwardly.*

† *Climbing shrubs with opposite leaves; (carpels one-seeded.)*

(1) **Clematis**—sepals 4–5, coloured and petal-like; petals none.

†† *Herbs with the leaves alternate or radical.*

‡ *Carpels several, short, one-seeded; (flowers regular).*

(2) **Thalictrum**—sepals 4–5 or more, often coloured and petal-like; petals none.

(3) **Adonis**—petals 5–10, conspicuous, usually red, without a nectariferous pore at their base.

(4) **Ranunculus**—petals five, rarely many, usually yellow or white, having a nectariferous pore at their base.

‡‡ *Carpels several, many-seeded.*

§ *Flowers regular.*

(5) **Trollius**—sepals 5–15, large, pale yellow, petal-like; petals as many, small, flat, linear.

§§ *Flowers irregular or spurred.*

(6) **Aquilegia**—sepals five, petaloid, flat, regular; petals five, funnel-shaped, each with a long horn-like basal spur.

(7) **Delphinium**—sepals five, the upper one with a long basal spur; petals combined into one, which is lengthened into a spur included in the spurred sepal.

(8) **Aconitum**—sepals five, the upper helmet-shaped, not spurred; two upper petals tubular, on long stalks concealed in the helmet-shaped sepal.

** *Anthers bursting inwardly.*

(9) **Pæonia**—sepals five, persistent; petals five or more, large, red or white.

2. Nymphæaceous Plants. NYMPHÆACEÆ.

(10) **Nymphæa**—sepals four, greenish outside; petals numerous, white.

(11) **Nuphar**—sepals 5–6, yellow; petals several, much smaller, yellow.

3. Papaveraceous Plants. PAPAVERACEÆ.

* *Fruit oblong or globular; stigmas radiate.*

(12) **Papaver**—petals four; stigmas four to twenty, sessile, connected, *i.e.* radiating on a flat sessile disk.

(13) **Meconopsis**—petals four; stigmas 5–6, free, supported by a short distinct style.

** *Fruit linear; stigmas two-lobed.*

(14) **Chelidonium**—petals four, small; stigmas two; capsule two-valved, one-celled, the placentas distinct; seeds crested.

(15) **Glaucium**—petals four, large; stigmas two, sessile; capsule two-valved, two-celled, the placentas connected by a spongy dissepiment; seeds not crested.

4. Cistaceous Plants. CISTACEÆ.

(16) **Helianthemum**—two outer sepals smaller; petals deciduous capsule three-valved.

5. Resedaceous Plants. RESEDACEÆ.

(17) **Reseda**—flowers in racemes; stamens indefinite, but not numerous, about 8–24, inserted on a glandular disk.

6. Cruciferous Plants. CRUCIFERÆ.

* *Fruit a silique or pod, linear or linear-lanceolate, usually much longer than broad.*

† *Pods two-valved, with a dissepiment.*

‡ *Style not forming a stout conical beak.*

§ *Calyx conspicuously bisaccate at the base.*

(18) **Matthiola**—pod round or compressed; stigma sessile, two-lobed, the lobes erect, gibbous or horned at the back.

(19) **Hesperis**—pod quadrangular or sub-compressed, the valves keeled, somewhat three-nerved; stigma nearly sessile, of two closely-converging erect elliptical obtuse lobes.

§§ *Calyx equal at the base, or very slightly gibbous.*

(20) **Erysimum**—pod tetragonal, the valves prominently keeled, with one longitudinal nerve; stigma obtuse, entire or slightly emarginate; seeds in a single row; calyx erect.

(21) **Sisymbrium**—pod terete, or angled; valves convex, with three longitudinal nerves; stigma entire; seeds in a single row, smooth; calyx spreading.

(22) **Nasturtium**—pod nearly cylindrical, short, the valves convex, almost nerveless; stigma capitate; seeds irregularly in two rows.

(23) **Barbarea**—pod tetragonal, somewhat compressed, the valves convex, with a prominent longitudinal nerve; stigma capitate; seeds in a single row.

L

(24) **Turritis**—pod compressed, the valves slightly convex, with a prominent longitudinal nerve; stigma capitate; seeds in two rows.

(25) **Arabis**—pod compressed, the valves nearly flat, with a prominent longitudinal nerve, or rarely nerveless with numerous longitudinal veins; stigma obtuse; seeds in a single row.

(26) **Cardamine**—pod compressed, the valves flat, nerveless; stigma capitate; seeds in a single row.

‡‡ *Style forming a stout, conical, often seed-bearing beak.*

(27) **Brassica**—calyx erect; pod terete or angular; seeds globose, in a single row.

(28) **Sinapis**—calyx spreading; pod terete or angular; seeds globose, in a single row.

(29) **Diplotaxis**—calyx spreading; pod compressed; seeds oval or oblong, in two rows; beak less distinct than in *Brassica* and *Sinapis*.

†† *Pods without valves or dissepiments.*

(30) **Raphanus**—pods linear or oblong, tapering upwards, divided transversely into several one-seeded joints, forming cells, the lowermost barren.

** *Fruit a silicule or pouch, scarcely more than one-half longer than broad.*

† *Pouch without valves, or one-celled, one-seeded.*

(31) **Cakile**—pouch two-jointed, the joints placed end to end, the upper one angular, deciduous, one-seeded, the lower sometimes sterile.

(32) **Crambe**—pouch two-jointed, the joints placed end to end, the upper globose, one-seeded, the lower barren, stalk-like.

(33) **Senebiera**—pouch somewhat kidney-shaped, almost two-lobed, of two cells placed side by side, not bursting; cells one-seeded.

(34) **Isatis**—pouch laterally compressed, one-celled, one-seeded, the valves keeled, scarcely separating.

†† *Pouch two-valved, with a dissepiment.*

‡ *Pouch laterally compressed, the dissepiment in the narrow diameter, the valves keeled or winged.*

(35) **Iberis**—pouch ovate or roundish, notched, the valves boat-shaped, winged at the back ; seeds one in each cell ; petals unequal, two outer ones much larger.

(36) **Lepidium**—pouch roundish or oblong, entire or notched, the valves compressed, keeled or winged at the back ; seeds one in each cell ; petals equal.

(37) **Capsella**—pouch triangular-obcordate, the valves compressed, keeled but not winged ; seeds numerous.

‡‡ *Pouch dorsally compressed, or globose, the dissepiment oval, in the broadest diameter.*

(38) **Cochlearia**—pouch globose, the valves very convex, one-nerved ; style permanent ; seeds many.

(39) **Armoracia**—pouch elliptical or globose, the valves very convex, nerveless ; style permanent ; seeds many.

(40) **Camelina**—pouch subovate, the valves ventricose, with a linear prolongation at the end, which is confluent with the persistent style ; seeds many.

7. Fumariaceous Plants. FUMARIACEÆ.

(41) **Fumaria**—upper petal spurred at the base ; fruit an inde-hiscent one-seeded nut.

(42) **Corydalis**—upper petal spurred at the base ; pod two-valved, many-seeded.

8. Frankeniaceous Plants. FRANKENIACEÆ.

(43) **Frankenia**—sepals combined into a tubular calyx, with

four or five teeth; petals four or five, with long claws and spreading laminas.

9. **Tamaricaceous Plants.** Tamaricaceæ.

(44) **Tamarix**—flowers in racemes; stamens as many or twice as many as the petals.

10. **Caryophyllaceous Plants.** Caryophyllaceæ.

   * *Sepals united into a cylindrical tube; stamens united below into a tube.*

     † *Calyx with two to four scaly bracts at the base, or sometimes covered by them.*

(45) **Dianthus**—calyx five-toothed; petals five, clawed; styles two; capsule one-celled, many-seeded, opening at the top with four valves.

     †† *Calyx without scales at the base.*

(46) **Silene**—calyx five-toothed; petals five, clawed; styles three; capsule more or less completely three-celled, opening at the top with six valves.

(47) **Lychnis**—calyx five-toothed; petals five, clawed; styles five; capsule one- or half five-celled, opening at the top with five or ten teeth.

    ** *Sepals distinct or only cohering at the base; stamens free.*
    † *Valves of the capsule equalling the styles.*

(48) **Sagina**—sepals four or five, spreading when in fruit; petals four or five or none; stamens four or five or twice as many; styles four or five; capsules four- or five-valved, many-seeded.

(49) **Alsine**—sepals five or four; petals five or four, entire or slightly emarginate; styles three; capsules opening with three valves.

†† *Valves or teeth of the capsule twice as many as the styles.*

(50) **Arenaria**—sepals five; petals five, entire or slightly emarginate; styles two or three; capsule opening with four to six valves.

(51) **Stellaria**—sepals five; petals five, bifid; styles three; capsule opening with six valves or teeth, many-seeded.

(52) **Malachium**—sepals five; petals five, bifid or entire; styles five; capsules opening with five bifid, not deeply-cleft, valves.

11. Illecebraceous Plants. ILLECEBRACEÆ.

* *Capsule one-celled, many-seeded.*

(53) **Spergula**—sepals and petals five, entire; stamens ten or five; styles five; capsule five-valved; leaves apparently whorled.

(54) **Spergularia**—sepals five, flattish; petals five, entire; stamens ten, sometimes only five; styles usually three; capsule three- to five-valved; leaves opposite.

(55) **Polycarpon**—sepals keeled on the back, hooded at the end; petals five, minute, emarginate; stamens three to five; styles three; capsule three-valved.

** *Capsule one-seeded.*

(56) **Corrigiola**—petals five, oblong, equalling the sepals; stamens five; stigmas three, sessile.

(57) **Illecebrum**—sepals thickened, horned at the back; petals five subulate, or wanting; stamens five; stigmas two.

12. Elatinaceous Plants. ELATINACEÆ.

(58) **Elatine**—capsule three- or four-celled, many-seeded.

13. Linaceous Plants. LINACEÆ.

(59) **Linum**—sepals, petals, stamens, and styles, five each; capsule ten-celled, ten-valved.

(60) **Radiola**—sepals four, connected below, deeply trifid; petals, stamens, and styles, four each; capsule eight-celled, eight-valved.

14. Geraniaceous Plants. GERANIACEÆ.

(61) **Geranium**—perfect stamens ten, the five alternate ones shorter.

(62) **Erodium**—perfect stamens five, with five rudimentary ones alternating with them.

15. Hypericaceous Plants. HYPERICINEÆ.

(63) **Hypericum**—capsule three- to five-celled, many-seeded.

16. Malvaceous Plants. MALVACEÆ.

   * *Outer bracts three, distinct, inserted on the calyx.*

(64) **Malva**—styles numerous; capsule orbicular, many-celled, the cells one-seeded.

   ** *Outer bracts united at the base into an involucre surrounding the proper calyx.*

(65) **Althæa**—styles numerous; involucre 5–9-lobed; capsule orbicular, many-celled, the cells one-seeded.

(66) **Lavatera**—styles numerous; involucre three-lobed; capsule orbicular, many-celled, the cells one-seeded.

17. Tiliaceous Plants. TILIACEÆ.

(67) **Tilia**—ovary five-celled, the cells with two ovules attached at the inner angle; fruit a small 1–2-seeded globular nut.

18. Balsaminaceous Plants. BALSAMINACEÆ.

(68) **Impatiens**—ovary five-celled, with several ovules in each cell; capsule bursting elastically into five valves, which roll inwards scattering the seeds.

19. Leguminous Plants. LEGUMINOSÆ.

* Stamens monadelphous.

(69) **Ononis**—calyx nearly equally five-cleft, campanulate, the segments narrow; lower petal or keel beaked; leaves simple or trifoliolate.

(70) **Anthyllis**—calyx tubular, inflated, five-cleft, segments unequal; keel not beaked; leaves pinnate.

** Stamens diadelphous.

† Leaves of three leaflets.

(71) **Medicago**—calyx five-toothed, nearly equal; keel of corolla obtuse; ovary curved; pod falcate or spirally twisted; leaves trifoliolate.

(72) **Melilotus**—calyx five-toothed, nearly equal; petals deciduous; keel obtuse; ovary straight; pod subglobose or oblong; leaves trifoliolate; flowers in long racemes.

(73) **Trigonella**—calyx five-toothed, nearly equal; petals distinct; keel obtuse; pod straight, or slightly curved, eight-seeded, much longer than the calyx; leaves trifoliolate.

(74) **Trifolium**—calyx five-toothed, unequal; petals persistent, cohering by their claws; keel obtuse; pod oval, 1–4-seeded, scarcely longer than the calyx; leaves trifoliolate.

(75) **Lotus**—calyx five-toothed, nearly equal; keel ascending, with a narrowed point or beak; style kneed at the base; pod linear, many-seeded, two-valved; leaves trifoliolate.

†† Leaves pinnate, or apparently simple, tendrilled; (pod one-celled, two-valved).

(76) **Vicia**—calyx five-cleft or five-toothed; style filiform, its upper part hairy all over, or bearded on the under side.

(77) **Lathyrus**—calyx five-cleft or five-toothed; style dilated upwards, plane on the upper side, and hairy below the stigma.

††† *Leaves pinnate, without tendrils.*
‡ *Pod bursting, imperfectly two-celled.*

(78) **Astragalus**—calyx five-toothed; keel obtuse; pods imperfectly two-celled, the cells formed by the inflexed margin of the lower suture.

(79) **Oxytropis**—calyx five-toothed; keel with a narrow straight point; pod imperfectly two-celled, the cells formed by the inflexed margin of the upper suture.

‡‡ *Pod indehiscent.*

(80) **Onobrychis**—calyx five-toothed, nearly equal; keel obliquely truncate longer than the wings; pod compressed, one-celled, one-seeded.

20. Rosaceous Plants. ROSACEÆ.

* *Petals four five or more.*
† *Fruit consisting of one or more follicles, containing 2–6 seeds, and invested by a dry calyx.*

(81) **Spiræa**—calyx five-cleft; stamens numerous, inserted along with the petals on a disk adhering to the calyx; follicles usually distinct.

†† *Fruit composed of numerous (rarely 1–2) small nuts or drupes (one-seeded) invested by a dry calyx.*
‡ *Fruit and receptacle both dry.*
§ *Calyx with external bracts, i.e. in two rows.*

(82) **Potentilla**—calyx 8–10-parted, in two rows; stamens numerous; styles withering; fruits numerous, placed upon a flattish receptacle.

(83) **Sibbaldia**—calyx ten-parted, in two rows; stamens five; styles withering; fruits five to ten.

(84) **Geum**—calyx ten-cleft, in two rows; stamens numerous; fruits numerous, tipped with the persistent jointed styles, hooked at the joint.

§§ *Calyx without external bracts.*

(85) **Dryas**—calyx 8–9-cleft, in one row; stamens numerous; fruits numerous, tipped with the persistent hairy styles, which are straight at the extremity.

(86) **Agrimonia**—calyx five-cleft, the tube turbinate, armed with hooked bristles above, and contracted at the throat; stamens fifteen, inserted with the petals into a glandular ring in the throat of the calyx; nuts two, included in the dry tube of the calyx; style terminal.

‡‡ *Fruit dry, receptacle succulent.*

(87) **Fragaria**—calyx ten-parted, in two rows; stamens numerous; receptacle large, succulent, pulpy, deciduous; style lateral near the base of the nut.

‡‡‡ *Fruit succulent, receptacle dry or spongy.*

(88) **Rubus**—calyx concave or flattish, five-parted; stamens numerous; styles nearly terminal; carpels numerous, succulent, drupaceous, placed upon a spongy receptacle.

††† *Fruit formed of numerous small dry nuts enclosed in the fleshy tube of the calyx, which is contracted at the orifice.*

(89) **Rosa**—calyx urceolate, contracted at the mouth, ultimately fleshy, five-cleft; stamens numerous, inserted with the petals on the rim of the tube of the calyx.

** *Petals none.*
† *Stamens four.*

(90) **Sanguisorba**—calyx four-cleft, with two or three external scales at its base, the tube quadrangular; stamens four, opposite to the segments of the calyx.

†† *Stamens from twenty to thirty.*

(91) **Poterium**—flowers monœcious or polygamous; calyx four-

cleft, with three external scales at its base, the tube quadrangular.

21. Crassulaceous Plants. CRASSULACEÆ.

* *Stamens as many as the petals, alternating with them.*

(92) Tillæa—sepals, petals, stamens, and carpels three or four.

** *Stamens twice as many as the petals, and opposite to them.*

(93) **Cotyledon**—sepals five; petals cohering in a tubular five-cleft corolla; stamens ten, inserted on the corolla; hypogynous scales concave.

(94) **Sedum**—sepals and petals five, rarely four, distinct; stamens ten, rarely eight; hypogynous scales entire.

(95) **Sempervivum**—sepals and petals 6–20, usually 10–12; stamens twice as many; hypogynous scales laciniated.

22. Saxifragaceous Plants. SAXIFRAGACEÆ.

* *Stamens ten.*

(96) **Saxifraga**—calyx five-cleft or five-parted; petals five; styles two, persistent, forming two beaks to the carpels.

** *Stamens five.*

(97) **Drosera**—calyx deeply five-cleft; petals five; styles 3–5, deeply bifid; capsule one-celled, 3–5-valved, many-seeded.

23. Lythraceous Plants. LYTHRACEÆ.

(98) **Lythrum**—calyx tubular, cylindrical, with 8–12 teeth; petals 4–6, longer than the calyx teeth; style filiform.

(99) **Peplis**—calyx campanulate, with twelve teeth; petals six, minute, fugacious; style very short.

24. Onagraceous Plants. ONAGRACEÆ.

* *Stamens eight; (flowers regular or nearly so).*

(100) **Epilobium**—calyx four-cleft, deciduous; seeds bearded; flowers purplish or white.

(101) **Œnothera**—calyx tubular below, four-lobed; seeds not bearded; flowers large, yellow.

** *Stamens two; (flowers very irregular).*

(102) **Circæa**—calyx deeply two-cleft; petals two.

25. Haloragaceous Plants. HALORAGACEÆ.

* *Stamens eight (sometimes four or six).*

(103) **Myriophyllum**—monœcious; stigmas four, sessile on the ovary; fruit tetragonal, separable into four nuts.

** *Stamens one only.*

(104) **Hippuris**—stigma one, sessile on the ovary; fruit a one-celled nut.

26. Cornaceous Plants. CORNACEÆ.

(105) **Cornus**—petals distinct; fruit a succulent drupe.

27. Umbelliferous Plants. UMBELLIFERÆ.

* *Umbels simple or imperfectly compound.*

† *Fruit without vittæ (oil-cysts); albumen solid (not furrowed).*

‡ *Fruit laterally compressed.*

(106) **Hydrocotyle**—fruit flat nearly orbicular, of two carpels, each with five filiform ridges; commissure linear; calyx teeth obsolete.

‡‡ *Transverse section of fruit nearly round.*

(107) **Sanicula**—fruit subglobose, covered with hooked spines; without ridges; calyx of five leafy teeth.

(108) **Eryngium**—fruit obovate, covered with chaffy scales without ridges; calyx of five leafy teeth.

†† *Fruit with vittæ; albumen furrowed or involute on the inner face.*

(131) **Torilis**—*see* page 158.

** *Umbels compound or perfect.*
† *Fruit not prickly.*
‡ *Albumen solid (not indented with a longitudinal furrow)*
§ *Fruit laterally compressed.*
‖ *Leaves compoundly divided.*
(a) *Calyx teeth foliaceous.*

(109) **Cicuta**—fruit roundish-cordate; carpels with five equal broad flattened ridges, the lateral marginal; interstices with single vittæ.

(b) *Calyx teeth small or obsolete; petals entire.*

(110) **Apium**—fruit roundish-ovate; carpels with five filiform equal ridges; vittæ single in the interstices; involucre none.

(111) **Helosciadium**—fruit ovate or oblong; carpels with five filiform prominent equal ridges; vittæ single in the interstices; involucre many-leaved.

(c) *Calyx teeth small or obsolete; petals obcordate or emarginate.*
1. *Vittæ short, clavate, single between the ribs.*

(112) **Sison**—fruit ovate; carpels with five filiform equal ridges.

2. *Vittæ elongated, linear.*

(113) **Bunium**—fruit oblong; carpels with five filiform equal ridges; the interstices with 1–3 vittæ.

(114) **Sium**—fruit ovoid or globose; carpels with five filiform obtuse equal ridges; vittæ three or more between the ribs; calyx of five small teeth.

(115) **Pimpinella**—fruit ovate; carpels with five filiform equal ridges; vittæ three or more between the slender ribs; calyx obsolete.

|||| *Leaves simple.*

(116) **Bupleurum**—fruit ovate-oblong; carpels with equal winged, or filiform sharp, or obsolete ridges.

§§ *Fruit ovate or elliptical, rounded, or slightly compressed dorsally.*
|| *Vittæ (of the carpels) single between the ribs.*

(117) **Œnanthe**—fruit ovoid-cylindrical, crowned with the long erect styles; carpels corky, with five blunt ridges.

(118) **Æthusa**—fruit shortly ovoid, crowned with the reflexed styles; carpels with five thick acutely-keeled edges.

|||| *Vittæ (of the carpels) two or more between the ribs.*

(a) *Seeds without vittæ, cohering with the carpels.*

(119) **Meum**—fruit elliptical, terete; carpels with five sharp somewhat winged ridges.

(b) *Seeds with many vittæ, loose from the carpels.*

(120) **Crithmum**—fruit elliptical, terete; carpels with five sharp elevated slightly-winged ridges.

§§§ *Fruit much compressed dorsally.*
|| *Petals entire.*

(121) **Pastinaca**—fruit with a flat dilated margin; carpels with three slender dorsal ridges, and two distant lateral ones near the edge; vittæ linear, solitary.

(122) **Heracleum**—fruit with margin and ridges as in *Pastinaca*; vittæ short, club-shaped.

(123) **Tordylium**—fruit with a thickened and wrinkled margin; vittæ 1–3 together.

§§§§ *Fruit globose, the carpels scarcely separating.*

(124) **Coriandrum**—fruit globose; carpels with the primary ridges obsolete, the four secondary conspicuous, prominent, keeled; interstices without vittæ.

‡‡ *Albumen furrowed, or involute on the inner face.*
§ *Fruit short, turgid.*

(125) **Conium**—fruit ovate, laterally compressed; carpels with five prominent wavy or crenate ridges, the lateral marginal; interstices striated; vittæ none.

(126) **Smyrnium**—fruit laterally compressed; carpels reniform-oblong, with three dorsal prominent sharp ridges and two lateral marginal nearly obsolete ones; interstices with many vittæ.

§§ *Fruit oblong.*
|| *Fruit with an evident beak; (vittæ none.)*

(127) **Scandix**—fruit with a very long beak; carpels with five obtuse ridges.

(128) **Anthriscus**—fruit with a short beak; carpels without ridges, the beak five-ridged.

||| *Fruit not beaked.*

(129) **Chærophyllum**—fruit not beaked; carpels with five equal obtuse ridges; interstices with single vittæ.

(130) **Myrrhis**—fruit not beaked; carpels covered with a double membrane, the outer with elevated keeled ridges hollow within, the inner close to the seed; vittæ none.

†† *Fruit prickly.*

(131) **Torilis**—fruit slightly laterally compressed; carpels with three dorsal bristly primary ridges, the secondary hidden by the numerous prickles which occupy the interstices.

(132) **Daucus**—fruit dorsally compressed ; carpels with three dorsal bristly primary ridges, the secondary ridges **equal**, winged, with one row of spines.

28. Cucurbitaceous Plants. CUCURBITACEÆ.

(133) **Bryonia**—diœcious ; styles three-cleft ; fruit a globose three-celled berry.

29. Caprifoliaceous Plants. CAPRIFOLIACEÆ.

* *Style single, filiform.*

(134) **Lonicera**—corolla with an irregular limb ; stamens five.

(135) **Linnæa**—corolla-limb nearly regular ; stamens four.

** *Stigmas three, sessile.*

(136) **Sambucus**—corolla rotate ; berry 3–4-seeded ; leaves pinnated.

(137) **Viburnum**—corolla shortly campanulate or funnel-shaped ; berry one-seeded ; leaves entire or lobed, not pinnated.

30. Galiaceous Plants. GALIACEÆ.

* *Corolla rotate, the tube very short.*

(138) **Rubia**—corolla usually five-lobed ; fruit fleshy.

(139) **Galium**—corolla usually four-lobed ; fruit dry, not crowned by the calyx.

** *Corolla-tube distinct, as long as the lobes.*

(140) **Sherardia**—corolla funnel-shaped ; fruit dry, crowned by the four calyx teeth.

31. Valerianaceous Plants. VALERIANACEÆ.

(141) **Centranthus**—corolla-tube spurred at the base ; stamen solitary.

(142) **Valeriana**—corolla-tube slightly swollen, but not spurred at the base ; stamens three ; fruit crowned by a feathery pappus.

32. Dipsacaceous Plants. Dipsacaceæ.

(143) **Dipsacus**—scales of the receptacle between the florets prickly or spinous.

(144) **Scabiosa**—scales of the receptacle between the florets not prickly or changed to hairs.

33. Composite Plants. Compositæ.

* *Florets all tubular, or those of the disk or centre tubular with the outer ones ligulate forming a ray, or slender and filiform; style of the perfect florets not swollen below its branches* (Corymbiferæ).

† *Pappus none, or consisting merely of a border or short scales or teeth.*

‡ *Receptacle scaly; (pappus none).*

(145) **Anthemis**—heads radiant; receptacle convex or conical; involucre imbricated, of few rows; fruit terete, or obtusely tetragonal, with a more or less prominent margin.

(146) **Achillea**—heads radiant; tube of the disk-florets plano-compressed, two-winged; receptacle narrow, slightly elevated; involucre ovate or oblong, imbricated; fruit compressed.

‡‡ *Receptacle without scales; (heads radiant).*

(147) **Chrysanthemum**—involucre hemispherical; receptacle plane or convex; fruit terete, without wings or slightly angular and somewhat winged; pappus none, or consisting of three minute teeth or an elevated border.

(148) **Matricaria**—involucre nearly flat; receptacle elongate-conical; fruit angular, not winged; pappus none.

†† *Pappus of two or three persistent awns.*

(149) **Bidens**—heads discoidal, sometimes radiant; receptacle flat; involucre of two rows, the outer spreading; fruit compressed, angular, rough at the edges.

††† *Pappus pilose, abundant.*

‡ *Involucral scales herbaceous.*

§ *Heads discoid or floscular.*

(150) **Eupatorium**—heads few-flowered; involucre imbricated, cylindrical or oblong, the scales oval-oblong; receptacle naked; fruit angular or striated.

§§ *Heads radiant.*

(151) **Erigeron**—involucre imbricated oblong, the scales linear acute; receptacle naked, foveolate; pappus in many rows; fruit compressed.

(152) **Solidago**—involucre imbricated; receptacle naked; pappus in one row; fruit terete.

(153) **Senecio**—involucre cylindrical or conical, of one row of equal scales, not membranous at the edge, with or without smaller scales at the base; receptacle naked; fruit cylindrical, with a pappus of simple soft hairs.

(154) **Inula**—involucre imbricated, in many rows; receptacle naked; pappus uniform, in one row; anthers with two bristles at the base.

(155) **Pulicaria**—involucre loosely imbricated in few rows; receptacle naked; pappus in two rows, the outer cup-like, membranous; anthers with two bristles at the base.

‡‡ *Involucral scales dry, scarious.*

(156) **Gnaphalium**—involucre hemispherical, imbricated, the scales unequal, blunt; receptacle flat, naked.

(157) **Filago**—involucre subconical, imbricated, the scales equal, acuminate; receptacle conical, scaly at the margin.

** *Florets all tubular; style swollen below its branches* (Cynarocephalæ).

† *Involucral scales hooked at the point.*

(158) **Arctium**—involucre globose, imbricated; receptacle flat,

with rigid subulate scales; fruit compressed, oblong; pappus short, pilose, distinct.

†† *Involucral scales not hooked.*

‡ *Pappus single, long, plumose, the hairs unequally united at the base; (achenes silky).*

(159) **Carlina**—involucre imbricated, the outer scales spinose, leafy, spreading at the points, the inner linear, scarious, coloured, resembling radiant florets; receptacle with cleft scales.

‡‡ *Pappus in many rows, equal, long, the hairs united at the base into a ring; (achenes smooth).*

(160) **Carduus**—involucre imbricated, with simple spinous pointed scales; receptacle with fimbriated scales; fruit compressed, oblong.

(161) **Silybum**—involucre imbricated, the scales leaf-like at the base, narrowed into a long spreading or recurved spinous point; receptacle scaly; fruit compressed, with a papillose ring at the end.

‡‡‡ *Pappus in many rows, unequal, equalling or shorter than the achenes, rarely none.*

(162) **Centaurea**—involucre imbricated, the scales leafy, scarious or spiny; receptacle with jagged scales; fruit attached laterally above the base.

‡‡‡‡ *Pappus in many rows, unequal, longer than the fruit.*

(163) **Serratula**—heads diœcious; involucre imbricated, the scales unarmed, acute; receptacle with scales split into linear bristles; fruit compressed.

*** *Florets all ligulate and perfect; style not swollen beneath its branches* (Cichoraceæ).

† *Pappus none.*

(164) **Lapsana**—heads 8–12-flowered; involucre with one row of

linear-lanceolate erect scales, and 4–5 external ones; receptacle naked; fruit compressed, striated.

†† *Pappus sessile, setaceous in the ray, feathery in the disk florets.*

(165) **Thrincia**—involucre oblong, in one row, with a few additional scales at the base; receptacle punctured; fruit beaked; pappus in two rows, the outer setaceous, deciduous, the inner longer, feathery; marginal row of fruits, scarcely beaked, with a short scaly fimbriated pappus.

(166) **Leontodon**—involucre imbricated, the exterior scales much smaller in 1–3 rows; receptacle punctured; fruit smooth, slightly beaked; pappus in two rows, the outer setaceous, persistent, the inner longer, feathery.

††† *Pappus sessile, uniform in the ray and disk florets.*

‡ *Pappus in one row.*

(167) **Oporinia**—involucre imbricated, the exterior scales much smaller, in several rows; receptacle not scaly; fruit attenuated, uniform; pappus feathery, dilated at the base.

(168) **Hieracium**—involucre imbricated, with many oblong scales; receptacle not scaly; fruit terete, ribbed, not beaked, with a very short crenulated margin; pappus pilose.

‡‡ *Pappus in two rows.*

(169) **Picris**—involucre of one row of equal scales, with unequal linear often spreading scales at the base; receptacle dotted; fruit terete, transversely striated, constricted or slightly beaked above; pappus feathery, the outer row subpilose.

(170) **Cichorium**—involucre in two rows, the outer of 5 shortish, the inner of 8–10 longer converging scales; receptacle naked, or slightly pilose; fruit crowned by a ring of two rows of minute erect scales.

‡‡‡ *Pappus in many rows.*

171) **Sonchus**—involucre imbricated, with two or three rows of

unequal scales; fruit plano-compressed, truncate above, not beaked.

(172) **Crepis**—involucre double, the inner of one row, the outer of short lax scales; fruit terete, narrowed upwards, or obscurely beaked.

†††† *Pappus stalked.*
‡ *Pappus in two rows.*

(173) **Hypochœris**—involucre oblong, imbricated; receptacle scaly; fruit glabrous, muricated, often beaked.

‡‡ *Pappus in many rows.*
§ *Pappus feathery.*

(174) **Helminthia**—involucre of eight equal scales, in one row, surrounded by three to five loose leafy bracts; receptacle dotted; fruit compressed, transversely rugose, rounded at the end, and with a slender beak longer than itself.

(175) **Tragopogon**—involucre simple, of eight to ten scales connected at the base; receptacle punctured; fruit longitudinally striated, with a long beak.

§§ *Pappus hair-like.*

(176) **Lactuca**—heads few-flowered; involucre imbricated, in two to four rows, the outer row shorter, the scales with a membranous margin; fruit plano-compressed, contracted and produced into a filiform beak.

(177) **Barkhausia**—heads many-flowered; involucre double, the inner of one row, the outer of short lax scales; fruit terete, all (or the inner ones only) gradually contracted into a long beak.

34. Lobeliaceous Plants. LOBELIACEÆ.

(178) **Lobelia**—corolla very irregular, split open on the upper side; anthers closely cohering.

35. Campanulaceous Plants. CAMPANULACEÆ.

* Corolla cut into deep narrow-linear segments.

(179) **Jasione**—corolla rotate, with five long linear segments; anthers cohering at their base; capsule two-celled, opening by a large pore at the end.

(180) **Phyteuma**—corolla rotate, with five long linear segments; anthers free, the filaments dilated at the base; capsule 2–3-celled, bursting at the sides.

** Corolla cut into short broad segments.

(181) **Campanula**—corolla bell-shaped, tubular, or rotate, with five broad segments; anthers free, the filaments dilated at the base; capsule 3–5-celled, crowned by the calyx-teeth, and opening laterally or at the top.

36. Vacciniaceous Plants. VACCINIACEÆ.

(182) **Oxycoccus**—corolla very deeply four-cleft, with reflexed segments; berry crowned by the persistent calyx-teeth.

37. Ericaceous Plants. ERICACEÆ.

* Anthers opening by pores.

(183) **Erica**—calyx four-parted; corolla campanulate or ovate, often ventricose, four-toothed; capsule four-celled.

(184) **Pyrola**—calyx five-parted; corolla of five petals; capsule five-celled, five-valved.

** Anthers opening by a transverse fissure.

(185) **Monotropa**—calyx 4–5-parted; corolla of 4–5 petals, each with a hooded nectariferous base; capsule five-celled, five-valved.

38. Primulaceous Plants. Primulaceæ.

  * *Ovary superior.*

   † *Calyx divided almost to the base; (submerged aquatics with divided leaves.)*

(186) Hottonia—calyx five-parted; corolla salver-shaped.

   †† *Calyx tubular or campanulate; (terrestrial plants with undivided leaves).*
   ‡ *Leaves radical.*

(187) **Primula**—calyx tubular; corolla salver-shaped; stamens five; flowers on radical peduncles.

   ‡‡ *Leaves opposite or whorled, cauline.*

(188) **Lysimachia**—calyx five-parted; corolla rotate, with a five-parted limb; stamens five; capsule opening at top.

(189) **Anagallis**—calyx five-parted; corolla rotate or funnel-shaped, with a five-parted limb; stamens five; capsule opening all round transversely.

  ** *Ovary half superior.*

(190) Samolus—calyx five-parted; corolla salver-shaped, the tube short; stamens five; flowers in a terminal raceme.

39. Plumbaginaceous Plants. Plumbaginaceæ.

(191) Statice—flowers panicled, arranged on one side of the branches in long rows, surrounded by scarious scales.

(192) Armeria—flowers in globular heads, surrounded by a common, imbricated, scarious involucre.

40. Plantaginaceous Plants. Plantaginaceæ.

(193) Plantago—flowers hermaphrodite, in terminal heads or spikes.

(194) Littorella—flowers unisexual, solitary or in pairs.

41. Lentibulariaceous Plants. LENTIBULARIACEÆ.

(195) **Pinguicula**—calyx 4–5-lobed ; leaves entire, radical.

(196) **Utricularia**—calyx two-lobed ; leaves floating, much divided.

42. Aquifoliaceous Plants. AQUIFOLIACEÆ.

(197) **Ilex**—flowers small, clustered in the axils of the alternate leaves.

43. Oleaceous Plants. OLEACEÆ.

(198) **Fraxinus**—calyx and corolla none; leaves pinnate; fruit dry, oblong, winged.

(199) **Ligustrum**—calyx and corolla four-parted ; leaves simple fruit a berry.

44. Convolvulaceous Plants. CONVOLVULACEÆ.

* *Stems furnished with leaves.*

(200) **Convolvulus**—calyx five-parted, with two small, distant, basal bracts ; corolla bell-shaped, five-plaited.

(201) **Calystegia**—calyx five-parted, enclosed in two foliaceous bracts; corolla bell-shaped, five-plaited.

** *Stems thread-like, leafless.*

(202) **Cuscuta**—calyx 4–5-cleft; corolla urceolate, or nearly globular.

45. Polemoniaceous Plants. POLEMONIACEÆ.

(203) **Polemonium**—corolla rotate, with five stamens.

46. Gentianaceous Plants. GENTIANACEÆ.

* *Leaves opposite ; (terrestrial herbs).*
† *Stamens four.*

(204) **Erythræa**—calyx divided to the base ; style simple ; flowers pink or red.

(205) **Gentiana**—calyx not divided below the middle ; style two-lobed ; flowers blue.

†† *Stamens eight.*

(206) **Chlora**—corolla rotate, yellow.

** *Leaves alternate ; (aquatic herbs).*

(207) **Villarsia**—leaves entire, orbicular ; flowers yellow.

47. Solanaceous Plants. SOLANACEÆ.

* *Fruit dry, capsular.*

(208) **Datura**—corolla funnel-shaped, with a long tube ; capsule prickly, four-valved.

(209) **Hyoscyamus**—corolla obliquely funnel-shaped ; capsule smooth, opening transversely with a lid.

** *Fruit fleshy, baccate.*

(210) **Solanum**—corolla rotate ; anthers porose, joined into a cone.

(211) **Atropa**—corolla bell-shaped ; anthers distinct.

48. Boraginaceous Plants. BORAGINACEÆ.

* *Corolla tube open or naked at the orifice.*
† *Stamens protruding ; corolla irregular.*

(212) **Echium**—corolla bell-shaped ; the throat dilated ; nuts wrinkled, attached by a flat triangular base.

†† *Stamens included ; corolla regular.*

(213) **Lithospermum**—corolla funnel-shaped; nuts smooth or tubercular, stony.

** *Corolla tube closed with scales at the orifice.*
  † *Scales of the corolla subulate.*

(214) **Borago**—corolla rotate; stamens exserted; filaments bifid, the inner fork bearing the anther; anthers connivent into a cone.

†† *Scales of the corolla obtuse, concave.*
  ‡ *Corolla funnel-shaped.*

(215) **Lycopsis**—tube of the corolla doubly curved ; the limb oblique.

(216) **Anchusa**—tube of the corolla straight.

‡‡ *Corolla hypocrateriform.*

(217) **Myosotis**—corolla salver-shaped, with a straight tube ; nuts ovate, compressed, smooth.

(218) **Cynoglossum**—corolla salver-shaped, with a straight tube; nuts roundish-ovate, depressed, muricated.

49. Labiate Plants. LABIATÆ.

* *Stamens two.*

(219) **Salvia**—calyx tubular, two-lipped ; corolla ringent ; anthers one-celled with a lengthened connective.

(220) **Lycopus**—calyx five-cleft; corolla nearly equal; anthers two-celled.

** *Stamens four.*
  † *Upper stamens longer than the lower.*

(221) **Nepeta**—calyx five-toothed ; corolla ringent, the upper lip flat, straight, emarginate or bifid.

†† *Stamens nearly equal.*
‡ *Corolla nearly equal.*

(222) **Mentha**—calyx equal, five-toothed; corolla four-lobed, the tube very short.

(223) **Origanum**—calyx with five equal teeth; corolla five-lobed obscurely two lipped; flowers in four-sided spikes, with imbricated bracts.

(224) **Thymus**—calyx two-lipped; corolla five-lobed, obscurely two-lipped; flowers whorled, axillary, or spiked.

††† *Two upper stamens shorter than the lower.*
‡ *Corolla distinctly two-lipped, the upper lip longer than the stamens.*
§ *Calyx regularly five-lobed.*
‖ *Stamens shorter than the corolla tube.*

(225) **Marrubium**—calyx tubular, the teeth nearly equal; upper lip of corolla straight, erect, cloven; nuts flatly truncate

‖‖ *Stamens longer than the corolla tube.*

(226) **Galeopsis**—calyx tubular, the teeth equal; upper lip of corolla arched; nuts rounded at the end.

(227) **Stachys**—calyx tubular-bellshaped, the teeth equal; upper lip of corolla concave; nuts obtuse and convex at the end.

(228) **Leonurus**—calyx tubular, the teeth nearly equal; upper lip of corolla nearly flat; nuts flatly truncate.

(229) **Ballota**—calyx funnel-shaped, the teeth equal; upper lip of the corolla erect, concave; nuts convex, and rounded at the end.

§§ *Calyx obviously two-lipped.*

(230) **Scutellaria**—calyx ultimately closed and compressed, both lips entire, upper with a concave scale on its back; corolla two-lipped, the upper lip concave.

(231) **Prunella**—calyx ultimately closed and compressed, the

upper lip flat, truncate, three-toothed, the lower bifid; corolla ringent, the upper lip concave.

(232) **Calamintha**—calyx two-lipped, the lower bifid; upper lip of corolla straight, nearly flat.

‡‡ *Corolla apparently one-lipped, the upper lip short or nearly wanting, shorter than the stamens.*

(233) **Teucrium**—calyx tubular, five-toothed; upper lip of corolla bipartite, the tube slit along the back.

50. Verbenaceous Plants. VERBENACEÆ.

(234) **Verbena**—calyx five-toothed; corolla with distinct tube, and unequally five-cleft spreading limb.

51. Scrophulariaceous Plants. SCROPHULARIACEÆ.

* *Corolla rotate, the tube very short.*
† *Stamens two.*

(235) **Veronica**—calyx 4–5-parted; corolla unequally four-lobed, the lower lobe smallest; capsule compressed.

†† *Stamens four.*

(236) **Sibthorpia**—calyx five-lobed; corolla irregularly five-cleft capsule compressed, orbicular.

††† *Stamens five.*

(237) **Verbascum**—calyx deeply five-cleft; corolla with five broad rounded lobes; stamens five, often woolly; capsule ovoid.

** *Corolla distinctly tubular.*
† *Corolla spurred at the base in front.*

(238) **Linaria**—calyx five-parted; corolla personate, the lower lip three-cleft, with a prominent palate closing the mouth.

†† Corolla gibbous at the base in front.

(239) **Antirrhinum**—calyx five-parted; corolla personate, the
lower lip three-cleft, with a prominent palate closing the
mouth.

††† Corolla equal at the base in front.
‡ Calyx five-cleft.

(240) **Scrophularia**—corolla globose, the limb minute, upper lip
two-lobed, lower three-lobed; rudiment of a fifth stamen
often present.

(241) **Digitalis**—corolla bell-shaped, with broad tube, and oblique,
4–5-cleft flat limb.

(242) **Pedicularis**—calyx inflated; corolla ringent, the upper lip
compressed laterally, the lower plane, three-lobed.

‡‡ Calyx four-cleft.

(243) **Bartsia**—calyx bell-shaped; corolla tubular, two-lipped, the
upper lip erect, arched, or concave, entire or notched;
seeds many, striated or furrowed.

(244) **Euphrasia**—calyx tubular or bell-shaped; corolla tubular, two-lipped, the upper lip two-lobed, spreading or
reflexed; seeds few, furrowed.

(245) **Rhinanthus**—calyx inflated, compressed, with a contracted
mouth; corolla ringent, the upper lip compressed laterally;
seeds compressed, with an orbicular margin.

(246) **Melampyrum**—calyx tubular; corolla ringent, upper lip
compressed laterally, with reflexed margins; seeds smooth.

52. Orobanchaceous Plants. OROBANCHACEÆ.

(247) **Orobanche**—calyx deeply divided into 2–4 pointed sepals;
corolla tubular or bell-shaped, often curved.

53. Chenopodiaceous Plants. CHENOPODIACEÆ.

* *Stem jointed, without leaves.*

(248) **Salicornia**—stamens 1–2; succulent maritime herbs.

* Stem continuous; leaves semicylindrical.

(249) **Suæda**—leaves succulent, not prickly; perianth segments remaining unaltered.

(250) **Salsola**—leaves prickly; perianth segments acquiring a dorsal appendage or wing.

*** Stem continuous; leaves flat.

(251) **Beta**—perianth segments succulent, becoming enlarged, with the lower half of the ovary adherent to them.

(252) **Chenopodium**—perianth segments herbaceous, scarcely enlarged, the ovary free.

(253) **Atriplex**—polygamous or monœcious; perianth of female flowers, consisting, when in fruit, of two much enlarged segments.

54. Ceratophyllaceous Plants. CERATOPHYLLACEÆ.

(254) **Ceratophyllum**—leaves whorled, 2–3 times furcately dissected.

(255) **Callitriche**—leaves opposite, entire.

55. Urticaceous Plants. URTICACEÆ.

(256) **Urtica**—perianth of female flowers two-sepaled; stamens four.

(257) **Parietaria**—perianth of female flowers four-cleft; stamens four.

(258) **Humulus**—perianth of female flowers a mere scale; stamens five.

56. Polygonaceous Plants. POLYGONACEÆ.

(259) **Rumex**—perianth six-parted, the three inner segments becoming enlarged.

(260) **Oxyria**—perianth four-parted, the two inner segments enlarged.

(261) **Polygonum**—perianth five-parted, the segments nearly equal.

57. Santalaceous Plants. SANTALACEÆ.

(262) **Thesium**—leaves alternate, entire, without stipules.

58. Euphorbiaceous Plants. EUPHORBIACEÆ.

(263) **Euphorbia**—flowers collected in heads, surrounded by a 4–5-toothed perianth-like involucre, enclosing several male flowers (each a single stamen, with jointed filament), and one female (a solitary three-celled stalked ovary).

(264) **Mercurialis**—herbs, with the male and female flowers separate, diœcious or monœcious.

59. Aristolochiaceous Plants. ARISTOLOCHIACEÆ.

(265) **Aristolochia**—perianth tubular, very oblique; stamens six.

60 Trilliaceous Plants. TRILLIACEÆ.

(266) **Paris**—perianth of 8–10 narrow, spreading or reflexed segments.

61. Dioscoreaceous Plants. DIOSCOREACEÆ.

(267) **Tamus**—perianth of six segments.

62. Orontiaceous Plants. ORONTIACEÆ.

(268) **Acorus**—spathe (flower sheath) a continuation of the scape, and similar to the leaves, not convolute.

63. Typhaceous Plants. TYPHACEÆ.

(269) **Typha**—spadix (flower spike) cylindrical.
(270) **Sparganium**—spadix spherical.

64. Pistiaceous Plants. PISTIACEÆ.

(271) **Lemna**—flowers monœcious; spathe urceolate; stamens 1-2.

65. Naiadaceous Plants. NAIADACEÆ.

(272) **Potamogeton**—flowers perfect, in pedunculated spikes or heads.

(273) **Zannichellia**—flowers monœcious, axillary and sessile.

66. Hydrocharidaceous Plants. HYDROCHARIDACEÆ.

(274) **Stratiotes**—stem scarcely any; leaves sword-shaped, tufted, radical, submerged; flowers polyandrous.

(275) **Hydrocharis**—stem root-like, bearing tufts of roundish floating leaves; flowers 9-12-androus.

(276) **Anacharis**—stems submerged, branched, with small opposite or whorled leaves; flowers minute, with nine stamens, or three styles.

67. Iridaceous Plants. IRIDACEÆ.

(277) **Iris**—perianth six-leaved, the three outer larger; stigmas three, petal-like, arching over the stamens.

(278) **Gladiolus**—perianth six-leaved, the divisions similar, with a somewhat oblique two-lipped arrangement.

68. Orchidaceous Plants. ORCHIDACEÆ.

* *Anther single, the pollen simple, or consisting of granules in a slight state of cohesion.*

(279) **Listera**—perianth ringent; lip deflexed, linear, or oblong, two-lobed; anther parallel with the stigma, enclosed in a hood.

(280) **Epipactis**—perianth patent; lip interrupted, the basal division concave, terminal one larger, with two projecting plates at its base above; anther terminal.

(281) **Cephalanthera**—perianth converging; lip interrupted, the basal division saccate, jointed to the recurved terminal one, which is without protuberances at the base; anther terminal.

  ** *Anther single, the pollen cohering in grains or masses, which are indefinite in number and waxy.*

  ‡ *Lip spurred.*

(282) **Orchis**—perianth ringent, hooded; lip three-lobed; pollen-masses with two glands enclosed in a common pouch; anther-cells converging at the base.

(283) **Gymnadenia**—as in *Orchis*, but the glands of the pollen-masses naked (without a pouch); anther-cells parallel at the base.

(284) **Habenaria**—perianth ringent, hooded; lip three-lobed or entire; glands of the stalks of the pollen-masses naked; anther-cells diverging at the base.

  ‡‡ *Lip not spurred.*

(285) **Aceras**—as in *Orchis*, except that the spur is wanting.

69. Liliaceous Plants. LILIACEÆ.

  * *Stem leafy and branching.*

(286) **Asparagus**—perianth bell-shaped, tubular below; stamens six; stigmas three, reflexed; flowers by abortion, diœcious.

  ** *Stem leafy and simple, or leafless.*
  † *Rootstock not bulbous.*
  ‡ *Flowers purplish.*

(287) **Simethis**—perianth-leaves spreading, deciduous; filaments bearded; flowers panicled.

  ‡‡ *Flowers yellow.*

(288) **Narthecium**—perianth-leaves persistent; filaments bearded; flowers racemose.

(289) **Tofieldia**—perianth-leaves persistent; flowers in short spikes.

†† *Rootstock bulbous.*

(290) **Allium**—perianth-leaves rather spreading; flowers umbellate.

70. Alismaceous Plants. ALISMACEÆ.

* *Perianth-segments six, large, coloured, nearly equal.*

(291) **Butomus**—perianth coloured, resembling a corolla; stamens nine.

** *Perianth-segments three large coloured, and three small herbaceous.*

(292) **Alisma**—flowers perfect; calycine and corolline segments three each; stamens six; carpels numerous, one-seeded.

(293) **Actinocarpus**—flowers perfect; calycine and corolline segments three each; stamens six; carpels 6–8, spreading in a radiant manner, two-seeded.

(294) **Sagittaria**—flowers monœcious; male flowers with numerous stamens; female flowers with numerous one-seeded compressed carpels upon a globose receptacle.

** *Perianth-segments all small, and but slightly coloured.*

(295) **Triglochin**—perianth of deciduous leaves; stamens six; carpels attached to an angular axis from which they at length separate at the base.

71. Juncaceous Plants. JUNCACEÆ.

(296) **Juncus**—perianth glumaceous, six-leaved; capsules three-celled, three-valved.

(297) **Luzula**—as in *Juncus*, but the capsules one-celled, three-valved.

N

72. Cyperaceous Plants. CYPERACEÆ.

* *Flowers unisexual, the stamens and ovaries within separate glumes.*

(298) **Carex**—flowers in imbricated spikes, each covered by a glume; ovaries enclosed in bottle-shaped utricles formed of the persistent perigones.

** *Flowers perfect, the stamens and ovaries within the same glume.*

   † *Glumes or bracts distichous or two-ranked.*

(299) **Schœnus**—spikelets 2–4-flowered; glumes 6–9, the lower ones smaller, empty.

   †† *Glumes or bracts regularly imbricated.*

    ‡ *Lower glumes empty or larger than the fertile ones, sometimes all fertile.*

    § *Bristles not so long as the glumes, or wanting.*

(300) **Blysmus**—bristles 3–6; nut plano-convex, tipped with the base of the style, which is not dilated; spikelets in two-ranked spikes.

(301) **Eleocharis**—bristles 3–6; nut compressed, crowned with the persistent dilated base of the style; spikelets solitary.

(302) **Scirpus**—bristles about six or none; nut plano-convex or trigonous, tipped with the filiform base of the style, which is not dilated; spikelets solitary clustered or panicled.

    §§ *Bristles at length much longer than the glumes.*

(303) **Eriophorum**—bristles (in the fruiting state) silky; nut trigonous; spikelets solitary or umbellate.

    ‡‡ *Lower glumes empty, or smaller than the fertile ones.*

(304) **Rhynchospora**—bristles about six; nut compressed, convex on both sides, crowned with the dilated base of the style.

73. Graminaceous Plants. GRAMINACEÆ.

* *Spikelets of one perfect terminal flower, with or without a perfect male flower below it.*

† *Flowering glumes of a firmer texture than the empty ones below* (Paniceæ).

(305) **Milium**—panicle loose; spikelets one-flowered; glumes two, flattish, equal; pales two, equal, awnless; stamens three; styles two.

†† *Two male or imperfect florets or minute rudimentary glumes below the perfect flower, besides the outer empty glumes* (Phalarideæ.)

('06) **Digraphis**—panicle somewhat contracted; glumes two, keeled, equal, each with a tuft of hairs at its base, representing an abortive floret; pales two, equal, awnless; stamens 3; styles 2.

** *Spikelets with one or more perfect flowers, the male or rudimentary, if any, terminal.*

† *Spikelets one-flowered, usually pedicellate* (Agrostideæ).

(307) **Phleum**—panicle spiked; glumes two, keeled, equal, longer than the pales, with an awn proceeding from the midrib; pales two, equal, awnless, membranous; stamens three; styles two.

(308) **Alopecurus**—panicle spiked; glumes two, equal, keeled, often connate at the base; pale single, with an awn arising from its base, ribbed; stamens three; styles two.

(309) **Lagurus**—panicle contracted, ovate; glumes two, equal, terminating in a long plumed awn; pales two, the lower with a dorsal twisted awn, bifid at the apex, the lobes long, sharp, awn-like; stamens three; styles two.

'310) **Polypogon**—panicle contracted, oblong; glumes two, equal, the lower awned under the apex, the upper from the apex; pales two, equal, convex, the lower awned beneath the apex; stamens three; styles two.

(311) **Agrostis**—panicle loose; glumes two, nearly equal, the upper smaller, acute, without awns; pales shorter, often bearing an awn below the middle, the inner sometimes wanting; stamens three; styles two.

(312) **Apera**—panicle loose; glumes two, acute, the lower smaller, nearly equal, without awns; outer pales with a hair-like awn, 3–4 times as long as the spikelet; stamens three; styles two.

(313) **Ammophila**—panicle contracted, cylindrical; glumes two, keeled, unequal, awnless, the lower smaller; pales two, the lower awned under the apex, with hairs at the base; stamens three; styles two.

(314) **Calamagrostis**—panicle somewhat spreading; glumes two, convex, nearly equal, keeled, pointed; pales two, unequal, membranous, ribbed, surrounded with hairs at the base, the lower awned; stamens three: styles two.

    †† *Spikelets two- or few-flowered, pedicellate, the flowering glumes (pales) usually shorter than the outer ones, their awns often bent or twisted* (Aveneæ).

(315) **Aira**—panicle loose; spikelets small; florets equally perfect; glumes two, nearly equal; pales two, the lower bifid, with a dorsal straight or slightly-twisted awn; stamens three; styles two.

(316) **Avena**—panicle loose; spikelets large; upper florets sterile and imperfect; glumes two, nearly equal; pales two, knee-bent, the lower bifid, with a twisted dorsal awn; stamens three; styles two.

(317) **Trisetum**—panicle oblong, somewhat spreading; spikelets small, crowded; florets equally perfect; glumes two; pales two, the lower deeply cleft, with a twisted, dorsal, knee-bent awn; stamens three; styles two.

(318) **Arrhenatherum**—panicle loose; upper floret hermaphrodite with a rudimentary prolongation of the axis, lower male; glumes two, nearly equal; pales two, the lower

emarginate, that of the male floret with a twisted awn at the base, that of the hermaphrodite floret with a small straight awn under the apex; stamens three; styles two.

(319) **Holcus**—panicle somewhat open; lower floret hermaphrodite, the upper male; glumes two, nearly equal, boat-shaped; pales two, the outer of the male florets awned · under the apex, of the hermaphrodite floret awnless; stamens three; styles two.

      ††† *Spikelets one-flowered, sessile along one side of the simple linear branches of the panicle* (Chlorideæ).

(320) **Spartina**—panicle of 2–4 erect, spike-like branches; spikelets in two rows, pressed close to the rachis; glumes two, unequal; pales two, nearly equal, both keeled; stamens three; styles two.

      †††† *Spikelets one- or several-flowered, sessile in the notches of a simple spike* (Hordeineæ).
    ‡ *Spikelets one-flowered.*

(321) **Lepturus**—spike slender, cylindrical; glumes two, hard, ribbed; pales two, thin, awnless, about as long as the glumes; stamens three; styles two.

(322) **Nardus**—spike slender, simple, one-sided; glumes none; pales two, the outer wrapping up the inner, which is flat; stamens three; style one.

(323) **Hordeum**—spikes cylindrical, dense; spikelets in threes, 1–3 usually neuter or barren; fertile ones with one perfect floret and a rudimentary neuter one; glumes two, collateral, awned; pales two; stamens three; styles two.

    ‡‡ *Spikelets several-flowered.*

(324) **Elymus**—spike cylindrical, rather dense; spikelets in pairs, each containing 2–4 florets; glumes two, collateral, awnless; pales two, less pointed than the glumes; stamens three; styles two.

(325) **Triticum**—spike simple; spikelets solitary, sessile, transverse to the rachis, many-flowered; glumes two, opposite, nearly equal, embracing the flowers; pales two, lanceolate, the outer acuminate or awned, the inner bifid.

(326) **Lolium**—spike elongated, simple, compressed; spikelets at right angles with the rachis, with a bract at the base of each; glumes two, lateral, often deficient; pales two, nearly equal; the outer sometimes awned under the apex; stamens three; styles two.

(327) **Brachypodium**—spike simple; spikelets transverse to the rachis, with a short pedicel; glumes two, unequal, shorter than the lower florets; pales two, the lower awned under the point, or awnless; stamens three; styles two.

    ††††† *Spikelets several-flowered, pedicellate; awns, if any, straight* (Festuceæ).

  ‡ *Pales awned.*

(328) **Bromus**—panicle loose; spikelets laterally compressed; glumes two, unequal, shorter than the lower florets; pales two, the lower awned under the apex, very seldom awnless; stamens three; styles two.

(329) **Festuca**—panicle loose; spikelets laterally compressed; glumes two, unequal, or nearly equal, acute; pales two, the lower mucronate or awned at the point; stamens three; styles two.

(330) **Dactylis**—panicle loose or somewhat contracted, the terminal ramifications always very short; spikelets clustered, horizontal; glumes two, unequal-sided; pales two, the lower awned under the apex; stamens three; styles two.

(331) **Cynosurus**—panicle contracted, spike-like, unilateral; spikelets two- or many-flowered, distichous, with a pectinated involucre at the base; glumes two, equal; pales two, the lower awned from the apex or mucronate; stamens three; styles two.

‡‡ *Pales awnless.*

(332) **Briza**—panicle loose; spikelets cordate; glumes two equal, convex, about as long as the lower florets; pales two, awnless, the outer boat-shaped; stamens three; styles two.

(333) **Poa**—panicle loose, seldom contracted; spikelets two- or many-flowered, ovate; glumes two, rather unequal, usually keeled; pales two, nearly equal, awnless, scarious at top; stamens three; styles two.

(334) **Glyceria**—panicle loose; spikelets cylindrical; glumes two, nearly equal, obtuse; pales two, nearly equal, awnless; stamens three; styles two.

(335) **Catabrosa**—panicle loose; spikelets two-flowered; glumes two, unequal, rounded or truncate, much shorter than the florets; pales two, nearly equal, truncate, awnless; stamens three; styles two.

(336) **Melica**—panicle loose; spikelets ovate; glumes two, nearly equal, about as long as the florets; pales two, unequal, awnless; stamens three; styles two.

(337) **Arundo**—panicle loose; spikelets many-flowered, the lower floret male and naked, the upper hermaphrodite and surrounded by hairs; glumes two, the lower smaller, the upper about as long as the florets; pales two, unequal, pointed; stamens three; styles two.

[III.—SPECIES AND VARIETIES.]

(1) Clematis. TRAVELLER'S JOY.

C. Vitalba: climbing shrub; leaves pinnate, with about five cordately-ovate stalked leaflets, the stalks forming tendrils; flowers greenish-white, in loose panicles; carpels with long

feathery awns.—Old Man's Beard.—Hedges and thickets.
Fl. July.

(2) **Thalictrum.** MEADOW RUE.

* *Panicle scarcely more than a simple raceme.*

T. alpinum: stem dwarf, simple, almost leafless; leaves
twice ternate, bearing small, roundish, crenate or lobed leaflets; flowers few, drooping; sepals four, small; stamens 10–20.
—Moist alpine pastures.   Fl. June and July.

** *Panicle compound, diffuse.*

T. minus: stem 1–3 feet high, zigzag branched; leaves
three or four times divided, with numerous, small, roundish,
or broadly wedge-shaped, trifid and toothed leaflets; flowers
drooping, pale greenish yellow, the sepals tinged with pink.—
Stony and bushy pastures in limestone districts.   Fl. June,
July.

A variable plant, the varieties distinguished by size, colour,
and pubescence, by luxuriance of foliage, or by the lower
leaves being fully developed or reduced to mere sheaths; there
are three or four British forms.

*** *Panicle compound, compact.*

T. flavum: stem stout, furrowed, 2–3 feet high; leaves
two or three times divided, bearing large obovate leaflets,
wedge-shaped at the base; flowers erect, decidedly yellow, the
panicles somewhat corymbose.—Moist meadows.   Fl. June,
July.

(3) **Adonis.** PHEASANT'S EYE.

A. autumnalis: annual, erect, one foot high, glabrous or
slightly downy; leaves triply pinnatifid, with narrow linear

segments; flowers bright scarlet, with a dark spot at the base of the petals; carpels numerous, collected in ovate or oblong heads.—Cornfields. Fl. July or October.

A variety with larger flowers is cultivated in flower-gardens, under the name of *Flos Adonis*.

(4) **Ranunculus.** CROWFOOT.

* *Flowers white; (aquatic plants).*

R. hederaceus: plant floating or spreading on mud; leaves roundish, ivy-like, with 3–5 broad, shallow lobes; flowers very small, white; carpels and receptacle glabrous.—Shallow water and mud. Fl. July, August.

** *Flowers yellow; (terrestrial plants).*
† *Leaves simple or undivided.*

R. Lingua: stem erect, hollow, 2–3 feet high; leaves long, lanceolate, smooth, entire, or with only a few small teeth at the edge; flowers in a loose panicle, large, the petals bright shining yellow; carpels with a short, broad, flat beak.—Great Spearwort.—Marshes and ditches. Fl. July.

R. Flammula: stems more or less reclining at the base, 1–1½ foot high, with a few loose branches; lowest leaves ovate, the remainder lanceolate or linear, nearly entire; flowers yellow, on long peduncles, seldom more than half an inch in diameter, often much smaller; carpels with a very short, usually hooked beak.—Spearwort. Marshes and wet pastures. Fl. June to August.

Var. *reptans*: stems slender, creeping; flowers much smaller.

†† *Leaves deeply cut.*
‡ *Carpels smooth, that is, having an even surface.*
§ *Leaves smooth, not hairy.*

R. sceleratus: annual, erect, branched, about one foot high, glabrous, the stem hollow; lower leaves stalked, three-parted with obtusely-lobed segments, the upper sessile, trifid with narrow segments; flowers small, numerous, pale yellow; carpels very small and numerous, in a dense oblong head.—Wet places, sides of pools, etc.　Fl. July to September.

§§ *Leaves hairy.*

R. acris : stems erect, six inches to 2–3 feet high, hairy; leaves stalked, palmately three-parted, the segments again cut into three acutely-toothed lobes, those of the upper ones narrower and fewer; flowers rather large, bright yellow, forming large loose panicles, the calyx lobes spreading horizontally; carpels ovate, glabrous.—Buttercups.—Meadows and pastures. Fl. June, July.

R. repens : distinguished by having creeping runners, and spreading calyx-lobes; and

R. bulbosus : known by the bulbous base of its stem, and by its reflexed calyx-lobes—both noticed at p. 43—are also met with in flower during summer.

‡‡ *Carpels slightly tuberculate near the edge.*

R. Philonotis (*hirsutus*) : annual, erect, much branched from the base, ½–1 foot high, hairy; leaves ternate, with stalked trifid leaflets; flowers numerous, rather small, pale yellow, the calyx-lobes reflexed; carpels marked with minute tubercles within the margin.—Moist fields and waste places.　Fl. July to October.

‡‡‡ *Carpels covered with prickles.*

R. arvensis : annual, erect, branching, nearly glabrous, pale green, ½–1½ feet high; leaves once or twice deeply three-cleft, with narrow segments; flowers small, pale yellow; carpels few, rather large, much flattened, covered on both sides with

conical prickles.—Cornfields; a troublesome weed. Fl. June.
[See also p. 43.]

(5) Trollius. GLOBE FLOWER.

T. europæus: stems erect, glabrous, 1–2 feet high, simple, or nearly so; root-leaves palmately divided into three or five segments, which are lobed and cut, those of the stem few, small, nearly sessile; flowers large, pale yellow, with ten to fifteen broad, concave, converging sepals, usually concealing the petals, stamens, and carpels.—Moist mountain pastures. Fl. June, July.

(6) Aquilegia. COLUMBINE.

A. vulgaris: stem 1½–3 feet high, leafy; radical leaves long-stalked, twice ternate, with broad, three-lobed, crenate, glaucous-green segments; flowers in a loose panicle, large, drooping, blue or dull purple, the petals with incurved spurs.— Coppices and open woods. Fl. June, July.

(7) Delphinium. LARKSPUR.

D. Consolida: annual, erect, about one foot high, branched, nearly glabrous; leaves stalked, deeply divided into linear segments, the upper ones sessile; flowers blue, sometimes reddish or white, in loose racemes, the spur of the calyx as long as the rest of the flower; petals two, their appendages united on the under side into an inner spur; carpel solitary.—Cornfields. Fl. June, July.

(8) Aconitum. MONK'S-HOOD.

A. Napellus: stem erect, 1½–2 feet high; leaves deeply 5–7-cleft, with linear pointed segments, stalked, or the upper ones nearly sessile, dark green; flowers, dense, racemose, large, dark blue, on erect stalks; spur of the small upper petals short, conical, and more or less bent downwards; carpels three.—Wolf's-bane.—Moist pastures. Fl. June, July.

(9) **Pæonia.** Pæony.

P. corallina: stem 1-2 feet high; leaves twice ternate, the segments ovate, entire, or divided into two or three deep lobes; flowers deep red; carpels large, thick, downy, and when ripe, more or less recurved.—Naturalized in the rocky clefts of the Steep Holme Island, in the Severn. Fl. May, June.

(10) **Nymphæa.** Water Lily.

N. alba: leaves deeply cordate, glabrous, 6–8 inches in diameter; flowers lying on the surface of the water, white, scentless, 3–4 inches in diameter.—Lakes and slow rivers. Fl. July.

(11) **Nuphar.** Yellow Water Lily.

N. lutea: leaves oblong-cordate; flowers yellow, raised two or three inches above the water, faintly scented, the concave sepals assuming a somewhat globular form; petals and stamens very numerous, scarcely more than half the length of the sepals.—Rivers and pools. Fl. July.

Var *pumila:* plant smaller, flowers with a more indented stigmatic disk.—Lakes of the north of Scotland.

(12) **Papaver.** Poppy.

* *Leaves glaucous, toothed or slightly lobed.*

P. somniferum: annual, erect, glabrous, or with a few hairs on the peduncles, scarcely branched, two feet high or more; leaves clasping the stem by their cordate base, oblong, irregularly toothed, slightly sinuate or lobed; flowers large, usually bluish-white, with a purple base; capsules large, globular, glabrous.—Opium Poppy.—Sandy wastes, especially near the sea, and in the fens of the eastern counties. Fl. July.

** *Leaves not glaucous, hairy, once or twice pinnate.*
† *Capsule smooth.*

P. Rhœas : annual, erect, branched, 1–2 feet high, with stiff spreading hairs; lower leaves stalked, once or twice pinnately divided, the lobes lanceolate, pointed, more or less cut; flowers large, rich scarlet, with a dark eye; capsule smooth, globular or slightly top-shaped.—A cornfield weed. Fl. June, July.

†† *Capsule hairy.*

P. hybridum : annual; stem 1–1½ feet high, branched, leafy; leaves bipinnatifid, with short segments, and few short hairs; flowers smaller, purplish red, with a dark spot in the centre; capsule nearly globular, covered with stiff, spreading bristles. —Sandy or chalky fields; rare. Fl. July.

P. Argemone : annual; stems slender, 1–1½ feet high, leafy, branched; leaves bipinnatifid, the segments few and narrow; flowers rather small, pale red, often with a dark spot; capsule oblong, contracted at the base, *i.e.* clavate, with a few stiff, erect bristles.—Sandy cornfields. Fl. June, July.

(13) **Meconopsis.**

M. cambrica : stems erect, one foot high; leaves long-stalked, pale green, slightly hairy, pinnate, the segments ovate or lanceolate, toothed or pinnately lobed; flowers rather large, pale yellow, on long stalks; capsules narrow ovate or oblong, glabrous.—Welsh Poppy.—Rocky shady places; rare. Fl. June.

(14) Chelidonium. CELANDINE.

C. majus : stems erect, slender, branching, 1–2 feet high, full of a yellow fetid juice ; leaves thin, glaucous underneath, once or twice pinnate, the segments ovate, coarsely toothed or lobed ; flowers small, yellow, in a loose umbel ; pod nearly cylindrical, glabrous, 1½–2 inches long.—Roadsides and waste places. Fl. June.

(15) Glaucium. HORNED POPPY.

G. luteum : annual or biennial ; stem 1–3 feet high, with hard spreading branches, everywhere glaucous ; leaves thick, pinnately lobed or divided, the lobes ovate or lanceolate, sinuate or lobed, rough with short thick hairs, the upper ones shorter, less divided ; flowers large, yellow, the petals very fugacious ; pods 6–12 inches long, crowned by the spreading lobes of the stigma.—Sandy sea-coasts. Fl. June to August.

(16) Helianthemum. ROCKCIST.

* *Annual herb.*

H. guttatum : annual, erect, hairy, often branched at the base, 2–3 inches to one foot high ; leaves narrow-oblong or lanceolate, or the lower ones obovate and very obtuse, upper ones more pointed ; racemes loose, with small flowers on slender pedicels ; petals very fugacious, yellow, with a dark spot at their base.—Warm sandy pastures. Fl. June, July.

** *Dwarf undershrubs.*

H. canum : stem decumbent, compact-growing ; leaves ovate or oblong, small, white underneath or on both sides ; racemes numerous and short ; flowers yellow, very small.—Limestone rocks ; rare. Fl. June.

H. vulgare : stems diffuse, much branched, with procumbent or ascending flowering branches, from a few inches to nearly a foot long; leaves mostly oblong, but varying from ovate to lanceolate, glabrous or slightly hairy, more or less hoary beneath; racemes loose; petals broadly spreading, bright yellow.—Rock Rose.—Dry hilly pastures. Fl. July, August.

Var. *surrejanum :* flowers much smaller; petals narrow, deeply cut.—Supposed to have been originally found near Croydon in Surrey.

(17) Reseda. MIGNONETTE.

* *Leaves undivided.*

R. Luteola : annual or biennial; stems 1–2 feet high, erect, glabrous; leaves linear or lanceolate, 2–3 inches long, slightly waved on the edges; flowers yellowish green, in long, stiff spikes; sepals four; petals 4–5, very unequal, the one or two lower ones entire, the upper ones divided into 2–5 lobes; capsules nearly globular.—Weld or Dyer's Rocket.—Waste places, in chalky districts.—Fl. July, August.

** *Leaves deeply cut.*

R. lutea : stems one foot high, branched; leaves variable, but always deeply divided, mostly once or twice trifid, but occasionally pinnatifid, with few oblong or linear segments, much waved on the margins; flowers in long racemes; sepals usually six, sometimes only five; petals as many, of a greenish-yellow, the lowest entire or two-cleft, the others irregularly divided into two, three, or four lobes; capsule oblong.—Waste places, in limestone districts. Fl. June to August.

(18) Matthiola. STOCK.

M. incana : stem 1–2 feet high, erect, usually perennial,

and almost woody at the base, but not of long duration, with
hard, slightly-spreading branches; leaves oblong-linear, obtuse, entire, soft and hoary on both sides; flowers purple or
reddish, rather large; pod 4–5 inches long, crowned by the
short stigmas.—Gilliflower.—Cliffs in the Isle of Wight. Fl.
May, June.

M. sinuata: stem one foot high, with spreading branches,
perennial, but of short duration; leaves deeply sinuate, covered
over with short hoary down; flowers purple; pods compressed,
usually more or less covered with glandular protuberances.—
Sandy sea-shores of the south and west coasts. Fl. August.

(19) **Hesperis.** Rocket.

H. matronalis: stems 2–3 feet high, slightly branched;
leaves shortly stalked, ovate-lanceolate or lanceolate, 2–3
inches long, the upper ones smaller; flowers lilac, usually
fragrant in the evening; pods 2–4 inches long, nearly cylindrical, much contracted between the seeds.—Dame's Violet.—
Hilly pastures, rare. Fl. June.

(20) **Erysimum.** Treacle Mustard.

E. cheiranthoides: annual; stem 1–2 feet high, stiff, erect,
slightly hoary with appressed hairs; leaves numerous, broadly
lanceolate, entire or slightly toothed, tapering into a stalk;
flowers small, pale yellow; pods numerous, seldom an inch
long.—Waste and cultivated places. Fl. July.

(21) **Sisymbrium.**

* *Leaves deeply pinnatifid.*

S. officinale: annual; stems a foot high or more, erect,
more or less downy, with rigid spreading branches; leaves

deeply pinnatifid, with few lanceolate toothed lobes, the terminal one 1–1½ inch long, the others often directed backwards
towards the stem, upper leaves sometimes undivided and
hastate; flowers very small, yellow; pods half an inch long,
tapering to the point, almost sessile, and closely pressed against
the axis in long, slender racemes, the midribs of the valves
prominent.—Hedge Mustard.—Waste places, by roadsides.
Fl. June, July.

S. Irio: annual; stem a foot high or more, erect, hard,
glabrous; leaves deeply pinnatifid or pinnate, the lobes or
segments lanceolate, more numerous and larger than in the
former; flowers small, yellow; pods 1½–2 inches long, forming dense racemes.—London Rocket.—Waste places, chiefly
about London and other towns. Fl. July, August.

** *Leaves 2–3 times pinnate.*

S. Sophia: annual; stem a foot high or more, erect, slender,
somewhat hoary with very short down; leaves divided into
numerous short linear segments; flowers small, yellow; pods
slender, glabrous, ¾–1 inch long, forming loose erect racemes.
—Flixweed.—Waste places. Fl. July to September.

(22) **Nasturtium.** Water-cress.

* *Flowers white.*

N. officinale: stem much branched, short and creeping, or
floating in shallow water; leaves pinnate, with distinct segments, the terminal one usually larger, ovate or orbicular;
flowers small, white, in short racemes; pods 6–8 lines long,
on spreading pedicels, slightly curved upwards.—The common edible Water-cress.—Brooks and rivulets. Fl. all the
Summer.

** Flowers yellow.
† Pods half an inch long or more.

N. sylvestre : stem creeping, the flowering branches erect
or ascending, a foot high or more ; leaves deeply pinnatifid or
almost pinnate, the lower lobes distinct and narrow, the terminal one often larger and broader; flowers small, yellow, the
petals twice as long as the calyx; pod slender.—River-banks
and wet places. Fl. June to September.

†† Pods not exceeding a quarter of an inch in length.

N. palustre : stem creeping, but weaker and not so tall as
the last ; leaves deeply pinnatifid, the lobes broader and more
toothed ; flowers small, yellow, the petals seldom exceeding
the calyx ; pod about three lines long, slightly curved.—Muddy
and watery places. Fl. June to September.

N. amphibium : stem 2–3 feet, erect ; leaves narrow lanceolate, 3–4 inches long, slightly toothed, more frequently
deeply toothed or lobed, sometimes divided to the midrib into
narrow segments ; flowers yellow, the petals longer than the
calyx : pod straight, elliptical, about two lines long.—Moist
meadows and watery places. Fl. June to August.

(23) **Barbarea.**   WINTER-CRESS.

B. stricta : biennial ; stem 1–2 feet long ; leaves lyrate, the
upper pair of lobes small, much narrower than the large
oblong-ovate terminal lobe, uppermost undivided, toothed ;
flowers small, yellow, the flowering raceme close ; pods straight,
adpressed.—Yellow Rocket.—Hedges. Fl. May to August.

B. præcox : biennial ; stem 1–2 feet long, slender ; leaves
lyrate, the upper pair of lobes as broad as the roundish subcordate terminal lobe, uppermost pinnatifid, with linear-oblong

entire lobes; flowers small, yellow, the raceme close; pods patent, straight.—Waste places. Fl. May to August. [See also p. 45.]

(24) Turritis. TOWER MUSTARD.

T. glabra: annual or biennial; stem two feet or more high, erect, glabrous except a few hairs at the base, usually glaucous; leaves spreading, obovate-oblong, sinuate or pinnately lobed, with a few forked hairs, the upper ones oblong-lanceolate, entire, clasping the stem by pointed auricles; flowers small, white or straw colour; pods long, narrow, erect, crowded in a long narrow raceme.—Banks and roadsides. Fl. June.

(25) Arabis. ROCK-CRESS.

* *Stem-leaves undivided, rounded or auricled at the base.*

A. hirsuta: annual or biennial; stem a foot high or more, rather stiff, erect, usually simple, rough with short hairs; leaves spreading, obovate or oblong, slightly toothed; stem-leaves erect, oblong or lanceolate, all, or the upper ones, clasping by short auricles; flowers small and white; pods slender, 1–2 inches long, erect, crowded in a long raceme.—Walls, banks, and rocks. Fl. May, June.

** *Stem-leaves narrowed at the base.*

A. petræa: stems branched at the base, loosely tufted, almost creeping, but seldom above six inches long; lower leaves obovate or oblong, stalked, mostly pinnately divided, with the terminal lobe largest, some of them nearly entire, the upper ones few, narrow, almost entire; flowers few, white or slightly purplish; pods spreading, rather more than half an inch long. —Moist mountainous places. Fl. July. [See also p. 45.]

(26) Cardamine. BITTER-CRESS.

C. impatiens: annual; stem 1½ foot high, stiff, erect, leafy, simple, or with a few erect branches; leaves pinnate, with numerous lanceolate or almost ovate segments, often deeply toothed; petals very minute, sometimes wanting; pods numerous, about an inch long, the valves rolling back at maturity with much elasticity.—Moist rocks and shady waste places. Fl. May, June. [See also p. 45.]

(27) Brassica.

* *Valves of the pod three-ribbed.*

B. monensis: biennial, or short-lived perennial; stem prostrate, glabrous, sometimes barely six inches high, the leaves mostly radical; leaves pinnatifid or pinnate, the lobes oblong with a few coarse teeth, upper ones more deeply divided, with narrower segments; flowers rather large, pale yellow; pods spreading 1½–2 inches long, terminating in a thick beak, which is from ⅓—½ of the whole length, and contains 1–3 seeds.—Sandy western coasts. Fl. June to August.

Var. *Cheiranthus :* more luxuriant, the stem 1–3 feet high, erect, leafy, hairy at the base.—South Wales and the Channel Islands.

** *Valves of the pod one-ribbed.*
† *Upper leaves stem-clasping, not auricled.*

B. oleracea: stock thick, almost woody, of 2–3 years' duration, branching into erect stems 1–2 feet high; leaves smooth, glaucous, the lower ones stalked, broad, sinuate, or lobed at the base, the upper oblong, clasping the stem by their broad base, but not auricled; flowers rather large, pale yellow; pod spreading, 1½ inches or more in length.—Maritime cliffs, probably escaped from cultivation. Fl. June to August.

The cultivated forms of this species include the garden *Cabbage, Cauliflower, Kale,* etc.

†† *Upper leaves stem-clasping, with auricles.*

B. campestris : annual; stem 1–2 feet high, erect, simple, or scarcely branched; leaves green, slightly glaucous, pinnately divided, with a large terminal lobe, rough with stiff hairs, which are rarely wanting, upper ones narrow-oblong or lanceolate, stem-clasping, with rounded projecting auricles; flowers bright yellow; pods resembling those of the last.—Borders of fields and waste places, a frequent weed of cultivation. Fl. June, July.

The cultivated varieties include the *Turnip* (*B. Napus*), the *Rape-seed* or *Colza* (*B. Rapa*), etc.

(28) Sinapis. MUSTARD.

* *Pods spreading from the axis of the inflorescence.*

S. alba : annual; stem 1–2 feet high, glabrous or with spreading hairs; leaves lyrate, *i.e.* pinnately lobed or divided, the terminal one largest, the lobes ovate or oblong, coarsely toothed; flowers rather large, yellow; pods spreading, knotty, hispid with stiff hairs, shorter than the sword-shaped beak.—Waste and cultivated places; often cultivated for salad. Fl. June.

S. arvensis : annual; stem 1–2 feet high, with a few spreading hairs; leaves rough, the lower ones with one large terminal oval-oblong, coarsely-toothed segment, and a few smaller ones, the upper often undivided; flowers yellow, rather large; pods spreading, about a third occupied by a stout awl-shaped beak, the valves glabrous, or rough with reflexed hairs.—Charlock. —A common weed of cultivation. Fl. June, July.

** *Pods erect, appressed to the axis of inflorescence.*

S. nigra : annual, less hairy than the preceding; stem two feet high; leaves deeply divided, with one large terminal ovate or oblong lobe and a few small lateral ones, the upper often small and entire ; flowers small, yellow ; pods closely pressed against the axis of the long slender racemes, glabrous, with a small beak.—Waste and cultivated places. Fl. June, July.

(29) **Diplotaxis.**

D. tenuifolia : stems 1-2 feet high, glabrous and somewhat glaucous, emitting a disagreeable smell when rubbed ; leaves irregularly pinnate, with few lanceolate or oblong, entire or coarsely-toothed segments, the upper ones often entire or nearly so ; flowers large, lemon-coloured ; pods in a loose raceme, about 1½ inch long, spreading, with numerous small seeds distinctly arranged in two rows.—Old walls, ruins, and waste places. Fl. June to October.

D. muralis : annual ; stem branching from the base, usually about six inches high ; leaves mostly radical, or crowded at the base, ovate-lanceolate, sinuately toothed, sometimes pinnatifid ; flowers much smaller than the last, the pods and seeds similar, but also smaller.—Fields and waste places in the south, near the sea. Fl. all summer.

(30) **Raphanus.** Radish.

R. Raphanistrum : annual ; stem erect or spreading, 1-2 feet high, much branched, with a few stiff hairs at the base ; leaves pinnately divided or lobed, the terminal segment large, obovate or oblong, rough with short hairs, the upper ones narrow and entire ; flowers white with coloured veins, or lilac ; pods beaded, often separating in joints between the seeds, shorter than the very long beak.—Jointed Charlock, a weed of cultivation. Fl. June, July.

Var. *maritimus :* biennial ; leaves more divided, with over-lapping lobes ; flowers yellow ; pods longer in proportion to the beak.—Sea-coasts, rare.

(31) Cakile. SEA ROCKET.

C. maritima : stems hard at the base, with loose, straggling, glabrous branches, a foot long or more; leaves few, thick, fleshy, with few distant oblong or linear lobes; flowers purplish, not unlike those of a Stock ; pods on short thick pedicels, distant, in long racemes, when young linear or lanceolate entire, when ripe separating transversely into two joints, the upper one mitre-shaped, the lower one persistent, not unlike the head of a pike, divided into two points, and containing a pendulous ovule, which seldom enlarges into a seed.— Sandy sea-coasts. Fl. June to September.

(32) Crambe. SEA KALE.

C. maritima : stems two feet high, branched ; leaves stalked, large, thick, smooth, glaucous, roundish, waved, coarsely-toothed, the upper ones few and smaller; panicle large, much branched, the flowers white ; pod 3–4 lines broad, the abortive joint short, within the calyx.—Sandy sea-coasts. Fl. June.

(33) Senebiera.

S. Coronopus : annual ; stems at first forming a close tuft, afterwards spreading , leaves once or twice pinnately divided, the segments linear or wedge-shaped, entire or toothed ; racemes forming close sessile heads, lengthening as the fruit ripens to 1–2 inches ; flowers inconspicuous ; pod small, marked with deep wrinkles, which form a kind of crest round the edge.—A common weed. Fl. June to September.

(34) Isatis. WOAD.

I. tinctoria : stem 1½–3 feet high, branched in the upper part, glabrous; leaves obovate or oblong, 2–4 inches long, coarsely toothed and stalked, the upper ones lanceolate, with prominent auricles; flowers small, yellow; pods hanging from slender pedicels, about 7–8 lines long and 2–2¼ broad, obovate or oblong, tapering to the base.—Cultivated fields, but scarcely naturalized. Fl. July.

(35) Iberis. CANDYTUFT.

I. amara : annual; stem erect, ¼–1 foot high, with a few erect branches, forming a terminal flat corymb; leaves oblong-lanceolate, with a few coarse teeth; flowers white; pod nearly orbicular, the long style projecting from the notch at the top. —Chalky cornfields. Fl. July.

(36) Lepidium. CRESS.

* Pods winged at top.

L. campestre : annual; stem about a foot high, hoary with minute hairs, erect or nearly so, branched in the upper part; leaves stalked, oblong, entire or pinnatifid, with a large ter-minal lobe, the upper ones oblong or lanceolate, clasping the stem with short pointed auricles; flowers very small, white; pods numerous, on spreading pedicels, broadly ovate, nearly surrounded by the wing, slightly notched at the top, with a short, often very minute style.—Mithridate Pepperwort.— Cultivated and waste places. Fl. July.

L. *Smithii* is very near this, but more or less perennial, the stems shorter and decumbent at the base; the foliage more hairy, the flowers not quite so small, and the style pro-minent.

** Pods not winged.

L. latifolium : stems stout, erect, two feet or more in height, glabrous, much branched in the upper part; leaves large, ovate, toothed, on long stalks, those of the stem oblong or broadly lanceolate, 2–3 inches long, the upper sessile, tapering at the base; flowers small, white; pods with the valves scarcely keeled, the style almost imperceptible.—Salt-marshes. Fl. July.

(37) Capsella.

C. Bursa-pastoris : annual; root-leaves spreading, pinnatifid; stem ¼–1 foot high, rather rough, with a few oblong or lanceolate, entire or toothed, clasping leaves, having projecting auricles; pods in a long, loose raceme, triangular, truncate at the top, with the angles slightly rounded.—Shepherd's Purse. —A common weed. Fl. nearly all the year.

(38) Cochlearia. Scurvy Grass.

C. officinalis : annual or biennial; stems seldom above six inches long, diffuse, glabrous, and somewhat fleshy; leaves stalked, orbicular or reniform, the upper ones ovate or oblong, sinuate, sessile; flowers white, in short racemes, the petals obovate, spreading; · pods globular or obovate.—Muddy or sandy soils on the sea-coasts. Fl. June to August.

Var. danica : leaves stalked, the lower cordate, lobed, the upper 3–5-lobed, subdeltoid ; pods ovate.—Sea-coast.

Var. anglica : leaves ovate-oblong, entire, the upper oblong often toothed, sessile ; flowers large ; pods ovate-oblong.—Seacoast.

(39) Armoracia. Horse-radish.

A. rusticana : stems 2–3 feet high, erect; leaves on long stalks, ½–1 foot long, and 4–6 inches broad, sinuate and toothed at the edges, glabrous but rough, the lower stemleaves often deeply toothed, almost pinnatifid ; flowers small, white, in numerous racemes, forming a terminal panicle;

pods ovoid or elliptical.—Waste places, naturalized near the sea. Fl. May, June. The thick roots are eaten as a condiment with meat.

(40) Camelina. GOLD OF PLEASURE.

C. sativa : annual ; stem simple or slightly branched, 1–2 feet high ; lowest leaves stalked, lanceolate, entire, 1–2 inches long, the upper ones sessile, clasping the stem with pointed auricles; pods obovate, in a long, loose raceme.—Corn and flax fields. Fl. June.

(41) Fumaria. FUMITORY.

F. officinalis : annual, tufted or diffuse, ¼–1 foot high ; leaves much divided into numerous flat segments, generally three-lobed, the lobes broadly lanceolate or oblong ; flowers in racemes of 1–2 inches long, at first dense, often lengthening out, red, the sepals ovate-lanceolate, narrower than the corolla tube ; nuts rugose, retusely globose.—Cultivated ground. Fl. May to September. The two following plants are often separated as species :—

Var. *micrantha*: leaf-segments usually small ; flowers pale red, smaller and in closer racemes, the sepals remarkably large, broader than the corolla tube.—Fields.

Var. *parviflora*: leaf-segments narrow ; flowers small, white or rarely red, the sepals very small, sometimes quite minute. —Fields.

F. capreolata : annual, climbing or diffuse ; stems two feet or more high ; leaves cut into numerous divisions, with broadish flat lobes ; flowers white or pale red, the sepals ovate toothed, as broad as the corolla tube ; nuts nearly orbicular.—Walls and Hedges. Fl. June to September.

(42) Corydalis.

C. lutea: stems 6–8 inches high; leaves pale green, much divided, the segments ovate or wedge-shaped, cut into two or three lobes; flowers in short racemes, pale yellow, with a short broad spur; pod three or four lines long.—Naturalized on old walls, rare. Fl. May to August.

C. claviculata: annual; stems slender, intricate, 1–2 feet long, climbing by means of the leaf-stalks, which usually terminate in delicate tendrils; leaf-segments small, ovate or oblong, and often toothed or cut; racemes short, compact; flowers small, white, with a slight yellow tinge, and a very short spur; pod two or three lines long.—Bushy shady places. Fl. June, July.

(43) Frankenia.

F. lævis: diffuse, much branched, spreading 6–8 inches, glabrous or nearly so ; leaves crowded in little opposite clusters along the branches, small, rather thick, and appearing linear from their edges being closely rolled down ; flowers few, sessile among the upper leaves, forming little terminal leafy heads or short spikes, the petals small, pink.—Sea Heath.—Maritime sands and salt-marshes. Fl. July.

(44) Tamarix. TAMARISK.

T. anglica: shrub, 3–6 feet high, with slender erect branches ; leaves crowded, scale-like ; flowers pink or whitish, small, crowded in spikes ½–1½ inch long, often forming branched terminal panicles.—South-west coast. Fl. July.

(45) Dianthus. PINK.

* *Flowers clustered, the scales as long as the calyx.*

D. prolifer: annual; stems ¼–1 foot high, stiff, erect, wiry, glabrous, simple or nearly so; leaves few, narrow, erect flowers small, in compact ovoid, terminal heads. Gravelly pastures. Fl. July.

D. Armeria: annual; stems about one foot high, erect, slightly branched, downy; leaves herbaceous, 1–3 inches long, obtuse, or the upper ones pointed; flowers small and scentless, in terminal clusters, pink with white dots, the petals crenate on the edge.—Gravelly pastures, rare. Fl. July, August.

　　** *Flowers few, distinct, the scales much shorter than the calyx.*

D. deltoides: diffuse, leafy, tufted, the flowering stems ascending, ½–1 foot long, usually forked above the middle; leaves seldom half an inch long, green, glabrous, obtuse, or the upper ones scarcely pointed; flowers scentless, pink or spotted with white, solitary or two together, on short peduncles; calyx-scales half the length of the tube.—Banks and open pastures. Fl. July to October.

D. plumarius: tufted; leaves linear-subulate, glaucous, crowded together on radical shoots; stem ½–1 foot high, 2–5-flowered, the flowers solitary, pale pink, rarely white, fragrant, the petals deeply digitate-multifid; calyx-scales four times shorter than the tube.—Established here and there on old walls and ruins. Fl. June.

The Cheddar Pink (*D. cæsius*) and the Clove Pink (*D. Caryophyllus*), the first smaller, the second larger than *D. plumarius*, occur sometimes in similar situations.

(46) **Silene.** CATCHFLY.

　* *Leaves smooth.*

S. acaulis: dense moss-like tufts, often many inches in diameter, much branched, the very short branches crowned by dense spreading clusters of short, linear, glabrous leaves;

flowers numerous, sessile or on one-flowered peduncles, which seldom attain an inch in length, reddish-purple, the petals obovate, slightly notched, with a small scale at the base.—Moss Campion.—High mountains. Fl. July, August.

S. inflata : stems ½–1 foot long, loosely branched, ascending, glaucous green; leaves ovate-oblong, pointed; flowers few, white, erect or slightly drooping, in loose terminal panicles; calyx at length almost globular, inflated, and much veined; petals two-cleft.—Banks and roadsides. Fl. July.

Var. *maritima :* stems short, diffuse; leaves lanceolate; flowers almost solitary, the petals two-cleft, with a scale at the base.—Sea-shores.

** *Leaves downy or hairy.*

S. anglica : annual; stems ½–1 foot high, hairy, slightly viscid, much branched, erect or decumbent at the base; leaves small, obovate, the upper ones narrow and pointed; flowers small, pale red or whitish, nearly sessile, generally all turned to one side, forming a simple or forked terminal spike, with a linear bract at the base of each; calyx very hairy, becoming ovoid.—Gravelly fields and waste places. Fl. June, July.

S. noctiflora : annual; stem 1–2 feet high, coarse, erect, hairy, viscid, simple or branched; leaves ovate or ovate-lanceolate, shortly stalked, the upper ones narrow and sessile; flowers largish, pale pink or nearly white, opening at night, 2–3, or sometimes several together, in a loose, terminal, dichotomous panicle.—Gravelly fields. Fl. July.

(47) Lychnis.

* *Calyx-lobes shorter than the petals.*

L. vespertina : biennial; stem 1–2 feet high, coarse, hairy, more or less viscid, loosely branched; leaves oval-oblong, usually pointed, the lower ones stalked; flowers few, opening in the evening, and then slightly scented, in loose panicles, rather large, white, or rarely pale pink, usually diœcious; calyx softly hairy.—Hedges, fields, and waste places. Fl. June to September.

** *Calyx-lobes much longer than the petals.*

L. Githago : annual; stem 2–3 feet high, erect, simple or slightly branched, clothed with long, whitish, appressed hairs; leaves long, narrow; flowers on long leafless peduncles, rather large, red, inodorous, remarkable for the long green linear lobes of the calyx, the petals broad, undivided, and without any scales at the base.—Corn Cockle.—Cornfields. Fl. June, July. [See also p. 48.]

(48) **Sagina.** PEARLWORT.

S. nodosa : tufted, often flowering the first year; stems numerous, decumbent, or nearly erect, 2–4 inches high, sparingly branched; leaves small, subulate, those of the stem much shorter, with little clusters of minute ones in their axils; flowers pedicellate, few on each stem, conspicuous, the white obovate petals being twice as long as the calyx; sepals obtuse; the parts of the flower usually in fives, with ten stamens.— Wet sandy places. Fl. July, August.

(49) **Alsine.**

A. tenuifolia : annual; stem 3-4 inches high, slender, erect, much branched, dichotomous, with the small white flowers in the forks, glabrous or minutely downy; leaves subulate; sepals narrow; petals obovate or oblong, shorter than the calyx. —Dry sandy fields. Fl. June.

(50) Arenaria. Sandwort.

* *Leaves fleshy; capsule three to five-valved.*

A. peploides : stems short, procumbent, usually forked from a creeping root-stock; leaves numerous, somewhat fleshy, ovate or elliptical, half an inch long, the upper smaller and broader; flowers few, white, in small, leafy, terminal cymes, more or less unisexual; petals scarcely longer than the sepals; capsule nearly globular, opening in three (sometimes 4–5) broad valves.—Sea Pimpernel, Sea Purslane, or Sea Chickweed.—Maritime sands. Fl. June, July.

** *Leaves thin; capsules six to ten-valved.*

A. serpyllifolia : annual; stems dichotomous, much branched, slender, slightly downy, seldom attaining six inches; high leaves small, ovate, pointed; flowers from the upper axils or forks, on long slender pedicels; petals usually much shorter than the sepals, obovate; capsule opening in six narrow valves. —Walls and dry sands. Fl. July. [See also p. 48.]

(51) Stellaria. Starwort.

S. graminea : stems diffuse or nearly erect, often 1–2 feet long, glabrous, slender, quadrangular; leaves sessile, linear-lanceolate, pointed; flowers small, white, in long loose panicles, which often become lateral as the flowering advances; sepals three-ribbed; petals narrow, deeply cleft, seldom exceeding the calyx.—Lesser Stitchwort.—Meadows and bushy pastures. Fl. June to August.

S. glauca : stems about one foot high, erect, weak, angular, smooth; leaves linear-lanceolate, acute, smooth, glaucous, sessile; flowers solitary or in a few-flowered lax corymb, white; sepals three-ribbed; petals deeply cleft.—Marshy places. Fl. June, July. [See also p. 49.]

(52) Malachium.

M. aquaticum : stems weak, often a foot or more in length ; lower leaves small, cordate-ovate, acuminate, stalked, the upper ones sessile or stem-clasping, 1–2 inches long; flowers white, in the forks of leafy cymes, the pedicels turned down after flowering; **styles** usually five; petals narrow, deeply cleft, about one-half longer than the calyx.—Wet places, along ditches and streams. Fl. June.

(53) Spergula. SPURRY.

S. arvensis : annual ; stems ½–1 foot high, slender, branching at the base, erect or ascending, glabrous or slightly downy ; leaves almost subulate, 1–2 inches long, 6–8 together in two opposite clusters, and spreading so as to appear whorled ; flowers small, white, on long slender pedicels, turned down after flowering, in terminal, forked cymes; stamens frequently ten or five in different flowers of the same plant ; seeds slightly flattened, with or without a narrow scarious border.—Cornfields. Fl. June to August.

(54) Spergularia. SAND-SPURRY.

S. rubra : annual or biennial; stems numerous, branching from the base, forming spreading or prostrate tufts, 3–4–6 inches long, glabrous ; leaves narrow-linear, the scarious stipules at the base conspicuous ; flowers usually pink, rarely nearly white, on short pedicels, in forked cymes, usually leafy at the base ; petals shorter, or rarely rather longer than the sepals; seeds more or less flattened, roughish, often surrounded by a narrow, scarious wing or border.—Sandy or gravelly waste places, chiefly maritime. Fl. July, August.

Var. *marina :* leaves thicker, somewhat fleshy; flowers

larger; the sepals 2–3 lines long; seeds usually bordered, smooth.—Sea-coast.

(55) Polycarpon.

P. tetraphyllum: annual; stems glabrous, much branched, 3–4 inches long, spreading or prostrate; leaves obovate or oblong, really opposite but placed two pairs so close together as to assume the appearance of a whorl of four; flowers very small and numerous, in loose, terminal cymes, the sepals barely a line long; petals much shorter.—South-west coasts. Fl. June, July.

(56) Corrigiola.

C. littoralis: annual; stems short, procumbent or ascending, slender, smooth; leaves linear or oblong, obtuse, tapered at the base; flowers small, white, crowded in little cymose heads at the ends of the branches.—Coasts of Devon and Cornwall. Fl. July, August.

(57) Illecebrum.

I. verticillatum: annual; stems prostrate, glabrous, much branched, the branches 1–3 inches high, ascending; leaves small, opposite, obovate or roundish; flowers white, in whorls in the leaf-axils.—Sandy marshes in Devon and Cornwall. Fl. July.

(58) Elatine.

E. hexandra: annual; forming small, matted, creeping tufts, often under water, the stems ½–2 inches long; leaves small, obovate or oblong, tapering at the base; flowers globular, with three rose-coloured petals scarcely longer than the

P

calyx; seeds numerous, appearing beautifully ribbed and transversely striated under the microscope.— Water Pepper.—Margins of pools under water. Fl. July, August.

E. Hydropiper: annual; differs from the last in having sessile flowers, with four sepals, petals, and styles, and eight stamens, a more deeply divided calyx, and fewer and larger seeds.—Margins of pools and in water. Fl. July, August.

(59) Linum. FLAX.

* *Leaves scattered ; flowers blue.*

L. usitatissimum: annual; stems 1–2 feet high, erect, glabrous, usually branched only at the top; leaves alternate, erect, narrow-lanceolate, pointed, entire; flowers rich blue, in a loose terminal corymb; sepals pointed; petals obovate; capsule globular or slightly depressed.—Linseed.—A weed of cultivation. Fl. July.

L. perenne: stem 1–2 feet long, erect or decumbent; very variable, sometimes resembling the last, but the stems usually more slender, the leaves smaller and narrower, and the sepals obtuse.—Dry limestone pastures. Fl. June, July.

L. angustifolium: stem 1–2 feet high, branching irregularly, decumbent; leaves narrow, linear-lanceolate; flowers smaller than the preceding, pale blue, the sepals pointed.—Limestone pastures and wastes. Fl. July.

** *Leaves opposite ; flowers white.*

L. catharticum: annual; stems 3–8 inches high, slender, erect, or slightly decumbent, glabrous; leaves small, opposite, obovate or oblong; flowers very small, pure white, on long slender pedicels.—Meadows and pastures. Fl. June to August.

(60) Radiola. ALLSEED.

R. Millegrana: annual; minute, erect, with numerous repeatedly-forked branches, forming dense corymbose tufts, 1–2 inches high; leaves small, opposite; flowers minute, globular, white; calyx-teeth 8–12; petals about the length of the calyx. —Sandy heaths and waste places. Fl. July, August.

(61) Geranium. CRANE'S-BILL.

* *Perennials with large flowers.*

G. sanguineum: stems numerous, a foot long, decumbent, rarely erect, with spreading hairs, leafy; leaves nearly orbicular, divided to the base in 5–7 segments, which are again cut into 3–5 narrow lobes; flowers large, dark purple, growing singly on long, slender peduncles; petals slightly notched, spreading.—Dry woods and pastures. Fl. June to September.

G. sylvaticum: stems 1–2 feet high or more, erect; leaves long-stalked, palmately divided almost to the base into 5–7 pointed lobes more or less cut and serrated; stem-leaves few; flowers in a rather dense, corymbose panicle, purplish; peduncles two-flowered; pedicels erect; petals obovate, slightly notched, scarcely twice as long as the calyx.—Moist thickets and mountain pastures. Fl. June, July.

G. pratense: this resembles the last, but differs chiefly in its more cut leaves, and larger bluish-purple flowers loosely panicled on longer peduncles, the pedicels always more or less spreading or reflexed after flowering.—Meadows and thickets. Fl. June, July.

** *Annuals, with small flowers.*
† *Petals entire.*

G. Robertianum: annual; stems ½–1 foot high, erect or

spreading, much branched, generally bearing a few soft hairs,
often bright red, smelling disagreeably when rubbed; leaves
divided into 3 pinnate or twice-pinnate segments; flowers
rather small, with hairy long pointed sepals; petals obovate,
entire, reddish-purple, rarely white, with glabrous erect claws;
carpels glabrous, with a few transverse wrinkles.—Herb Ro-
bert.—Waste places, woods, etc. Fl. May to October.

G. lucidum: annual; stems ¾-1 foot high, often turning
red, glabrous, shining; leaves orbicular, palmately lobed, with
broad segments usually obtuse, or rarely slightly pointed;
flowers as in the last, but smaller; calyx pyramidal, the edges
of the erect sepals forming very projecting angles.—Waste
places, old walls, etc. Fl. May to August.

†† *Petals notched.*

G. pusillum: annual; stems prostrate, downy; leaves reniform, palmate, with 5–7 trifid lobes; flowers small, bluish
purple, on two-flowered peduncles; petals but slightly notched;
carpels not wrinkled, but hairy; seeds smooth.—Waste and
cultivated places. Fl. June to August.

G. dissectum: annual; stems diffuse, hairy, branched;
leaves downy, deeply divided into 5–9 narrow segments, which
are again deeply trifid or lobed; peduncles very short, bearing
two small purple flowers, the petals slightly notched; carpels
hairy, without wrinkles; seeds beautifully and minutely reticulated or dotted.—Dry pastures and cultivated places. Fl.
June to August.

G. columbinum: annual; stems slender, decumbent, slightly
hairy; leaves deeply divided into 5–9 narrow deeply-lacini-
ated segments, the alternate lobes mostly linear; peduncles
and pedicels long and slender; flowers small, rose-coloured,
the petals notched; carpels slightly hairy, or glabrous, not
wrinkled; seeds dotted.—Dry pastures. Fl. June, July

(62) Erodium.

E. cicutarium: annual; forming a dense hairy tuft, the stems short, or sometimes ½–1 foot long; leaves mostly radical, on long stalks, pinnate, the leaflets deeply pinnatifid, with narrow more or less cut lobes; peduncles erect, bearing an umbel of 10–12 small purple or pink flowers; carpels slightly hairy.—Waste and cultivated lands. Fl. June to September.

E. moschatum is a larger and coarser plant, with a strong smell of musk; it has ovate leaflets, and umbels of numerous bluish-purple flowers, and is found in sandy, waste places, especially near the sea.

E. maritimum is distinguished by having simple, toothed, not pinnate leaves, and 1–2-flowered peduncles ; the flowers reddish-purple; it grows in maritime sands.

(63) Hypericum. St. John's-wort.

* *Stamens pentadelphous,* i.e. *in five groups or clusters.*

H. Androsæmum: undershrub; stems numerous, erect, 1½– 2 feet high, simple or slightly branched ; leaves sessile, ovate, obtuse, cordate at the base, glabrous, with minute pellucid dots ; flowers few, in small corymbs, yellow, the petals scarcely longer than the sepals; stamens numerous, connected at the very base into five clusters; styles three; capsule globular, slightly succulent before it is ripe.—Tutsan.—Open woods. Fl. July, August.

** *Stamens triadelphous,* i.e. *in three groups or clusters.*

  † *Stems erect.*

    ‡ *Sepals entire.*

H. perforatum: stems 1–1½ feet high, branching above, cylindrical or with two slightly prominent opposite angles, glabrous; leaves sessile, oblong, marked with pellucid dots, and occasionally a few black ones beneath; flowers bright yellow, in a handsome terminal corymb; sepals with a few glandular lines or dots; petals twice as long, marked with black dots; stamens shortly united; styles 3.——Woods, hedges, and thickets. Fl. July and August.

H. quadrangulum: stem 1-2 feet high, with four prominent angles; leaves ovate, clasping the stem at the base, with numerous pellucid dots, and a few black ones round the margin beneath; flowers pale yellow; sepals lanceolate, pointed; petals with or without a few black dots.——Moist pastures. Fl. July, August.

‡‡ *Sepals fringed.*

H. pulchrum: stems 1-2 feet high, with short lateral branches, glabrous; leaves broadly cordate, clasping, those of the branches smaller, much narrower, all marked with pellucid dots; flowers in an oblong or pyramidal panicle, golden yellow; sepals broad, obtuse, fringed at the top with black, glandular teeth.——Dry woods, heaths, and wastes. Fl. July.

H. hirsutum: stem 1½-2 feet high, nearly simple, downy or hairy; leaves oblong or elliptical, narrowed into a short stalk, hairy underneath on the veins, marked with numerous pellucid dots; flowers in an oblong or pyramidal panicle, pale yellow, the sepals narrow, fringed with rather long glandular teeth.——Woods and thickets. Fl. July, August.

†† *Stems prostrate.*

H. humifusum: stems decumbent, much branched, almost trailing, 2-6 inches long, sometimes forming dense spreading tufts; leaves oval-oblong, obtuse; flowers few, small, pale yellow, in short, loose, leafy cymes, the sepals oblong, unequal, entire or with a few glandular teeth, generally bordered by black dots; petals with few black dots.—Stony heaths, pastures, and bogs. Fl. July.

H. Elodes: stems diffuse, 6-12 inches long, covered with loose, woolly, whitish hairs; leaves orbicular, stem-clasping, woolly; flowers pale yellow, few together in a leafless cyme, at first terminal, but becoming lateral; sepals small, ovate, copiously fringed with glandular teeth; petals with a small fringed appendage at their base.—Spongy and watery bogs. Fl. July, August.

(64) Malva. MALLOW.

* *Stems decumbent or spreading at the base.*

M. rotundifolia: annual; stems procumbent, $\frac{1}{2}$-1 foot long, tough, slightly downy; leaves long-stalked, orbicular, cordate at the base, with 5-7 short and broad crenate lobes; flowers clustered in the axils, small, pale purplish; carpels about fifteen, downy, rounded on the back.—Roadsides and waste places. Fl. June to September.

** *Stems erect or ascending.*

M. sylvestris: biennial; stems 1-3 feet high; leaves long-stalked, orbicular, slightly cordate at the base, with 5-7 broad, short, deep lobes; flowers in axillary clusters, reddish purple; carpels about ten, flat on the back, with angular edges.— Waste places, roadsides, etc. Fl. June to September.

M. moschata: stems erect, simple, or slightly branched, $1\frac{1}{2}$ foot high; radical leaves orbicular, with short, broad lobes, those of the stem deeply divided into linear or wedge-shaped segments, which are again pinnatifid or three-lobed; flowers large, rose-coloured, rarely white, crowded at the summits of the branches; carpels rounded on the back, very hairy.— Hedge-banks and gravelly pastures. Fl. July, August.

(65) Althæa.

A. officinalis : stems erect, branched, 2–3 feet high, covered, as well as the foliage, with soft, velvety down ; leaves stalked, broadly ovate, undivided or three-lobed ; flowers pale rose-colour, on short pedicels in the upper axils, or forming almost leafless terminal spikes ; carpels rounded on the back.—Marsh Mallow.—Marshes, especially near the sea. Fl. July to September.

(66) Lavatera.

L. arborea : undershrub ; stem woody at the base, 1–5 feet high ; leaves long-stalked, broadly orbicular, palmately divided into 5–9 broad, short, crenate lobes, softly downy on both sides ; flowers numerous, pale purple-red, collected into clusters, forming a long terminal raceme or narrow panicle.— Maritime rocks, rare. Fl. July to September.

(67) Tilia. LIME TREE.

T. europæa : tree ; leaves stalked, broadly heart-shaped or nearly orbicular, often oblique, pointed, serrate on the edge, glabrous above, downy underneath, especially in the angles of the principal veins ; peduncles hanging amongst the leaves, bordered or winged half-way up by the long, narrow, leaf-like bract ; flowers sweet-scented, of a pale whitish-green.— Linden.—Woods, much planted. Fl. July.

(68) Impatiens.

I. Noli-me-tangere : annual ; stems 1–2 feet high, erect, glabrous, branching, rather succulent, and swollen at the nodes ; leaves stalked, ovate, pointed, toothed, pale green, flaccid ; peduncles axillary, slender, bearing one or two perfect flowers, which are large and showy, yellow, spotted with orange, pods chiefly produced by minute, imperfect flowers, of which there are several with the perfect ones.—Touch-me-not.—Moist woods and shady places. Fl. July to September.

I. **fulva**: annual; closely resembling the last, but the flowers are deeper orange-colour, spotted with reddish-brown, and the spur is very closely bent back upon the calyx.—On the Wey, and some other streams in Surrey. Fl. July, August.

[*Euonymus* and *Rhamnus*, late spring or early summer-flowering genera described at p. 90, will be found included in the summary at p. 55.]

(69) **Ononis**.

O. **arvensis** : undershrub, low, spreading, much branched, sometimes nearly erect, one foot high or more, with soft spreading hairs, glutinous; leaves trifoliolate, the leaflets obovate or oblong; flowers stalked, solitary, pink, the standard streaked with deeper pink.—Barren pastures. Fl. June to September.

Var. *campestris* or *antiquorum* : glabrous, more erect, and usually thorny.—Barren pastures.

(70) **Anthyllis**. KIDNEY VETCH.

A. **Vulneraria**: stems spreading or ascending, $\frac{1}{2}$–1 foot long, clothed with short, silky hairs; leaves pinnate; leaflets narrow and entire; flower-heads usually in pairs at the ends of the branches, each surrounded by a digitate, leafy bract; flowers numerous, closely sessile, varying from pale or bright yellow to deep red.—Lady's Fingers.—Dry pastures and stony places. Fl. June to August.

(71) Medicago. MEDICK.

* *Flowers purple.*

M. sativa: stem erect, 1–1½ foot high; leaves trifoliolate;
leaflets obovate-oblong, dentate above; flowers almost always
violet or blue; pod spirally twisted so as to form 2–3 complete
rings or coils.—Lucerne.—Borders of fields, scarcely naturalized. Fl. June, July.

** *Flowers yellow.*

M. lupulina: annual; stems 1–2 feet long, branching at
the base, spreading, hairy; stipules broad, shortly toothed;
leaflets obovate; peduncles longer than the leaves, bearing a
compact raceme or oblong head of very small bright yellow
flowers; pods small, black when ripe, curved almost into a
complete spire, containing a single seed.—Nonsuch.—Pastures
and waste places; cultivated. Fl. May to August.

M. maculata: annual; stems ½–1 foot long, spreading,
branched at the base, glabrous; stipules toothed; leaflets
triangular-obcordate, with usually a dark spot in the centre;
flowers 1–4 in the raceme; pod with 3–4 spires, edged with
two rows of fine curved prickles.—Cultivated and waste
places. Fl. May to August.

(72) Melilotus. MELILOT.

M. officinalis: annual or biennial; stems erect, 2–4 feet
high, branched, glabrous; leaves distant, on long leafstalks;
the leaflets obovate or nearly orbicular, those of the upper
leaves narrower, often linear; flowers numerous, bright yellow,
in long, axillary racemes; pod oval.—Roadsides, banks, and
bushy places. Fl. June to August.

M. arvensis has the leaflets rather broader, and the racemes
looser, with fewer flowers; probably introduced.

M. alba has harder and more wiry stems, narrower leaflets, and white flowers, and is also occasionally found, but probably introduced with corn or ballast.

(73) Trigonella. TRIGONEL.

T. ornithopodioides: annual; stems thickly matted, spreading, rarely more than 2–3 inches long, glabrous; leaflets inserted close together at the summit of the stalk, obovate or obcordate, toothed; flowers small, nearly white, solitary or two or three together in each axil; petals remaining round the pod as in the *Clovers*; pod slightly curved, glabrous, containing 6–8 seeds.—Dry sandy pastures, chiefly near the sea. Fl. June, July.

(74) Trifolium. CLOVER.

* Flowers purple or red.

T. pratense: stems decumbent or nearly erect, 1–2 feet long, hairy; stipules large, ovate, bristle-pointed; leaflets obovate or obcordate; flowers reddish-purple, in dense terminal ovoid or globular heads, with two sessile, trifoliolate leaves close at their base.—Meadows and pastures. Fl. May to September.

T. medium: stems ascending zigzag; stipules linear-lanceolate; leaflets elliptical or lanceolate; flower-heads always more or less pedunculate above the last floral leaves; the corolla larger than in the last, brighter and richer coloured.— Open woods and bushy pastures. Fl. June to September.

T. striatum: annual; small, tufted, spreading, covered with 'soft hairs; stipules ovate, ending in a fine point; leaflets obovate; flower-heads small, ovoid or globular, chiefly terminal, and closely sessile within the last leaves; calyx softly hairy, with short, subulate teeth, which remain erect after flowering; corolla very small and pale red.—Dry pastures, banks, and waste places. Fl. June, July.

T. fragiferum: stem creeping; leaflets obovate, emarginate, finely serrated; stipules ovate, with a long point; flower-head globose on long axillary peduncles, very compact, often assuming a pink tint, so as to have been compared to a strawberry; corolla small and red; calyx, after flowering, much inflated.— Dry meadows and pastures. Fl. July, August.

** *Flowers white or whitish.*

T. arvense: annual; stems slender, branching, erect, seldom reaching a foot in height, clothed with short soft hairs; stipules and leaflets narrow; flowers small, in pedunculate heads, which are at first nearly globular but soon become oblong or cylindrical, appearing very soft and feathery owing to the fine hairy teeth of the calyx projecting beyond the corolla.— Sandy cornfields. Fl. July to September.

T. scabrum: stems procumbent, less hairy than *T. striatum*; leaflets obovate; stipules ovate, with a fine point; flower-heads ovate, sessile, terminal, and lateral; the flowers small, whitish; calyx-teeth lanceolate, spreading or recurved after flowering. —Dry pastures and waste places. Fl. May to July.

T. repens; stems creeping and rooting, glabrous; stipules small; leaflets obovate, distinctly toothed, usually marked in the centre, the leafstalks often very long; peduncles axillary, long, and erect, bearing a globular head or umbel of white flowers, the pedicels recurved after flowering.—Dutch Clover. —Meadows and pastures. Fl. May to September.

*** *Flowers yellow.*

T. agrarium (*procumbens* of authors): annual; stems

slender, much branched at the base, glabrous or slightly downy, procumbent or nearly erect, ½–1 foot long; stipules broad, pointed; leaflets obovate or obcordate; flower-heads loosely globular or ovoid, on rather long axillary peduncles, containing 30–50 small yellow flowers, which in fading become reflexed, and turn pale brown.—Dry pastures, borders of fields, etc. Fl. June to August.

T. procumbens (*minus* of authors): annual; resembling the last, but more slender and procumbent; flowers smaller, 12–20 in a head, paler yellow.—Dry pastures. Fl. June, July.

T. filiforme is a still more slender plant, with the stems decumbent, ascending, or erect, seldom six inches long; the leaflets narrower; and the flowers 2–3, rarely 5–6, in each head.—Sandy or stony pastures in south-eastern England, but rare. [See also p. 51.]

(75) Lotus. BIRDS'-FOOT TREFOIL.

L. corniculatus: stems decumbent or ascending, ½–2 feet long; leaflets obovate, pointed, the stipules ovate; peduncles much longer than the leaves; umbels of 5–10 bright yellow flowers, the standard often red on the outside; calyx-teeth about the length of the tube, the two upper ones converging. —Meadows and pastures. Fl. June to September. A very variable plant. Mr. Bentham places the following as a mere variety of it:

L. major: stem 1–3 feet high, ascending or nearly erect, glabrous or slightly hairy, luxuriant in all its parts; leaflets obovate; stipules roundish-ovate; flowers 6–12 in the umbel; calyx-teeth spreading like a star, the upper ones diverging.— Moist meadows and bushy places. Fl. July, August.

(76) **Vicia.** VETCH.

* Peduncles short, few-flowered; calyx gibbous at the base on one side.

V. sepium: stems 1–2 feet high, weak, straggling, but scarcely climbing; stipules small, entire, or larger and toothed; leaflets 4–6 pairs, ovate or oblong, the leafstalk tendrilled; flowers smallish, light reddish-purple, 2–4 in the axils of the upper leaves, forming a sessile cluster or very short raceme; style with a dense tuft of hairs under the stigma on the outer side, with a few short hairs on the opposite side; pod glabrous; seeds few.—Woods, shady places, and hedges. Fl. June to August.

** Peduncles elongated, many-flowered.
† Calyx gibbous at the base.

V. Cracca: stems weak, climbing by means of branched tendrils, to the length of 2–3 feet or more, hairy, or nearly glabrous; stipules narrow and entire; leaflets numerous, oblong or linear; flowers numerous, in one-sided racemes, on peduncles rather longer than the leaves, of a fine bluish-purple; style hairy all round below the stigma; pod flattened, glabrous.—Hedges and bushy places. Fl. June to August.

V. sylvatica: stems 6–8 feet long, glabrous, climbing; stipules deeply divided at their base; leaflets usually 8–10 pairs, oblong, or the lower ones ovate, obtuse or notched at the top; flowers considerably longer than in the last, white with bluish streaks, drooping in long racemes; pod glabrous, broad.—Woods and bushy places. Fl. June to August.

†† Calyx equal at the base.

V. tetrasperma: annual; stems glabrous or nearly so, weak, often climbing, ½–2 feet long; leaflets narrow, the lower ones obtuse, 3–6 pairs, the tendrils simple or branched; peduncles slender, with 1–7 small pale bluish flowers; calyx-teeth much shorter than the standard; pod flat, containing 4, sometimes 5–6 seeds.—Smooth Tare.—Fields, hedges, and waste places. Fl. June to August.

V. hirsuta: annual; stems hairy, slender, 1–3 feet long, often climbing, the tendrils branched; stipules small, narrow, often divided; leaflets small, oblong, 6–8 pairs; peduncles slender, with very few, usually 2–3, insignificant, pale blue flowers, the calyx-teeth almost as long as the standard; pod flat, hairy, containing two seeds.—Hairy Tare.—Hedges, cornfields, and waste places. Fl. June to August. [See also p. 51.]

(77) Lathyrus. Pea or Vetchling.

* *Leafstalks without real leaflets.*

† *Flowers red.*

L. Nissolia: annual; stems erect, glabrous, branching from the base, about one foot high; leaves reduced to linear, grass-like, flattened leafstalks ending in a fine point, without leaflets or stipules; peduncles long, bearing one rarely two small pale red flowers; pod long, narrow, and straight.—Grass Vetch.—Bushy places and stony pastures. Fl. June.

†† *Flowers yellow.*

L. Aphaca: annual; stems weak, branching, glabrous, about one foot long; leaves reduced to slender branched tendrils between the pairs of large broadly heart-shaped or sagittate stipules, which appear like simple opposite leaves; peduncles long and slender, with one or rarely two small yellow flowers; pod flattened, glabrous.—Cultivated places, occasionally. Fl. June to August.

** Leaves with one pair of leaflets.
　† *Flowers yellow.*

L. pratensis : stems much branched, glabrous, straggling
or half climbing to the length of 1–2 feet; stipules large,
broadly-lanceolate, sagittate; leaflets narrow-lanceolate or
linear; peduncles elongated, with a short raceme of about
6–10 yellow flowers; pod glabrous.—Moist meadows and
pastures. Fl. July, August.

　†† *Flowers red.*

L. sylvestris : stems glabrous, straggling or climbing, 3–6
feet high, the angles expanded into narrow green wings; leafstalks flattened or winged; leaflets long-lanceolate; stipules
narrow; flowers in loose racemes, large, pale reddish-purple,
the standard broad, with a green spot on the back.—Hedges
thickets, and bushy places. Fl. July to September.

The *Everlasting Pea* of our gardens is a broad-leaved variety, with larger and more richly-coloured flowers.

　*** Leaves with 2–4 pairs of leaflets.

L. macrorrhiza : tuberous; stems glabrous, erect, simple
or nearly so, ½–1 foot high; leaves without tendrils; leaflets
usually 2 sometimes 3–4 pairs, oblong-lanceolate or linear;
peduncles slender, bearing a loose raceme of 2–4 bright
reddish-purple flowers; pod glabrous.—Thickets and open
woods. Fl. June, July.

(78) **Astragalus.** MILK VETCH.

　* *Flowers bluish-purple.*

A. hypoglottis : stem prostrate, branching at the base, 2–6
inches long, slightly hairy; stipules free from the leafstalk,
more or less united together on the opposite side of the stem;

leaflets usually 10–12 pairs, with an odd one; flowers bluish-purple, in short spikes, on long axillary peduncles; calyx erect, downy, with short black hairs; pod ovoid, hairy.—Dry hilly pastures. Fl. June, July.

** *Flowers dull yellow.*

A. **glycyphyllos**: glabrous; stems zigzag, spreading along the ground to the length of two feet or more; stipules free; leaflets in 5–6 pairs, ovate; flowers dingy yellow, in racemes rather shorter than the leaves; pods erect, curved, glabrous. —Dry, open woods, and bushy places. Fl. June.

(79) **Oxytropis.**

* *Flowers yellow.*

O. **campestris**: stem short, tufted, scarcely lengthening into shortly ascending branches; leaflets 10–15 pairs, with an odd one, oblong or lanceolate, hairy; flowers in short spikes, pale yellow tinged with purple; calyx hairy; pod erect, ovoid, covered with short, usually black, hairs.—Clova mountains, rare. Fl. July.

** *Flowers purple.*

O. **uralensis**: stem short; the foliage, inflorescence, and pod as in the last, but more densely covered with soft, silky hairs; flowers bright purple.—Dry hilly pastures in Scotland. Fl. July.

(80) **Onobrychis.** SAINTFOIN.

O. **sativa**: stems ascending, 1–2 feet long; stipules brown, finely pointed; leaflets numerous, oblong, glabrous above; peduncles longer than the leaves, bearing a spike of pale pink flowers, at first closely packed, but afterwards lengthening out; pod twice as long as the calyx, the upper edge nearly straight, the lower semicircular, bordered with short, sometimes prickly teeth.—Limestone districts, cultivated for forage. Fl. June, July.

(81) Spiræa.

S. Ulmaria : stems erect, rather stout, 2–3 feet high, glabrous, reddish; leaves large, interruptedly pinnate, with 5–9 ovate or broadly-lanceolate, irregularly-toothed segments, green above, soft and whitish beneath, the terminal one deeply divided into three, besides several smaller segments along the common stalk; stipules broad, toothed; flowers small, yellowish-white, sweet-scented, numerous, in compound corymbose cymes at the summit of the stems; capsules 5–8, very small, more or less spirally twisted.—Meadow-sweet.—Meadows and banks of streams. Fl. June to August.

S. Filipendula : stems erect, 1–2 feet high; leaves chiefly radical, 3–5 inches long, with numerous small, oval, oblong or lanceolate segments, deeply toothed or pinnately lobed, green, glabrous; stipules broad, adhering to the leafstalk nearly their whole length; flowers creamy white, often tipped with red; carpels 6–12, not twisted.—Dropwort.—Meadows, pastures, and open woods. Fl. June, July.

(82) Potentilla. POTENTIL.

* *Leaves digitate.*
  † *Petals usually four.*

P. Tormentilla : stems erect, or procumbent from a thick woody rootstock, forked, silky-hairy; lower leaves often shortly stalked, upper ones always sessile, consisting of three, rarely five, deeply-toothed leaflets; peduncles in the forks of the stem, or in the axils of the upper leaves, forming a loose, leafy, terminal cyme; flowers small, bright yellow, mostly with four petals; the first one, however, of each stem has occasionally five.—Heaths, moors, and pastures. Fl. June to August.

The *Tormentilla reptans* of authors is a more procumbent variety, occasionally creeping at the base, with rather larger flowers, more frequently furnished with five petals.

†† *Petals usually five.*

P. reptans : stems slender, prostrate, often rooting at the nodes; stipules ovate, mostly entire; leaves stalked, with five obovate, coarsely-toothed leaflets; flowers large, yellow, solitary, on long peduncles, axillary.—Cinquefoil.—Pastures, borders of woods, and hedges. Fl. June to August.

P. argentea : stems decumbent at the base, ascending; leaves long-stalked, of five wedge-shaped, deeply-toothed leaflets, the upper nearly sessile, all clothed beneath with close white down; flowers in a loosely-forked leafy corymb or panicle, rather small, yellow.—Gravelly pastures, wastes, and roadsides. Fl. June, July.

** *Leaves pinnate.*
† *Flowers yellow.*

P. anserina : stem forming long creeping runners rooting at the nodes; leaves with numerous, oblong, deeply-toothed leaflets, shining silver-white with silky down beneath; peduncles long, solitary, bearing a rather large yellow flower.—Silverweed.—Roadsides and waste places. Fl. June and July.

†† *Flowers purple.*

P. Comarum : stems decumbent and rooting at the base, 1–1½ feet high; stipules not distinct from the enlarged base of the leafstalk; leaflets 5–7, oblong, toothed, often hoary beneath; flowers few, in a loose, irregular corymb, dingy purple; inner segments of the calyx broad; petals shorter than the calyx; carpels numerous, small, on a somewhat enlarged, rather spongy receptacle.—Marsh Cinquefoil.—Marshes, and peat-bogs. Fl. June, July.

[See also p. 53.]

(83) Sibbaldia.

S. procumbens: stem short, dense, spreading, tufted; leaves ternate, the leaflets obovate or wedge-shaped, three-toothed at the end; flower-stems short, almost leafless, bearing a cyme of small flowers, of which the green calyx is the most conspicuous part, the petals being very small and pale yellow.— Scotch Highlands. Fl. July.

(84) Geum. Avens.

G. urbanum: stem erect, slightly branched, 1–2 feet high; stipules large and leaflike, coarsely toothed; leaves interruptedly piunate, with several large segments intermixed with small ones, the upper ones ternate, all coarsely-toothed; flowers erect, small, yellow; carpels in a close, sessile head, covered with silky hairs, the awn curved downwards, with a minute hook at the tip.—Herb Bennet.—Roadsides, banks, and margins of woods. Fl. June to August.

G. rivale: stems erect or ascending, usually simple; leaves mostly radical, with one large, orbicular, terminal segment, coarsely toothed or lobed, and a few very small segments below; flowers few, drooping, dull purple; carpels very hairy, in a shortly-stalked globular head.—Marshes and wet ditches.— Fl. June, July.

(85) Dryas.

D. octopetala : stems short, much branched, prostrate or creeping, forming dense spreading tufts ; leaves crowded, oblong, deeply and regularly crenate, shining green above, white and downy beneath ; peduncles erect, 2-3 inches long, with rather large white solitary flowers ; awn of the carpels above an inch long, feathery.—Limestone mountain districts. Fl. July, August.

(86) Agrimonia. AGRIMONY.

A. Eupatoria : stems 2-3 feet high, clothed, like the leaves, with soft hairs ; leaves pinnate, with 5-9 ovate, coarsely-toothed leaflets, intermixed with smaller ones ; spike long, leafless, each flower in the axil of a small three-cleft bract ; flowers, small, yellow ; tube of the calyx turned downwards after flowering, forming a small burr.—Roadsides, waste places, and borders of fields. Fl. June, July.

(87) Fragaria. STRAWBERRY.

F. elatior : stem bearing few runners, which root and form new plants at the nodes, as in *F. vesca*, but taller, with fewer runners and flowers, usually entirely or partially unisexual ; leaves ternate, the leaflets oblong, plaited, coarsely-toothed, hairy ; fruit perfumed.—Hautbois Strawberry.—Woods in the south ; rare. Fl. June to September.

(88) Rubus.

R. Idæus : stems biennial, erect, 3-4 feet high, armed with weak prickles ; leaves pinnate, leaflets five in the lower, three in the upper ones, ovate or oblong, pointed, coarsely toothed, whitish underneath ; flowers white, in panicles at the ends of

the short branches; petals small, narrow, whitish; fruit red,
separating from the receptacle when ripe.—Raspberry.—
Thickets and woods. Fl. June.

R. fruticosus: stems biennial, or of few years' duration,
erect, or more frequently arched straggling or prostrate,
armed with prickles stiff hairs or glandular bristles; leaves
digitate, the leaflets 3–5, rather large, coarse, ovate, toothed,
the midribs and stalks armed with hooked prickles; flowers
white or pink, in panicles at the ends of the branches;
fruit black, or dull-red.—Bramble, or Blackberry.—Hedges,
thickets, woods, and waste places. Fl. June and July.

A large number of Brambles, often considered as species,
occur in Britain, but for these we must refer to more technical
books, just mentioning a few of the most distinct:—

R. fruticosus (type) has the leaflets covered underneath with
a close, white down; flowers usually numerous.—Hedges and
thickets.

R. corylifolius has the leaflets green underneath, usually
large and broad; flowers not so numerous as in the last.—
Hedges and thickets, flowering earlier.

R. carpinifolius has the leaflets green underneath, but not
so broad, and more pointed than in the last, the stems more
hairy; flowers not numerous.—Woods.

R. glandulosus has the leaflets as in the last, or sometimes
broader, the stems with numerous stiff, glandular hairs mixed
with the prickles.—Shady woods.

R. suberectus has the leaflets green, or slightly hoary underneath; stems shorter and more erect than in the common
forms; flowers usually few, and the fruit not so black.—Wet
woods and thickets.

R. cœsius: stems slender, more or less glaucous when
young, spreading, or creeping along the ground, seldom arched;

leaves ternate; flowers few, in small, loose panicles; fruit covered with glaucous bloom when ripe.—Dewberry.—Open fields and stony wastes. Fl. June to August.

R. saxatilis: stems ascending, simple, seldom above one foot high, slender, downy, with few prickles; leaves ternate, the leaflets obovate, coarsely-serrate; flowers on slender pedicels, two or three together in the axils of the upper leaves, forming short racemes or corymbs, dirty white or greenish yellow; berries red, with few large carpels. — Stony mountainous places.—Fl. June.

R. Chamæmorus: stems simple, herbaceous, unarmed, 6–10 inches high; leaves few, large, simple, broadly orbicular or reniform, deeply-cut into 5–9 broad lobes; flowers white, rather large, solitary on terminal peduncles; fruit orange-red. —Cloudberry.—Turfy Alpine bogs. Fl. June.

(89) **Rosa.** Rose.

* *Branches bearing glandular spiny hairs (setæ).*

† *Prickles straight, slender, scarcely dilated at the base.*

R. pimpinellifolia (*spinosissima* of authors) : shrub erect, branched, 1–2 feet high, with numerous, unequal, straight, slender prickles, intermixed with glandular hairs; leaflets small, 7–9, with simple teeth; flowers small, white or pink, solitary at the end of the short branches; calyx-segments lanceolate, almost always entire; fruit black, rarely red, globular or nearly so, crowned by the persistent segments of the calyx.—Sandy heaths.—Fl. May, June. The origin of our garden Scotch Roses.

†† *Prickles hooked, very unequal, much dilated at the base.*

R. rubiginosa: shrub; bushy, somewhat slender, the prickles of the stems curved and intermixed with a few setæ; leaflets small, usually doubly-toothed, glandular, scented;

flowers pink, usually solitary; fruit ovoid or oblong, the primordial ones pear-shaped, smooth or rarely bearing a very few small prickles.—Sweetbriar.—Hedges and thickets. Fl. June, July. There are two or three forms of this Rose.

    ** *Stems without glandular spiny hairs (setæ).*
     † *Leaves glandulose on their disk or surface.*

    R. villosa: shrub; erect, bushy, 3–6 feet high, the prickles of the stem straight or but slightly curved; leaflets softly downy on both sides, almost always doubly-toothed; flowers white or pale pink; calyx-segments long, and often expanded near the top, sometimes entire, sometimes pinnately lobed; fruit red, globular, covered with small fine prickles.—Hedges and thickets. Fl. June, July. There are several forms.

    †† *Leaves without glands on their disk.*
     ‡ *Styles distinct.*

    R. canina: shrub, stems of several years' duration, the first year erect, simple, 3–4 feet high, the flowering stems of two or more years branched, rather weak and straggling, 6–8 feet long, glabrous, without glands, armed with curved or hooked prickles; leaflets five, sometimes seven, ovate, usually glabrous and simply toothed, or downy on the under side and then often doubly-toothed; flowers pink or white, sweet-scented, solitary or 3–4 together at the ends of the branches; fruit ovoid, without bristles, the calyx-lobes pinnate, deciduous before the fruit is ripe.—Hedges and thickets; the commonest of our Roses, and variable. Fl. June, July.

    ‡‡ *Styles united into a column.*

    R. arvensis: shrub; stems long, trailing, often extending many feet, with slender branches; foliage and prickles nearly as in the last, the prickles usually small and much hooked, and the leaflets simply serrate; flowers white, scentless, usually 3–4 together at the ends of the branches, rarely solitary; fruit globular or nearly so, without bristles, the calyx-divisions mostly entire, and falling off before the fruit is ripe; styles united into a column protruding from the orifice of the calyx-tube.—Hedges and thickets. Fl. June to August.

(90) Sanguisorba. GREAT BURNET.

S. officinalis: stem glabrous, erect, about two feet high; leaves chiefly radical, pinnate, with 9–13 ovate or oblong toothed leaflets; upper part of the stem almost leafless, divided into a few long peduncles, each terminated by a single flower-head; flowers crowded, dark purple, the heads at first globular, becoming ovoid or oblong.—Moist meadows. Fl. June, July.

(91) Poterium. LESSER OR SALAD BURNET.

P. Sanguisorba: stem glabrous, seldom above a foot high; leaves pinnate; leaflets many, small, ovate, deeply-toothed; flower-heads small, almost globular, light green, seldom acquiring a purplish tinge, the lower flowers all male with the numerous stamens projecting in hanging tufts, the upper female with a long style ending in a purple tufted stigma.— Dry pastures and limestone rocks. Fl. July.

(92) Tillæa.

T. muscosa: stems 1–2 inches high, much branched, reddish, slender, succulent, crowded with flowers; leaves narrow-lanceolate or linear; flowers solitary in each axil, or several together in little clusters; petals minute, subulate, white tipped with red.—Moist barren sandy heaths and wastes. Fl. June, July.

(93) Cotyledon. NAVELWORT.

C. Umbilicus : stems erect, ½–1 foot high, simple or slightly
branched ; radical and lower leaves on long stalks, fleshy, orbicular, broadly crenate, more or less peltate ; flowers in long
racemes, pendulous, yellowish-green; corolla cylindrical.—
Pennywort.—Rocks, walls, and old buildings.   Fl. June,
July.

(94) Sedum.

* *Leaves with a flat expanded surface.*

S. Rhodiola : stems erect, stout, simple, ½–¾ foot high, leafy
to the top; leaves alternate, sessile, obovate or oblong, slightly
toothed ; flowers diœcious, yellow or rarely purplish, forming
dense cymes, surrounded by the upper leaves, the males with
eight stamens longer than the petals, the females with four
carpels.—Rosewort.—Clefts of alpine rocks.   Fl. May, June.
The smell of the rootstock, when drying, has been compared
to that of roses.   It is sometimes called *Rhodiola rosea.*

S. Telephium : stems hard, erect, simple, 1–2 feet high or
more ; leaves scattered, obovate or oblong, coarsely-toothed ;
flowers numerous, purple, forming a handsome terminal corymb ; stamens shorter than the petals.—Orpine.—Borders of
fields, hedge-banks, and bushy places.   Fl. July, August.

** *Leaves tumid, as thick as broad.*
  † *Flowers white or red.*

S. anglicum : stems glabrous, decumbent, three inches high,
much branched at the base ; leaves short, thick, almost globular, crowded on the short barren branches, more loosely scattered and occasionally opposite on the flowering ones ; flowers
white, occasionally tinged with pink, in a short irregular cyme.
—Rocky or stony places.   Fl. July.

S. villosum : annual; stems erect, nearly simple, 3–4 inches high, the upper part of the plant with short viscid hairs; leaves linear, flat above, alternate or scattered; flowers few, pale dingy rose colour, in a small, loose, terminal cyme.— Mountain bogs and stony rills. Fl. June, July.

†† *Flowers yellow.*

S. acre : stems numerous, short, tufted, glabrous, procumbent, the erect ascending flowering branches 1–3 inches high; leaves small, thick, ovoid, or sometimes nearly globular; flowers bright yellow, in small terminal cymes.—Wall Pepper. —Walls and rocks. Fl. June.

S. rupestre : tufted, with numerous short barren shoots, 1–3 inches long, the terminal flowering stems ascending $\frac{1}{2}$–1 foot high; leaves narrow, cylindrical, shortly spurred at the base; flowers large, yellow, forming a terminal cyme of 4–8 recurved branches, each bearing 3–6 sessile flowers.—Old walls and stony places. Fl. July.

(95) **Sempervivum.** HOUSELEEK.

S. tectorum : barren shoots forming globular tufts; flowering stems stout, succulent, one foot high; leaves very thick, fleshy, ending in a short point, fringed; flowers pink, sessile along the spreading or recurved viscid branches of the cyme; petals linear.—Cottage roofs and old walls. Fl. July.

(96) **Saxifraga.** SAXIFRAGE.

* *Flowers yellow.*

S. aizoides : flowering stems ascending about six inches high; leaves alternate, narrow, smooth, shining, entire, rarely with 1–2 teeth; flowers yellow, in a loose panicle of 3–12 or more; calyx segments almost as yellow as the petals.—Wet rocks in mountainous districts. Fl. June to September.

S. Hirculus : flowering stem ascending, about six inches high, terminated by a single, rather large yellow flower ; leaves alternate, narrow-oblong or linear, entire.—Wet moors at high elevations. Fl. August.

   ** *Flowers white.*

    † *Calyx adherent at the base.*

S. hypnoides : barren shoots, numerous, decumbent, sometimes 2–3 inches long, sometimes contracted into a short dense tuft ; leaves mostly entire, narrow-linear, pointed, some of the larger ones often 3–5-lobed, glabrous or more or less ciliated ; stems 3–6 inches high, with few leaves, and 1–8 large white flowers.—Rather moist rocky situations. Fl. May to July. Very variable in the development of its stems, leaves, and flowers, and in the more or less pointed or almost obtuse segments of the leaves and calyx.

   †† *Calyx free.*

S. stellaris : tufted, and when luxuriant elongated into leafy branches, 1–2 inches long ; leaves spreading, thin, varying from oblong to obovate, with a few coarse teeth, tapering at the base ; stems erect, 3–6 inches high ; flowers from 2–3 to 8–10, forming a loose terminal panicle, small, white, star-like, on slender, spreading pedicels, each with a small leafy bract. —Wet rocks in mountain ranges. Fl. June, July.

   *** *Flowers pink.*

S. umbrosa : root-leaves collected in dense rosulate tufts, spreading, thick and leathery, glabrous, obovate, bordered with cartilaginous crenatures or coarse teeth, and narrowed at the base into a short flattened stalk ; flowering stems erect, leafless, ½-1 foot high ; flowers in a loose, slender panicle, small, pink, elegantly spotted with dark crimson.—London

Pride or St. Patrick's Cabbage.—Mountains of south-western Ireland, and Yorkshire. Fl. June.

S. Geum : resembling the last, but the leaves are orbicular, usually notched or cordate at the base, with long hairy stalks, the leaves also having a few scattered hairs on both surfaces. —West of Ireland. Fl. June.

(97) Drosera. SUNDEW.

D. rotundifolia : leaves on long stalks, nearly orbicular, covered on the upper surface with long, red, viscid hairs, each bearing a small gland at the top ; flower-stems slender, erect, glabrous, 2–6 inches high, the upper portion, consisting of a simple or once-forked unilateral raceme, rolled back when young, but straightening as the flowers expand ; petals white, expanding in sunshine.—Bogs, and wet heathy ground. Fl. July, August.

D. longifolia : resembling the last, but the leaves much more erect, not half so broad as long, and gradually tapering into the footstalk ; flowering stem shorter and stouter.—Bogs. Fl. July, August.

(98) Lythrum. LOOSESTRIFE.

L. Salicaria : stems erect, 2–4 feet high, slightly branched, glabrous or softly downy ; leaves opposite, or in threes, sessile, clasping the stem, lanceolate, entire ; flowers reddish-purple or pink, in rather dense whorls, forming handsome terminal spikes, more or less leafy at the base, the upper floral leaves reduced to bracts.—Wet ditches and marshy places. Fl. July, August.

(99) Peplis. WATER PURSLANE.

P. Portula : annual ; stems slightly branched, creeping and

rooting at the base, seldom more than 2–3 inches high, often tufted; leaves obovate or oblong, tapering into a stalk; flowers minute, sessile in the axils of nearly all the leaves.—Ditches, and moist watery places. Fl. July, August.

(100) Epilobium. WILLOW HERB.

* *Flowers somewhat irregular or unequal-petaled.*

E. angustifolium: stems simple or scarcely branched, 2–4 feet high, glabrous; leaves shortly stalked, lanceolate, entire; flowers large, purplish-red, in long terminal, sometimes branched, racemes, the petals slightly unequal, entire, spreading from the base, the stamens and styles inclined downwards; pod 1–2 inches long.—French Willow, or Rosebay.—Moist open woods. Fl. July, August.

** *Flowers regular, the petals equal.*
  † *Stigma deeply four-lobed.*

E. hirsutum: stems stout, branched, 3–5 feet high, softly hairy; leaves lanceolate, hairy, clasping the stem, bordered with small teeth; flowers large, handsome, deep rose-colour; the petals erect, deeply notched; pod very long, quadrangular, hairy.—Codlins and Cream.—Sides of ditches and rivers. Fl. July.

E. montanum: stems erect, simple or slightly branched, ½–1 foot or more high, cylindrical, without decurrent lines or angles, glabrous or slightly hoary; leaves shortly stalked, or almost sessile, ovate or broadly-lanceolate, toothed; flower-buds erect or slightly nodding, the petals pink, deeply-notched; pod slender, 2–3 inches long.—Waste and cultivated places, roadsides, woods, etc. Fl. July.

E. parviflorum differs chiefly in its softly hairy stem, narrower short-stalked leaves, and rather larger flowers. It is not unlike *E. hirsutum* on a smaller scale, and is distinguishable from it by its much smaller flowers.

†† *Stigma club-shaped, entire.*

E. tetragonum: stems erect, often branched, 1–2 feet high, glabrous or hoary with a very short down, more or less angular from raised lines descending on each side from the margins of the leaves; leaves sessile, narrow, and toothed; flowers small, in terminal leafy racemes, the buds erect, the petals pale pink, deeply notched; pod often very long.—Wet ditches and watery places. Fl. July.

E. palustre: stems round, ½–1½ feet high; leaves narrow, lance-shaped, sessile, entire or not much toothed; flowers small, pale-coloured, in short terminal racemes, the buds nodding.—Boggy places, and watery ditches. Fl. July.

E. alsinifolium: stems frequently branched, seldom more than six inches high; leaves very shortly stalked, ovate, toothed; flowers large, purplish, in very short, leafy racemes, the buds nodding.—Alpine rivulets and springs. Fl. July.

E. alpinum: stems slender, 2–5 inches high, decumbent and much branched, glabrous; leaves stalked, small, ovate or lanceolate, usually obtuse and entire; flowers rather large, growing in the axils of the upper leaves, and forming short leafy racemes, the buds nodding, the petals notched; pod 1–2 inches long, narrowed at the base into a long stalk.—Alpine rills, and wet places in mountain ranges. Fl. July.

(101) **Œnothera.** EVENING PRIMROSE.

Œ. biennis: biennial; stems almost simple, 2–3 feet high, hairy; leaves ovate-lanceolate or lanceolate, slightly toothed, hoary or downy; flowers yellow, large, fragrant, opening in the evening, forming long terminal spikes often leafy at the

base, the petals broad, spreading.—Naturalized on sandy
wastes in Lancashire. Fl. July to September.

(102) Circæa. ENCHANTER'S NIGHTSHADE.

C. Lutetiana : stems erect or shortly decumbent, rooting at
the base, 1–1½ feet high, clothed with short whitish hairs ;
leaves thin, stalked, broadly ovate, rather coarsely toothed ;
flowers white or pink, in slightly-branched, leafless, terminal
racemes ; capsule pear-shaped, forming a small burr.—Woods
and shady situations. Fl. June, July.

C. alpina : resembling the last, smaller in all its parts, glabrous, seldom above six inches high ; leaves heart-shaped,
glossy.—Woods, and stony places in mountain districts. Fl.
July, August.

(103) Myriophyllum. WATER MILFOIL.

M. spicatum : aquatic ; stems immersed, ascending to the
surface, more or less branched ; leaves whorled along the
whole length of the stem, the numerous capillary segments
entire ; flower-spike terminal, slender, 2–3 inches long, protruding from the water, bearing minute flowers arranged in
little whorls, the upper flowers usually males, and the lower
ones females ; bracts small, entire.—Ditches and ponds. Fl.
July, August.

M. verticillatum : aquatic ; like the last, but the flowers in
the axils of the upper leaves, immersed, and rarely forming a
spike protruding above the water ; bracts or floral leaves
longer than the flowers, and pinnate like the stem-leaves.—
Ditches and ponds. Fl. July.

(104) Hippuris. MARESTAIL.

H. vulgaris : aquatic ; stems erect, simple, the upper part

projecting 8–10 inches out of the water, crowded by whorls of 8–12 linear entire leaves, the submerged ones more elongated and flaccid; flowers minute, sessile in the axils of the upper leaves, consisting of a minute ovary, crowned by a scarcely perceptible border representing a calyx, and bearing a single small stamen, and a short thread-like style.—Shallow ponds and ditches. Fl. June, July.

(105) Cornus. CORNEL.

C. sanguinea : shrub, 5–6 feet high, erect; leaves opposite, broadly ovate, stalked, hoary or silky, with closely appressed hairs when young; glabrous when full grown; flowers numerous, forming terminal cymes of 1½–2 inches across; dull white; drupes globular, almost black, very bitter.—Dogwood. —Hedges and thickets. Fl. June.

(106) Hydrocotyle. PENNYWORT.

H. vulgaris : stems slender, creeping, rooting at every node, and emitting small tufts of leaves and flowers; leaves obicular, crenate or slightly lobed, attached by the centre to a rather long stalk; peduncles shorter than the leafstalks, with a single terminal head or 2–3 whorls of minute white almost sessile flowers; fruits small, flat, glabrous.—Bogs, marshes, and edges of ponds. Fl. June to August.

(107) Sanicula. SANICLE.

S. europæa : leaves on long stalks, palmately 3–5-lobed, the lobes obovate or wedge-shaped, serrated; stems 1–1½ feet high, bearing several capitate umbels in an irregular, slightly umbellate panicle; fruit ripening into little burrs in which the prickles almost conceal the calyx-teeth.—Woods. Fl. June, July.

R

(108) **Eryngium.** ERYNGO.

* Scales of the receptacle three-lobed.

E. maritimum : stems stiff, erect, much branched, nearly a foot high, glabrous, bluish or glaucous; leaves stiff, broad, sinuate, divided into three broad, short lobes, elegantly veined, bordered by coarse prickly teeth, the radical ones stalked, the rest stem-clasping; flower-heads nearly globular, pale blue, with an involucre of 5–8 small narrow leaves resembling those of the stem.—Sea Holly.—Sandy sea-coasts. Fl. July, August.

** Scales of the receptacle entire.

E. campestre : stems erect, a foot high, more slender and branched than in the last; leaves 2–3 times pinnatifid, with lanceolate lobes, waved, coarsely toothed, bordered and terminated by strong prickles; flower-heads more numerous, smaller, the involucral leaves purplish, pinnately toothed.—Ballast-hills near Plymouth, by the Tyne, etc. Fl. July, August.

(109) **Cicuta.** COWBANE.

C. virosa : stem hollow, somewhat branched, 3–4 feet high; leaves twice ternate, with large narrow-lanceolate acute unequal-toothed segments; general umbels of 10–15 or more rays, the partial involucres with numerous subulate bracts.—Water Hemlock.—Ditches and edges of lakes and rivers. Fl. August.

(110) **Apium.** CELERY.

A. graveolens : stems rather slender, branched, furrowed, glabrous, 1–2 feet high; leaves pinnate, with 3–5 broad crenate or three-lobed segments; umbels small, nearly sessile on the upper branches opposite the leaves, or on very short terminal peduncles, of 3–6 rays, bearing numerous small white flowers on short pedicels; fruits very small.—Marshy places near the sea. Fl. June to August. The *Celery* of our gardens is a cultivated variety, with enlarged leafstalks.

(111) Helosciadium.

H. nodiflorum: stems procumbent, rooting, the flowering ones ascending or nearly erect, 1–2 feet high, glabrous; leaves pinnate, with 3–10 or more pairs of ovate-lanceolate toothed segments; umbels nearly sessile, either opposite to the leaves or between the upper branches, with 5–6 (rarely 8 or 4) rays; general involucre usually wanting; partial involucre of several small, lanceolate bracts.—Marshy meadows and ditches. Fl. July, August. It varies much in size and foliage.

H. repens is a smaller, much branched, more creeping form, with 3–5 small broad leaf-segments, and longer peduncles to the umbels, occasioned by growing in drier situations.

H. inundatum: stems glabrous, creeping and rooting at the base like *H. nodiflorum*, but smaller, and more slender, often partly immersed in water, the submerged leaves divided into capillary segments; flowering stems 6–8 inches high, with small ternate or pinnate leaves, the segments three-toothed or three-lobed, each lobe often again three-toothed; umbels on short peduncles opposite the leaves, generally of 2–3 rays without involucre, the partial umbels of 5–6 small flowers, with 2–3 minute bracts.—Swamps and shallow pools. Fl. June, July.

(112) Sison.

S. Amomum: annual or biennial; stems erect, glabrous, two feet high or more, with many stiff, slender branches;

R 2

leaves pinnate, the segments ovate or oblong, toothed or lobed ;
the upper leaves much smaller, with small narrow deeply
three-lobed segments; umbels slender, of 3–5 rays, with few
white flowers on short pedicels; involucres of few, linear bracts,
those of the partial umbels smaller, and often turned to one
side.—Honewort.—Hedges and thickets.   Fl. August.

(113) Bunium.   EARTH-NUT, OR PIG-NUT.

B. flexuosum: tuberous stems erect, slender, glabrous, 1–2
feet high, with a few forked branches; leaves few decaying
early, thrice ternate, the divisions short, narrow, pointed, entire
or three-lobed; stem-leaves few, with narrow-linear divisions,
the central lobe of each segment much longer than the lateral
ones; umbels terminal, or one opposite the last leaf, of 6–10
rays with general involucres of 1–3 bracts, the partial ones
more numerous.—Woods and pastures.   Fl. May, June.

(114) Sium.   WATER PARSNIP.

S. latifolium: stems glabrous, stout, erect, angular, 2–4
feet high; leaves pinnate, the lower very long, with 6–10
pairs of large ovate-lanceolate toothed segments; the upper
shorter, with fewer and smaller segments; umbels large, of
15–20 rays, all terminal, general and partial involucres of
several lanceolate often toothed bracts.—Ditches and edges of
streams.   Fl. July, August.

S. angustifolium: stems erect, branched, leafy, round,
striated 1–3 feet high ; leaves pinnate, with 8–10 pairs of un-
equally-lobed ovate segments, deeply and sharply toothed;
umbels numerous, mostly lateral, with 8–15, rarely more,
rays; involucral bracts lanceolate, often toothed.—Ditches
and shallow streams.   Fl. August.

(115) Pimpinella. BURNET SAXIFRAGE.

P. Saxifraga: stems erect, 1–2 feet high, slightly branched, glabrous; leaves usually pinnate, with 7–9 pairs of broadly ovate or orbicular toothed segments, those of the stem bipinnatifid with linear segments; umbels terminal, with 10–15 rather slender rays; flowers white.—Pastures, roadsides, etc. Fl. July, August.

P. magna: larger and stouter than the last; the stems often two feet high or more; leaf-segments usually undivided, ovate or lanceolate, with more pointed teeth; flowers pink.— Shady places. Fl. July, August.

(116) Bupleurum.

B. rotundifolium: annual; stems erect, stiff, glabrous, one foot or more high, branched above; leaves broadly ovate, the upper ones embracing the stem, and joined round the back of it, so that they appear *perfoliate*, the lowest ones tapering to a stalk; umbels terminal of 3–6 short rays, without any general involucre, the partial involucres very much longer than the flowers, consisting of 4–6 broadly ovate, yellowish bracts.— Hare's-ear, or Thorow-wax.—Cornfields in chalky soils. Fl. July.

B. aristatum, and B. tenuissimum, are very rare dwarf annual species with grassy leaves, the first with non-granulated, the second with granulated fruits.

(117) Œnanthe.

* Segments of the upper leaves few, long and linear

Œ. fistulosa: stems thick, hollow, erect, 2–3 feet high, slightly branched; radical leaves twice pinnate, with small cuneate segments divided into 3–5 lobes, those of the stem with long hollow stalks bearing a few pinnate segments with linear lobes; umbels terminal, the central main one with three rays, each supporting numerous sessile fertile flowers, and few or no pedicellate barren ones, while those which terminate the branches have usually five rays, and only pedicellate, barren flowers; partial involucres of a few small narrow bracts, the general one either entirely wanting or reduced to a single bract; fruits in compact globular heads.—Water Dropwort.— Wet meadows and marshes. Fl. July, August.

Œ. pimpinelloides: stems erect, firm, almost solid, 1–2 feet high or more, with a few long branches; leaves much more divided than in the last; the upper ones usually with long, narrow segments, those of the radical ones much shorter and broader, and sometimes very numerous; umbels of 8–15 rather short rays; general involucre of a few small, linear bracts, sometimes wanting, partial ones of several small, linear bracts. —Meadows, pastures, and marshes. Fl. July.

Œ. Lachenalii and Œ. silaifolia are closely allied plants found in salt marshes.

** Segments of stem-leaves numerous, broadly wedge-shaped or oblong.

Œ. crocata: stem stout, branched, 3–5 feet high; leaves twice or thrice pinnate, the segments broadly cuneate or rounded, deeply cut into 3–5 lobes; umbels on long, terminal peduncles, with 15–20 rays, the bracts of the involucres small and linear, several in the partial ones, few or none under the general umbel; fruit somewhat corky.—Ditches, and sides of streams. Fl. June, July.

Œ. Phellandrium: stem rooting at the base, erect, elongated and creeping, or floating, the flowering branches erect or ascending; stem-leaves twice or thrice pinnate, with small oblong entire or cuneate, lobed segments; when under water, the lobes are long, narrow, sometimes capillary; umbels on short peduncles, opposite to the leaves or in the forks of the branches, the rays seldom above twelve; partial involucre of small, narrow bracts.—Ditches, ponds, and streams. Fl. July.

(118) **Æthusa.** FOOL'S PARSLEY.

Æ. Cynapium: annual; stems erect, glabrous, leafy, 1–2 feet high, with forked branches, emitting a nauseous smell when rubbed; leaves twice or thrice pinnate, the segments ovate-lanceolate, more or less deeply cut into narrow lobes; umbels on long peduncles, either terminal or opposite to the leaves, of 8–12 rays, usually without general involucres; partial involucres of 2–3 long linear bracts, towards the outside of the umbels, and turned downwards.—A common weed in fields and gardens. Fl. July, August.

(119) **Meum.** SPIGNEL.

M. Athamanticum: stems 1–3 feet high, leaves mostly radical, tufted, bipinnate, their segments deeply cut into numerous very fine, short, bristle-like lobes, and appearing whorled or clustered along the common stalk; the stems bear a few smaller, less divided leaves; umbels terminal, of 10–15 rays, with one or two narrow bracts to the general involucre, and a small number of short, slender bracts, to the partial one, —Bald-money.—Mountain pastures. Fl. June, July.

(120) **Crithmum.** SAMPHIRE.

C. maritimum: stems glabrous, seldom above a foot high, almost woody at the base, the young branches, foliage, and umbels, thick and fleshy; leaves twice or thrice ternate, with

few lanceolate linear segments; umbel of 15–20 or more rays,
with an involucre of several small linear or lanceolate bracts.—
Clefts of rocks by the sea. Fl. August.

(121) Pastinaca. PARSNIP.

P. sativa: annual or biennial; stem erect, 2–3 feet high;
leaves pinnate, coarse, downy, with 5–9 sharply-toothed, and
more or less lobed segments; upper leaves small and less
divided; umbels of 8–12 rays, usually without involucres;
fruits flat and oval.—Pastures and thickets, on calcareous
soils. Fl. July.

(122) Heracleum. COW PARSNIP.

H. Sphondylium: stems four feet high, coarse, rough with
short stiff hairs; leaves very large, pinnate, with 3–7 large
broad segments, which are three-lobed or pinnatifid and
toothed; umbels large, flattish, of about twenty rays; outer
petals much larger than the rest; carpels nearly orbicular,
flat.—Hogweed.—Pastures, hedges, and thickets. Fl. July.

(123) Tordylium. HARTWORT.

T. maximum: annual; stems erect, two feet or more high,
rough with short, stiff hairs; leaves pinnate, with 5–9 lanceolate, almost ovate, coarsely toothed segments; umbels terminal, of 8–10 short rays, with a few rather long, narrow
bracts to the involucres; petals small pink.—Waste and cultivated lands; rare. Fl. June, July.

(124) Coriandrum. CORIANDER.

C. sativum: annual; stems erect, branching, glabrous,
1–1½ feet high, emitting a very disagreeable smell when

rubbed; lowest leaves once or twice pinnate, with broadly-ovate or cuneate, deeply cut segments, the others more divided, with linear segments; umbels terminal, of 5–8 rays, without general involucre, and only a few small slender bracts to the partial ones; flowers white, the outer petals larger.—A weed of cultivation. Fl. June.

(125) **Conium.** HEMLOCK.

C. maculatum : annual or biennial; stem erect, branching, 3–5 feet high, smooth, glaucous, spotted, emitting a nauseous smell when bruised; leaves large, much divided into numerous small ovate or lanceolate deeply-cut segments; umbels terminal, of 10–15 rays; bracts short, lanceolate, those of the general involucre variable in number, those of the partial ones almost always three, turned to the outside of the umbel.— Hedges and borders of fields, etc. Fl. June, July.

(126) **Smyrnium.** ALEXANDERS.

S. Olusatrum : annual or biennial; stems coarse, erect, 2–4 feet high, glabrous; lower leaves twice or thrice, upper ones once ternate, the segments broadly ovate, coarsely toothed or three-lobed; umbels terminal, of 8–12 rays; flowers greenish yellow, crowded in the partial umbels; fruits black, aromatic. —Meadows and waste places. Fl. May, June.

(127) **Scandix.** SHEPHERD'S-NEEDLE.

S. Pecten : annual; stem branching, erect, or spreading, ½–1 foot high, hairy; leaves twice or thrice pinnate, with short segments cut into linear lobes; umbels terminal of 2–3 rays, without general involucres; partial involucres of several lanceolate bracts, often 2–3 lobed at the top; flowers small

and white, with a few large outer petals; carpels cylindrical
and ribbed, terminated in a stiff, flattened beak three times
their length, sometimes compared to the tooth of a comb.—
Venus's Comb.—A cornfield weed.　Fl. June to August.

(128) Anthriscus. CHERVIL.

A. sylvestris: stems hairy, erect, branched, 2–3 feet high;
lower leaves on long stalks, twice pinnate, with ovate-lanceolate, pointed, deeply pinnatifid, toothed segments, upper leaves
smaller, all hairy; umbels numerous, of 8–10 rays, without
general involucre, the partial ones of several bracts; fruits
smooth, shining, narrowed at the top into a short beak.—
Hedges, borders of fields, etc.　Fl. May, June.

A. vulgaris: annual; stems erect, branched, hairy, two
feet high; leaves twice or thrice pinnate, with ovate or ovate-
lanceolate, pinnately-lobed and toothed segments; umbels
small, on short peduncles, opposite to the leaves, of 3–7 rays,
without general involucres, and but few bracts to the partial
ones; fruits covered with short hooked bristles, narrowed at
top into a short smooth beak.—A weed of cultivation.　Fl.
May, June.

(129) Chærophyllum. CHERVIL.

C. temulum: biennial; stem erect, 2–3 feet high, rough
with short reflexed hairs; leaves twice pinnate or ternate, with
ovate or wedge-shaped pinnatifid or toothed segments, more
or less hairy; umbels of few rays, without a general involucre,
the partial involucres of 5–6 broadly lanceolate bracts.—
Hedges and thickets.　Fl. June, July.

(130) Myrrhis. CICELY.

M. odorata: stem erect, branching, hairy, 2–3 feet high, highly aromatic; leaves large, twice or thrice pinnate, with numerous lanceolate, deeply pinnatifid, and toothed segments; umbels terminal, not large, of 8–10 rays, without general involucre, the bracts of the partial ones numerous, lanceolate, acuminate; fruits, when ripe, nearly an inch long.—Mountain pastures, indigenous or naturalized. Fl. May, June.

(131) Torilis. HEDGE PARSLEY.

T. nodosa : annual; stems procumbent or spreading, scarcely a foot long; leaves twice pinnate, with small, narrow, pointed segments; umbels forming little heads, closely sessile, opposite to the leaves, composed of 2–3 exceedingly short, scarcely distinct rays, sometimes of a simple cluster; fruits covered with short, straight or hooked bristles.—Roadsides and waste places. Fl. May, June.

T. Anthriscus : annual; stem erect, 2–3 feet high, with slender, wiry branches; leaves once or twice pinnate, the segments lanceolate, pinnatifid, or coarsely toothed; umbels on long slender peduncles, of 3–8 rays; involucres, both general and partial, of small subulate bracts, one close under each ray; fruit a small burr.—Hedge Parsley.—Hedges, roadsides, and waste places. Fl. July, August.

T. infesta : annual; smaller and more spreading than the last; general involucre entirely wanting or reduced to a single bract, often lanceolate; bristles of the fruit less curved, with a minute hook at the top.—Cultivated and waste places. Fl. July, August.

(132) Daucus. CARROT.

D. Carota : annual or biennial; stem 1–3 feet high; leaves twice or thrice pinnate, with deeply three-lobed or pinnatifid, lanceolate or linear segments; umbels terminal, large, with numerous crowded rays, the inner ones very short, the outer much longer, and closing over after flowering; bracts of both involucres usually divided into 3–5 long linear lobes; fruit covered with prickles.—Fields, pastures, and waste places, especially near the sea. Fl. June, July.

Var. *maritimus :* leaves somewhat fleshy with shorter segments; umbels more spreading, and prickles more flattened.— Sea coasts.

(133) Bryonia. BRYONY.

B. dioica: tuberous; stems climbing to a great length by means of spirally-twisted tendrils; leaves palmately 5–7-lobed, the lobes toothed; flowers diœcious, the males broadly-campanulate pale yellow in stalked racemes, the females smaller nearly rotate; berries red.—Red Bryony.—Hedges and thickets. Fl. June to September

(134) Lonicera. HONEYSUCKLE.

L. Periclymenum : shrub; stems scrambling over bushes and trees to a considerable height; leaves ovate or oblong, the lower ones contracted, stalked, the upper ones rounded and closely sessile, but not united; flowers closely sessile, in terminal stalked heads, pale yellow, reddish outside.—Woodbine.—Woods and hedges. Fl. June to September.

(135) Linnæa.

L. borealis : evergreen; stems creeping and trailing to the length of one foot or more; leaves opposite, small, broadly-ovate; flowering branches short, erect, with 2–3 pairs of leaves, two-flowered; flowers bell-shaped, gracefully drooping,

fragrant, pale pink.—Woods: chiefly fir woods, in Scotland.—
Fl. July.

(136) Sambucus. ELDER.

S. nigra : small tree or shrub ; leaves pinnate, with 5–7 ovate,
pointed, sharply-toothed segments ; flowers white or cream-
coloured, in broad cymes of about five main branches, many
times divided ; fruits black.—Woods, coppices, and waste
places. Fl. June.

S. Ebulus : stems erect, thick, pithy, 2–3 feet high ; leaf-
segments 7–11, lanceolate, serrated, with a smaller one on
each side of the leafstalk, looking like stipules ; cymes with
three primary branches ; flowers white, or tinted outside with
purple ; fruits black.—Danewort.—Roadsides and rubbishy
wastes. Fl. August.

(137) Viburnum.

V. Lantana : shrub, much branched, covered with soft,
mealy down ; leaves ovate, cordate at the base, with small
pointed teeth, whitish and downy beneath ; flowers small,
white, in dense cymes, 2–3 inches across ; berries oblong, pur-
plish-black.—Wayfaring Tree.—Woods and hedges. Fl.
May.

V. Opulus : shrub, everywhere glabrous ; leaves divided
into 3–5 broad, angular, pointed, coarsely-toothed lobes, the
slender leafstalks beset with glands ; flowers white, in cymes,
the outer ones becoming enlarged, and perfectly barren ; berries globular, blackish-red.—Guelder Rose.—Hedges and coppices. Fl. June, July. The *Guelder Rose* seen in shrubberies
is a variety, in which all the flowers are enlarged and barren,
and the cyme becomes globular.

(138) Rubia. MADDER.

R. peregrina : stems straggling, dwarf, or trailing over bushes to the length of several feet, clinging by means of short recurved prickles; leaves 4–6, in a whorl, ovate-oblong or lanceolate, nearly sessile ; flowers small, greenish, in loose axillary or terminal panicles, rather longer than the leaves; fruit a small black two-lobed berry.—Dry woods and stony places. Fl. June to August.

(139) Galium.

* *Flowers yellow.*

G. cruciatum : stems prostrate or creeping, the flowering ones erect or ascending, $\frac{1}{2}$–$1\frac{1}{2}$ foot long, hairy ; leaves in whorls of four, ovate, hairy on both sides; flowers small, yellow, in little leafy axillary cymes or clusters, the fertile ones few and often five-lobed ; fruits small, smooth, almost succulent.—Crosswort.—Hedge-banks and bushy places.   Fl. May, June.

G. verum : glabrous, or with only a slight asperity on the edges of the leaves; stems $\frac{1}{2}$–1 foot high, much branched at the base, decumbent or ascending ; leaves small, linear, in whorls of 6–8 ; flowers small, yellow, numerous, in an oblong terminal panicle ; fruits small, smooth.—Ladies' Bedstraw.— Banks and pastures.   Fl. July, August.

** *Flowers white.*
 † *Fruits smooth or granulated.*
  ‡ *Leaves in whorls of four.*

G. palustre : stems one foot or more long, glabrous, with few spreading branches, almost always rough on the angles ; leaves mostly four in a whorl, linear or oblong, obtuse, usually rough on the edges ; flowers small, white, not numerous, in

spreading panicles; fruit rather small, slightly granulated.—
Marshes and wet places. Fl. July.

G. *uliginosum* differs chiefly from this in having the leaves
six or eight in a whorl, narrower, and terminated by a fine
point.

‡‡ *Leaves in whorls of six or eight.*

G. saxatile : stems much branched, often tufted at the base,
the flowering ones numerous, weak, six inches rarely a foot
high, smooth or nearly so on the angles; leaves 6–8 in a
whorl, occasionally on the barren shoots only 4–5, the lower
ones small and obovate, the upper narrow, all with a little
point at the tip, the edges smooth or rough; flowers, numerous, white, in short terminal panicles, their lobes scarcely
pointed; fruits small, more or less granulated.—Open heaths
and pastures.—Fl. June to August.

G. Mollugo : stems 1–3 feet long, smooth, shining, more or
less branched; leaves usually eight in a whorl, obovate, oblong,
or linear, more or less rough on the edges, always terminated
by a little point; flowers white, numerous, in large terminal
panicles, their lobes ending in a small point; fruit small,
smooth or slightly granulated.—Hedges, thickets, and rich
pastures. Fl. July, August.

A more erect narrower-leaved form closely allied to this, is
called G. *erectum.*

†† *Fruits bristly.*

G. Aparine : annual; stems several feet in length, scrambling over bushes, to which they cling by means of recurved
prickles on their angles and on the edges and midribs of
the leaves; leaves 6–8 in a whorl, linear or linear-lanceolate;
peduncles opposite and axillary, rather longer than the leaves,
bearing a loose cyme of 3–10 small, greenish-white flowers;

fruits covered with hooked bristles, forming small adhesive
burrs.—Cleavers, or Goosegrass.—Hedges and thickets. Fl.
June, July.

Allied plants, with more slender, shorter, and less hispid
stems, and smaller fruits, are sometimes called G. *Vaillantii*
and G. *spurium*. Another related plant, a cornfield species,
is G. *tricorne*, but it is altogether smaller, with shorter leaves,
and 1–3-flowered peduncles, and has the fruit granulated only,
and not bristly.

(140) Sherardia. FIELD MADDER.

S. arvensis : annual; stems decumbent, branched, seldom
over six inches high; leaves about six in a whorl, the lower
ones obovate, the upper linear or lanceolate, acute, rough-
edged; flowers small, blue or pink, in little terminal heads,
surrounded by a broad, leafy, eight-lobed involucre, calyx-
teeth enlarged after flowering, forming a little leafy crown at
the top of the fruit.—Cultivated and waste places. Fl. June
to August.

(141) Centranthus.

C. ruber : stems stout, tufted, glabrous, somewhat glaucous, 1–2 feet high; leaves ovate-lanceolate, entire; flowers
numerous, red, rarely white, in dense cymes, forming a handsome oblong, terminal panicle; border of the calyx unrolling
in the ripe fruit into a bell-shaped feathery pappus.—Red
Valerian.—Naturalized on chalk cliffs and old walls. Fl.
June to September.

(142) Valeriana. VALERIAN.

* *Lower leaves undivided.*

V. dioica : stem emitting creeping runners, the flowering

ones erect, 6–8 inches high; radical leaves on long stalks, ovate, entire, the stem-leaves few, mostly pinnate, with a large oval or oblong terminal segment; flowers small, pale rose, in terminal corymbs, mostly unisexual, the corolla tube short.— Boggy meadows. Fl. June.

** *Leaves all pinnate.*

V. **officinalis**: stems erect, 2–4 feet high, nearly simple, hairy at the base; leaves pinnate, with 9–21 or more large, lanceolate segments, with a few coarse teeth; flowers small, white or tinged with pink, in broad terminal corymbs.—All-heal.—Damp woods, sides of ditches and streams, etc Fl June.

(143) Dipsacus. TEASEL.

D. **sylvestris**: biennial; stems erect, 4–5 feet high, prickly, as well as the midribs, peduncles, and involucres; leaves sessile, long, lanceolate, entire or coarsely-toothed, the upper ones broadly-connate at the base; heads of flowers at first ovoid, but gradually becoming cylindrical, nearly three inches long, the involucre of 8–12 long, unequal, stiff, linear, prickly bracts, usually curved upwards; flowers pale lilac.— Roadsides and waste places. Fl. July.

The Fullers' Teasel (*D. fullonum*) is believed to be a cultivated variety, differing only in the scales of the receptacle being hooked.

D. **pilosus**: biennial; stems 2–4 feet high, branching, with stiff spreading hairs or bristles; leaves with a large ovate coarsely-toothed terminal segment, and 1–2 pairs of smaller ones; flowers white, forming globular, hispid heads, barely an inch in diameter, on long peduncles.—Moist hedges, thickets, and banks. Fl. August.

(144) Scabiosa. SCABIOUS.

* *Florets five-lobed.*

S. Columbaria : stems 1–2 feet high, glabrous or hoary ; leaves pinnate, with an ovate or oblong terminal segment, and several smaller ones, the stem-leaves few, with linear segments, entire or pinnatifid ; flower-heads pale purplish-blue, the outer florets larger and more oblique ; involucres short ; involucel with a scarious, cup-shaped border, in the centre of which appears the summit of the fruit crowned by the five bristles of the calyx.—Pastures and waste places.   Fl. June to August.

** *Florets four-lobed.*

S. arvensis : stems erect, hairy near the base, 1–3 feet high ; leaves variable, the radical ones lanceolate, stalked, the upper ones broader at the base, sessile, all coarsely toothed or slightly lobed, or deeply-cut or pinnate ; flower-heads large, pale lilac-purple, on long peduncles, the outer florets much larger and more oblique than the central ones ; involucre short ; fruit crowned by the 8–10 radiating bristles of the calyx.—Pastures and cornfields.   Fl. July, August.

(145) Anthemis. CHAMOMILE.

* *Ray-florets without a style.*

A. Cotula : annual ; stems erect, branching, one foot high or more, glabrous, with a disagreeable smell ; leaves bipinnatifid, with narrow linear, pointed, entire or divided lobes ; flower-heads in a loose terminal corymb ; involucre slightly cottony, the inner bracts scarious at the top ; receptacle convex, lengthening out as the flowering advances, with linear-setaceous scales ; ray-florets white, without any trace of style ; achenes rough with glandular dots, without any border.— Waste places.—A common weed.   Fl. June, July.

** *Ray-florets having a style.*

A. arvensis: annual; stem 1-2 feet high, downy, much branched, sometimes decumbent, the leafy branches terminating in single flower-heads; leaves bipinnatifid, hairy, the segments linear-lanceolate; flower-heads solitary, the ray-florets white, always having a style; receptacle conical, with lanceolate scales beween the florets.—Cultivated fields. Fl. June, July.

Var. *maritima*, with a stem of more spreading habit, thicker leaves, a flat receptacle, and subulate scales, has been called *A. anglica.*—North-east coast.

A. nobilis: stem procumbent or creeping, branched, the flowering branches short, ascending, leafy; leaves bipinnate, the leaflets linear-subulate, somewhat downy, pleasantly aromatic; flower-heads on terminal peduncles, with white rays; inner involucral bracts scarious at the top; scales of the receptacle broad, obtuse.—Commons and sandy pastures. Fl. July, August.

(146) Achillea.

A. Ptarmica: stems erect, glabrous, 1-2 feet high, nearly simple; leaves broadly linear, regularly serrate; flower-heads few, in a loose terminal corymb; florets of the ray 10-15, short, broad, white.—Sneezewort.—Moist hilly pastures. Fl. July, August.

A. Millefolium: stems numerous, short, leafy, the flowering ones erect, almost simple, about one foot high; leaves oblong or linear, finely cut into a multitude of short, narrow, deeply pinnatifid segments; flower-heads numerous, small, ovoid, in a dense terminal corymb; the ray-florets 5-6 in each head, white or pink.—Milfoil or Yarrow.—Pastures, meadows, and waste places. Fl. June to August.

(147) Chrysanthemum.

* *Ray-florets white.*

C. Leucanthemum : stems erect, simple or slightly branched, 1–2 feet high ; leaves obovate, coarsely-toothed, on long stalks, those of the stem narrow, sessile ; flower-heads solitary on long terminal peduncles, rather large ; involucral bracts bordered by a brown, scarious edge.—Oxeye Daisy.—Pastures, banks, etc.　Fl. June to August.

C. Parthenium : biennial ; stems erect, branching, a foot or more high ; leaves pinnate, the segments ovate or oblong, pinnatifid, toothed ; flower-heads numerous, about half an inch in diameter, in a terminal corymb ; achenes crowned by a minute toothed border.—Feverfew.—Roadsides, and in waste places.　Fl. June, July.　This plant is often called *Pyrethrum Parthenium*, and is a strongly scented aromatic medical herb, with bitter tonic properties.

C. inodorum : annual ; stem erect or spreading, branched, 1–1½ feet high ; leaves twice or thrice pinnate, with numerous narrow-linear, almost capillary lobes ; flower-heads rather large, on terminal peduncles ; involucral bracts with a brown scarious edge ; receptacle convex or hemispherical, not conical ; achenes ribbed, crowned with a minute entire or four-toothed border.—Fields and waste places.　Fl. July, August.

Var. *maritimum :* stems diffuse ; leaves rather fleshy ; flowers smaller.—Sea coast.

** *Ray-florets yellow.*

C. segetum : annual ; stem glabrous, erect, one foot high or more, with spreading branches ; lower leaves obovate and stalked, the upper ones narrow and stem-clasping, with a few teeth at the top ; flower-heads rather large, on terminal peduncles; involucral bracts broadly scarious; florets of the ray and disk yellow.—Corn Marigold.—A cornfield weed. Fl. June to August.

(148) Matricaria. WILD CHAMOMILE. ·

M. Chamomilla: annual, closely resembling *Anthemis Cotula*; stems erect, branching; leaves bipinnatifid, with short narrow linear segments; flower-heads rather large, on terminal peduncles, the involucral bracts nearly all of the same length, with scarious edges; ray-florets white; receptacle much elongated as the flowering advances, scaly between the florets; achenes without any border.—Fields and waste places. Fl. June, July.

(149) Bidens.

B. cernua: annual; stems erect, 1–2 feet high, with spreading branches; leaves sessile, lanceolate, serrated; flower-heads drooping, on terminal peduncles, the florets yellow, usually all tubular, but occasionally a few of the outer ones become ligulate; awns of the achenes usually 2–3, rarely 4.—Bur Marigold.—Wet ditches and marshes. Fl. July to September.

B. tripartita differs chiefly in the leaves being stalked, and deeply cut into 3–5 lanceolate segments, in the flower-heads being rather less drooping, and in the achenes usually having only two awns or bristles.

(150) Eupatorium.

E. cannabinum: stems erect, 3–4 feet high; leaves divided into three broadly lanceolate, coarsely-toothed lobes, slightly downy; flower-heads numerous, in compact terminal corymbs, pale reddish-purple; involucres cylindrical, of few unequal bracts, usually containing five florets.—Hemp Agrimony.— Banks and moist bushy places. Fl. July, August.

(151) Erigeron.

E. acris: annual or biennial; stems erect, ½–1 foot high, slightly branched, rough with short hairs; leaves linear or lanceolate, entire; flower-heads rather small, forming a short loose panicle; florets numerous, mostly filiform and short, the outer ones pale purple, projecting slightly beyond the involucre.—Fleabane.—Pastures, roadsides, and waste places. Fl. July, August.

(152) Solidago. GOLDEN ROD.

S. Virgaurea: stems erect, stiff, nearly simple, ½–2 feet high; leaves obovate, stalked, those of the stem oblong or lanceolate; flower-heads crowded in a narrow-oblong terminal panicle, with a spreading ray of 10–12 bright yellow florets.—Woods. Fl. July to September.

(153) Senecio. GROUNDSEL.

* *Ray-florets small, revolute.*

S. sylvaticus: annual; stems 1–2 feet high, erect, branched, downy; leaves deeply pinnatifid, downy with oblong unequally-toothed segments; flower-heads numerous, in a loose corymb; the outer florets ligulate, but small and rolled back.—Banks, waste places, and borders of woods. Fl. July to August.

S. viscosus resembles this, but is coarser, and covered with a short, viscous, strong-smelling down.

** *Ray-florets spreading, conspicuous.*

S. Jacobæa: stems 2–4 feet high, erect, scarcely branched, except at top; lower leaves oblong-ovate, lyrate pinnatifid, those of the stem sessile, bipinnatifid, with oblong deeply-toothed segments; flower-heads rather large, bright yellow, in a compact terminal corymb; involucral bracts tipped with black ; ray-florets 12–15, linear-oblong ; achenes of the disk covered with short hairs, those of the ray glabrous.—Ragwort. —Roadsides, waste places, and bushy pastures. Fl. July to September. It sometimes assumes the spreading inflorescence of *S. aquaticus*, in which the achenes both of the disk and ray are however always smooth.

S. erucæfolius resembles this, but the leaves are more regularly divided into narrower segments, the terminal ones not very different from the others; and the achenes of the ray-florets are hairy like those of the disk. [See also p. 58.]

(154) Inula.

I. Helenium : stems stout, erect, scarcely branched, 3–5 feet high; leaves often a foot long, oblong, narrowed into a stalk ; the upper ones clasping the stem, downy beneath ; flower-heads very large, the ray-florets numerous, linear, yellow ; involucral bracts broadly ovate, softly hairy.—Elecampane.—Moist pastures. Fl. July, August.

I. Conyza : biennial ; stems hard, erect, 2–3 feet high, roughly downy ; leaves ovate-lanceolate, downy, the upper sessile ; flower-heads numerous, in a terminal corymb; the ray-florets numerous but very small.—Ploughman's Spikenard. —Hedges, banks, and roadsides. Fl. July to September.

(155) Pulicaria. FLEABANE.

P. dysenterica : stems erect, 1–2 feet high, loosely branched, downy or woolly ; leaves oblong, waved, clasping the stem with rounded auricles, woolly ; flower-heads pedunculate, axillary or terminal, corymbose, with a ray of very numerous, linear, spreading bright yellow florets.—Wet pastures, ditches, and roadsides. Fl. August.

P. **vulgaris** resembles this, but has narrower, less woolly leaves, and the flower-heads are much smaller, the florets of the ray being very short; the minute outer scales of the pappus are also distinct, not forming a little cup as in *P. dysenterica.*

(156) Gnaphalium.

* *Flower-heads diœcious.*

G. **dioicum**: flowering stems almost simple, 2–5 inches high; leaves obovate or oblong, the upper ones linear, cottony beneath; flower-heads 3–4 together, in compact, terminal corymbs, diœcious; inner bracts of the involucre in the males with broad, white, petal-like tips, spreading like ligulate florets: in the females narrow, white-tipped, not spreading.— Cat's-ear.—Mountain pastures. Fl. June, July.

** *Flower-heads hermaphrodite.*

G. **sylvaticum**: stems simple, erect, 2–8 inches high; leaves linear, cottony on the under side; flower-heads small, cylindrical or ovoid, solitary or in little clusters in the axils of the upper leaves, forming a long leafy spike; the involucres scarcely cottony, with brown, shining bracts.—Woods and heaths. Fl. July to September.

G. **uliginosum**: annual; stems much branched, cottony six inches high; leaves linear or narrow-oblong, the upper ones waved; flower-heads small and clustered, many together, within a tuft of rather long leaves at the extremity of the branches; involucral bracts brown and scarious.—Wet sandy places. Fl. July, August.

(157) Filago. CUDWEED.

F. **germanica**: annual; stems erect, cottony, 6–8 inches high, simple or branched at the base, each branch terminated by a single globular cluster of flower-heads, or throwing out immediately under it 2–3 branches, each ending in a similar cluster; leaves erect, lanceolate or linear, pointed or obtuse, sometimes slightly spathulate; flower-heads very small, 20–30 in each cluster, the involucres ovoid-conical, more or less angular, pale yellow or brown, the bracts usually acute.—Dry pastures, and stony or sandy wastes. Fl. July, August.

F. apiculata and *F. spathulata* are sometimes distinguished from the above, by their shorter or longer, and more or less obtuse or acute floral leaves, by the quantity of cotton on their involucres, and by their obtuse or acute bracts.

F. minima: annual; more slender and smaller than the last, which it resembles; stems more irregularly branched at top, leaves smaller, clusters of flower-heads smaller and more numerous, consisting of 3–10 minute conical heads.—Fields, and stony or sandy wastes. Fl. June to September.

(158) **Arctium.** BURDOCK.

A. Lappa: biennial; stems stout, erect, branching, 3–5 feet high; leaves heart-shaped, large, sometimes 1½ feet in length, the upper ones broadly ovate, green above, white and cottony beneath; flower-heads in terminal panicles; involucres nearly globular, the bracts numerous with spiny hooked points, glabrous or covered with loose, white, cottony wool; florets purple, equal.—Waste places, roadsides, etc. Fl. July, August. This varies much in the size of the flower-heads and in some other points, and is sometimes divided into five species.

(159) **Carlina.** CARLINE THISTLE.

C. vulgaris: biennial; stem erect, corymbose, 6–8 inches high; leaves oblong-lanceolate, toothed or pinnatifid, very prickly; flower-heads hemispherical, 3–4 in a small terminal corymb, the outer involucral bracts broadly lanceolate, bordered with very prickly teeth or lobes, the inner ones linear, entire, with smooth and shining, horizontally-spreading tips. —Dry, hilly pastures and fields. Fl. July to September.

(160) Carduus. THISTLE.

* *Pappus of simple hairs.*

C. nutans: biennial; stems erect, cottony, 2–3 feet high; leaves deeply pinnatifid, very prickly, their edges decurrent along the stem, forming narrow very prickly wings; flower-heads large and drooping, crimson, solitary or 3–4 in a loose corymb; involucral bracts numerous, with a stiff, narrow-lanceolate appendage, ending in a spreading prickle.—Musk Thistle.—Waste places. Fl. July, August.

C. acanthoides much resembles this, but is usually taller and rather more branched; the leaves narrower and more prickly; the stem more thickly covered with prickly appendages, decurrent from the base of the leaves; the flower-heads smaller, globular, drooping; and the numerous narrow involucral bracts ending in a linear, spreading prickle.

** *Pappus of feathery hairs.*
† *Leaves decurrent, forming prickly wings to the stem.*

C. lanceolatus; biennial; stems stout, 3–4 feet high, winged and prickly; leaves waved pinnatifid, with short narrow lobes, the terminal one longer and lanceolate, all ending in a stiff prickle, rough above, with short almost prickly hairs, white and cottony beneath; flower-heads few, ovoid, the involucral bracts lanceolate, cottony, ending in a stiff, spreading prickle; florets purple.—Fields, pastures, and waste places. Fl. June to September.

C. palustris: annual or biennial; stems stiff, scarcely branched, 4–5 feet high, quite covered with the prickly de-current margins of the leaves; leaves long, narrow, pinnatifid with numerous ovate wavy prickly lobes, and a few rough hairs scattered on both surfaces; flower-heads numerous, small, ovoid, usually collected in clusters, forming an irregular ter-minal corymb; involucral bracts numerous, with very small somewhat prickly points, the inner ones often coloured; florets purple.—Wet fields and meadows. Fl. July, August.

†† *Leaves sessile or very partially decurrent.*

C. arvensis: stems erect, 3–4 feet high; leaves narrow, pinnatifid, very prickly, either embracing the stem with prickly auricles or shortly decurrent; flower-heads in loose terminal corymbs, dioecious; the males nearly globular, with project-ing purple florets, the females with much longer involucres and shorter florets; involucral bracts numerous, appressed, with small prickly points.—Cultivated and waste places. Fl. July.

C. heterophyllus: stems stout, 3–4 feet high, deeply fur-rowed, with a little loose cottony wool, not prickly; leaves clasping the stem, with scarcely decurrent auricles, lanceolate, glabrous above, cottony white beneath, bordered with small, bristly but scarcely prickly teeth, sometimes slightly lobed; flower-heads large, growing singly on long peduncles; involu-cral bracts glabrous, lanceolate, obtuse, or with a very minute not prickly point.—Mountain pastures. Fl. July, August.

C. pratensis: stems 1–2 feet high, usually simple, with a single ovoid flower-head, or occasionally divided into 2–3 long one-headed branches; leaves sinuate or shortly pinnatifid, the stem-leaves lanceolate, bordered with short, slightly prickly teeth; heads somewhat cottony; involucral bracts lanceolate, attenuate, appressed.—Wet meadows and pastures. Fl. June. This plant is sometimes regarded as a variety of *C. tuberosus*. Luxuriant specimens, with more divided leaves, sometimes slightly decurrent, have been named *C. Forsteri*.

(161) Silybum. Milk Thistle.

S. Marianus: annual or biennial; stems 2–3 feet high, slightly branched, nearly glabrous; leaves smooth, shining above and variegated by white veins, the lower ones deeply pinnatifid with broad very prickly lobes, the upper ones clasping the stem by prickly auricles scarcely decurrent; flowerheads large, drooping, solitary at the ends of the branches, with purple florets; involucral bracts, broad at the base, with a stiff, spreading, leafy appendage, ending in a long prickle; hairs of the pappus simple.—Waste places. Fl. June, July.

(162) Centaurea.

* *Involucral bracts not prickly.*

C. nigra: stems erect, hard, branched, 1–2 feet high; leaves linear-lanceolate or oblong, the upper ones entire or nearly so, the lower with a few coarse teeth; involucres globular, on terminal peduncles, the bracts closely imbricate, so as only to show their appendages, which are brown or black, and deeply fringed, except on the innermost, where they are shining and jagged; florets purple, all equal or the outer row much larger and neuter; fruits crowned by a ring of very minute, scaly bristles, occasionally intermixed with a few longer, very deciduous ones.—Knapweed.—Meadows, pastures, and waysides. Fl. June to August.

Var. *Jacea*: appendages of the involucral scales much paler, with a much shorter fringe, or jagged.—Sussex, rare.

C. **Scabiosa** : stem 2–3 feet high, stoutish, branched at the base ; leaves deeply pinnatifid, with **linear** or lanceolate lobes, often coarsely-toothed or lobed ; flower-heads large, **with** purple florets, the outer ones always enlarged and neuter ; involucral bracts broad, bordered only with a black appressed fringe, leaving the green centre exposed ; pappus of stiff hairs or bristles nearly as long as the achene.—Pastures, waste places, roadsides, etc. Fl. July to September.

C. **Cyanus** : annual ; stems erect, branching, two feet high, cottony ; leaves narrowish, the lower ones toothed, the upper linear and entire ; involucres solitary, on long terminal peduncles, ovoid, the bracts appressed, bordered by a fringe of very small teeth ; central florets bluish purple, the outer ones much larger, bright blue.—Cornflower or Bluebottle.—A cornfield weed. Fl. June to August. In our gardens where this plant is cultivated as an annual flower, it yields various colours.

** *Involucral bracts ending in a long stout prickle.*

C. **Calcitrapa** : annual ; stems somewhat cottony, spreading or prostrate, seldom above a foot high ; leaves pinnatifid, with a few long linear or lanceolate lobes ; flower-heads sessile among the upper leaves or in the forks of the branches, not large, the florets purple ; involucral bracts ending in stiff spreading spines, with 1–2 smaller prickles at their base ; achenes without any pappus.—Waste places and roadsides, most abundant near the sea. Fl. July, August.

(163) **Serratula.** SAWWORT.

S. **tinctoria** : stem stiff, erect, scarcely branched, 1–3 feet high ; leaves pinnate, with lanceolate, pointed, finely-toothed segments, the terminal one largest ; upper leaves toothed only, or with a few lobes at their base ; flower-heads in a terminal corymb, partially diœcious; florets purple.—Open woods, thickets, and bushy pastures. Fl. August.

(164) Lapsana. Nipplewort.

L. communis: annual; stem 1–3 feet high, with stiff hairs at the base, branched and glabrous upwards; leaves hairy, lyrate, the terminal lobe ovate, coarsely-toothed, upper ones narrow, entire; flower-heads on slender peduncles, in a loose panicle or corymb, yellow; involucre of about eight nearly equal scales.—A common weed. Fl. July, August.

(165) Thrincia.

T. hirta: leaves oblong or linear, coarsely-toothed, sinuate or shortly pinnatifid, hispid with forked hairs; peduncles six inches high, with a single small head of bright yellow flowers; involucres glabrous, of 10–12 nearly equal bracts, with several small imbricated ones at the base; achenes of the outer row curved, slightly tapering at the top, with a very short, scaly pappus.—Dry open pastures and waste places. Fl. July to September.

(166) Leontodon. Hawkbit.

L. hispidum: leaves long, narrow, coarsely-toothed or pinnatifid; the whole plant more or less hispid with erect, stiff, short hairs; peduncles ½–1 foot or more long, slightly swollen at the top, with a single rather large yellow flower-head; involucre-bracts narrow, hispid, the inner row much longer than the outer ones; achenes long, striate, and transversely rugose; pappus of about a dozen brown, feathery hairs, as long as the achene, surrounded by 5–6 others not a quarter that length. —Meadows and pastures. Fl. July.

(167) Oporinia.

O. autumnalis: leaves all radical, long, narrow, pinnatifid with a few narrow lobes ; flower-stems erect, with 1–2 single-headed branches; flower-heads yellow; involucres oblong, tapering at the base into the enlarged summit of the peduncle, with closely-appressed imbricated bracts, nearly always hairy ; achenes long, striate, transversely wrinkled ; pappus brown, feathery.—Meadows,pastures, and waste places. Fl. August.

Var. *Taraxaci :* dwarf; flower-stems often simple ; flower-heads rather larger ; involucre more or less covered with black hairs.—Scotch Highlands.

(168) Hieracium. HAWKWEED.

* *Peduncles radical, one-flowered.*

H. Pilosella : leaves oblong or lanceolate, entire, tapering at the base, white beneath with short stellate down ; peduncles radical, with a single head of lemon-coloured flowers, often tinged with red on the outside ; achenes short in proportion to the pappus.—Dry pastures, banks, and roadsides.—Fl. May to July.

** *Stems several-flowered or many-flowered.*

H. murorum : leaves spreading, rather large, ovate or oblong, stalked, sometimes obtuse and nearly entire, more frequently pointed and coarsely-toothed, persistent at the time of flowering ; flower-stems erect, 1–2 feet high, with 1–2 leaves near the base ; flower-heads large, yellow, usually 3–4, sometimes 20–30, in a loose terminal corymb ; involucres and peduncles more or less clothed with black glandular hairs, intermixed with a shorter rusty-coloured down, the outer scales few and much shorter than the inner.—Banks, old walls, meadows, pastures, and open woods. Fl. July to August.

Mr. Bentham includes as a variety, *H. vulgatum* (*sylvati-cum*), which has a much more leafy stem.

H. umbellatum: stems leafy, erect, rigid, 1–3 feet high, hairy at the base; leaves narrow-lanceolate or oblong, the lower ones stalked, all tapering at the base; flower-heads numerous, on rather short lateral branches towards the summit of the stem, several of which start from so nearly the same point as to form an irregular umbel; involucres and peduncles glabrous or shortly downy; scales of the involucre regularly imbricated, the outer ones usually spreading at the tips. —Woods, stony places, and banks. Fl. July to September.

H. sabaudum: stems 1–3 feet high, stout, equally tall with *H. umbellatum*, but without radical leaves at the time of flowering, less rigid and more hairy; leaves larger, broader, and more toothed, the upper ones shorter, rounded at the base, sometimes almost stem-clasping; flowering branches forming a loose corymb, never an umbel.—Woods, hedges, and shady places. Fl. August and September.

(169) Picris.

P. hieracioides: biennial; stem 1–3 feet high, covered with short, rough, minutely-hooked hairs; leaves lanceolate, the lower tapering into a stalk, the upper stem-clasping; peduncles rather long and stiff; pappus dirty white, the hairs usually very feathery, except a few of the outer ones of each achene.—Roadsides, borders of fields, and waste places. Fl. July to September.

(170) Cichorium. CHICORY.

C. Intybus: stems hispid, 1–3 feet high; leaves spreading on the ground, hairy, runcinately pinnatifid, with a large terminal lobe and smaller lateral ones, all pointed and coarsely-toothed, the upper leaves small, less cut, embracing the stem by pointed auricles; flower-heads in closely-sessile clusters of two or three along the stiff, spreading branches, the florets large, bright blue.—Succory.—Dry wastes, roadsides, and borders of fields. Fl. July, August.

(171) Sonchus. SOWTHISTLE.

S. arvensis : stems 2–3 feet high; leaves long, runcinately-pinnatifid or sinuate, the lobes lanceolate or triangular, bordered by small prickly teeth, the upper ones clasping the stem with short broad auricles; flower-heads large, bright yellow, in loose terminal panicles, the branches, peduncles, and involucres more or less hispid with brown or black glandular hairs. —A cornfield weed. Fl. August.

S. oleraceus: annual; stem thick, hollow, 1–4 feet high, glabrous; leaves thin, pinnatifid, with a broad, heart-shaped or triangular terminal lobe, bordered with irregular, pointed or prickly teeth, and a few smaller lobes or coarse teeth along the broad leafstalk, the upper leaves narrow and clasping the stem with short auricles; flower-heads rather small, pale yellow, in a short corymbose panicle, sometimes almost umbellate. —A weed of cultivation. Fl. June to August.

Var. *aspera :* leaves darker, less divided, but much more closely bordered with prickly teeth; clasping auricles broader, rounded, and more prickly toothed.

(172) Crepis. HAWK'S-BEARD.

C. virens: annual or biennial; stems erect or ascending, 1–3 feet high, glabrous or nearly so; leaves linear or lanceolate, toothed or pinnatifid, with triangular or narrow, but short

T

lobes, the upper leaves clasping the stem by pointed spreading auricles; flower-heads small, yellow, in loose, often leafy panicles; involucres often slightly hispid, becoming conical after flowering, the outer bracts narrow-linear, and rather close.— Pastures, dry banks, roadsides, and waste places. Fl. June to September.

C. paludosa: stems erect, scarcely branched, nearly glabrous, 1–2 feet high; leaves ovate, coarsely-toothed, with a few small lobes along the stalk, those of the stem broadly-oblong or lanceolate, pointed, toothed, and clasping the stem by large, pointed auricles; flower-heads yellow, rather large, in corymbs of 8–10, the involucres more or less hairy, with black spreading hairs.—Moist, shady situations. Fl. July to September.

(173) Hypochœris.

H. glabra : annual; stem branched, leafless, glabrous, 3–10 inches high; leaves oblong, dentate-sinuate; flower-heads small, yellow, the florets scarcely longer than the involucres; pappus feathery, sessile on the wrinkled achenes of the outer florets, supported on a slender beak in the central ones.— Sandy situations. Fl. July, August.

H. radicata : leaves all radical, spreading, narrow, more or less toothed or pinnately lobed, hispid on both sides; stems erect, leafless, 1–2 feet high, divided into two or three long branches, each bearing a few small scales, and terminated by a rather large head of flowers; involucres nearly an inch long, shorter than the florets; achenes transversely wrinkled, all narrowed into a long, slender beak, with a feathery pappus.— Cat's-ear.—Meadows, pastures, waysides, and waste places. Fl. July.

(174) **Helminthia.** Ox-tongue.

H. echioides: annual or biennial; stem coarse, erect, 1–3 feet high, rough with short, stiff, almost prickly, often hooked hairs; leaves lanceolate, sinuate or coarsely-toothed, rough, the upper ones clasping the stem, or shortly decurrent; flower-heads small, crowded, on short peduncles, forming an irregular terminal corymb.—Hedge-banks, fields, and waste places.—Fl. July to September.

(175) **Tragopogon.** Salsify.

T. pratense: stem erect, slightly branched, 1–2 feet high; radical and lower leaves linear, keeled, dilated at the base, glabrous, slightly glaucous; peduncles long, thickened at the summit, each with a single head of yellow flowers; involucral bracts narrow-lanceolate; florets sometimes not half as long; achenes long, striate, the slender beak as long as the achene itself, the hairs of the pappus long and very feathery.—Yellow Goat's-beard.—Meadows and rich pastures. Fl. June.

T. porrifolium resembles the last, but is generally of more luxuriant growth, the peduncles more thickened at the top, the involucres longer in proportion to the florets, and the beak of the achenes and pappus longer. It also has deep violet or purple flowers.

(176) **Lactuca.** Lettuce.

* *Leaves smooth-keeled, on long stalks.*

L. muralis: annual or biennial; stems glabrous, erect, two feet high, with slender branches, forming a loose terminal panicle; leaves few, thin, rather large, with a broadly-triangular, toothed or lobed terminal segment, and a few irregular smaller ones below, the upper ones small, narrow, entire or

T 2

toothed; flower-heads small, yellow, on slender pedicels; in-volucres of five equal linear bracts, with 1–3 very small outer ones, containing 4–5 florets; beak of the black achenes much shorter than the achene itself.—Woods and shrubby places. Fl. July to August.

** *Leaves sessile, with a bristly keel.*

L. **virosa**: annual or biennial; stem erect, stiff, 2–4 feet high, with short, spreading branches, glabrous, except a few stiff bristles or small prickles on the edges and midrib of the leaves; leaves spreading, broadly-oblong, toothed; flower-heads in a leafy panicle; florets 6–12, pale yellow; achenes much flattened, obovate-oblong, striated, nearly black, with a slender beak about their own length.—Dry, stony wastes, banks, and roadsides. Fl. July, August.

L. **Scariola**, which is allied to this, has upright leaves, arrow-shaped at the base, and the achenes are pale-coloured.

(177) Barkhausia.

B. **taraxacifolia**: biennial; leaves chiefly radical, runcinately pinnatifid, with a large, terminal, coarsely-toothed lobe; stems erect, 1–2 feet high, bearing a few small, narrow leaves; flower-heads small, forming a loose, terminal, flat corymb; florets yellow, purple beneath; achenes all terminated by a slender beak about the length of the achene itself.—Dry pastures and waste places, chiefly in limestone districts. Fl. June, July.

B. **fœtida**, an allied plant, growing about a foot high, has irregularly-pinnatifid leaves, few flower-heads on long pedun-cles, and the beak of the outer achenes very short, often scarcely distinct, whilst that of the inner ones is long and slender, carrying up the whole pappus above the tips of the involucral bracts.

(178) Lobelia.

L. Dortmanna: aquatic; leaves tufted, nearly cylindrical, hollow, forming a dense green carpet at the bottom of the water; flowering-stems erect and simple, rising 6–8 inches above the surface of the water, almost leafless; flowers pale blue, drooping, in a simple terminal raceme.—Common in the lakes of Scotland, Ireland, and west of England. Fl. July.

(179) Jasione.

J. montana: annual or biennial; stems short, decumbent or ascending, sometimes nearly erect, one foot high, with spreading branches; leaves linear or lanceolate, waved on the edges, hairy; flower-heads on long terminal peduncles, the involucral bracts broadly ovate, the flowers or florets small, pale blue.—Sheep's-bit.—Heathy pastures, banks, etc. Fl. June, July.

(180) Phyteuma. RAMPION.

P. orbiculare: stems simple, erect or slightly decumbent, $\frac{1}{2}$–$1\frac{1}{2}$ foot high; radical leaves ovate, cordate, on long stalks, those of the stem narrow-oblong or lanceolate, the upper ones sessile; flowers deep blue, in a globular terminal head, surrounded by a few short, broadly lanceolate bracts.—Chalk downs of southern England. Fl. August.

(181) Campanula. BELL-FLOWER.

* *Calyx-tube and capsule short, broad.*
   † *Capsule opening at the sides.*
     ‡ *Capsule sessile, erect.*

C. glomerata: stem firm, erect, a foot high or more, hairy; lower leaves stalked, the rest sessile, broadly lanceolate, clasping the stem by their cordate base, roughly hairy; flowers

sessile, deep blue, in small clusters in the upper leaves, the
uppermost ones forming a compact leafy head.—Dry pastures.
Fl. July, August. In very dry soils it sometimes becomes
much dwarfed, sometimes reduced to one or two inches in
height, and single-flowered.

‡‡ *Capsule stalked, nodding.*

C. **Trachelium**: stem 2–3 feet high, erect, simple, leafy;
leaves on long stalks, broadly heart-shaped coarsely-toothed,
the upper ones small, ovate-lanceolate; flowers large, deep
blue, 2–3 together in short leafy racemes in the upper axils
or at the summit of the stem, sometimes solitary; calyx with
broad lanceolate segments.—Hedges and thickets. Fl. July,
August.

C. **latifolia**: stems 2–3 feet high, erect, simple, leafy; leaves
ovate-lanceolate, pointed, toothed, narrowed at the base, the
lower ones stalked; flowers large, blue or white, solitary in
the axils of the upper leaves, forming a leafy raceme, the
uppermost exceeding the leaves; calyx with long lanceolate
segments.—Woods and thickets. Fl. July, August.

C. **rotundifolia**: radical leaves orbicular or heart-shaped,
those of the stem narrow-lanceolate or linear, entire; stems
ascending or erect, 6–18 inches high, often branched, with a
few elegantly drooping blue, bell-shaped flowers, in a loose
raceme or panicle, sometimes solitary.—Harebell.—Hilly
pastures, heaths, banks, and roadsides. Fl. July, August.

†† *Capsules opening at top.*

C. **hederacea**: stems slender, prostrate, branched, thread-
like; leaves small, delicate, mostly orbicular or broadly heart-
shaped, with a few broad, angular teeth; flowers on long, filiform peduncles, drooping in the bud, nearly erect when fully
out, and often drooping again as the fruit ripens: corolla

narrow-bellshaped, pale bluish-purple; capsule almost globular, opening in three valves between the calycine teeth.—
Moist shady pastures and woods. Fl. July, August. Sometimes called *Wahlenbergia*.

** *Calyx tube and capsule long, narrow, prismatical.*

C. hybrida: annual; stems erect or decumbent, branched at the base, 6–8 inches high, hairy; leaves oblong, waved; flowers blue, sessile in the axils of the upper leaves, remarkable for their long, narrow, triangular ovary and capsule, crowned by the linear or oblong leafy segments of the calyx; capsule opening by short clefts close under the segments of the calyx.—A cornfield weed. Fl. July, August. Sometimes called *Specularia*.

(182) Oxycoccus.

O. palustris: shrub; stem creeping, wiry; leaves small, evergreen, ovate or lanceolate, the edges rolled back, glaucous beneath; flowers drooping, on long, slender peduncles; the corolla deeply divided into four lobes, which are turned back, exposing the stamens; berry globular, red.—Cranberry.—
Peat-bogs. Fl. June, July.

(183) Erica. HEATH.

* *Corolla shorter than the calyx.*

E. vulgaris: shrub; stems tufted, about a foot high; leaves small, short, opposite, linear, rigid; flowers small, purplish-pink or white, on short pedicels along the upper branches, forming irregular, leafy racemes; calyx coloured like the corolla, which is concealed by it, and is deeply four-lobed. ·
—Ling.—Dry heaths; abundant. Fl. June, July. Now generally separated under the name of *Calluna*.

** *Corolla longer than the calyx.*

E. cinerea: shrub; stems one foot or more high, bushy; leaves linear, pointed, three in a whorl, with clusters of small leaves in their axils; flowers numerous, reddish purple, in very showy, dense terminal racemes, the corolla ovoid.— Scotch Heather.—Scotch, Irish, Welsh, and English moors, covering immense tracts. Fl. July to October.

E. Tetralix: shrub; stems about a foot high, bushy at the base, with short, erect flowering branches; leaves in fours, linear, ciliate with short stiff hairs, the branches and upper leaves clothed with a short, whitish down; flowers rose-coloured, ovoid, forming little terminal clusters or close umbels.—Boggy heaths. Fl. July, August.

E. Mackaiana, a plant with shorter and broader leaves of a darker green, from Cunnamara, in Ireland, which is closely related to *E. Tetralix,* is sometimes regarded as distinct, sometimes as a mere variety.

(184) Pyrola. WINTER-GREEN.

P. minor: leaves stalked, broadly ovate or orbicular, entire or slightly crenated, with a minute tooth or gland in each notch; peduncle erect, 4–8 inches high, leafless; flowers drooping, white or pinkish, in a short, loose raceme, the petals ovate or orbicular, free, concave closing over the stamens.— Woods and moist shady places. Fl. July.

P. rotundifolia is a somewhat larger plant, with larger and whiter flowers, chiefly distinguished by having a long protruding much curved style.

P. media is intermediate in size, and is distinguished by having a long protruding nearly straight style.

Both the latter plants are found in woods and moist shady places, but are more rare than *P. minor.*

(185) Monotropa.

M. Hypopitys: stem 6–8 inches high, bearing oblong or ovate concave scales instead of leaves; flowers few, in a short terminal raceme; the whole plant pale yellowish-brown.—Yellow Bird's-Nest.—Fir, birch, and beech woods. Fl. June.

(186) Hottonia. WATER VIOLET.

H. palustris: aquatic; branches whorled, leafy, entirely submerged; leaves alternate deeply pinnatifid, with narrow-linear lobes; flower-stem erect, leafless from the centre of the whorl, rising out of the water, bearing whorls of 3–6 handsome, pale purplish-lilac flowers.—Featherfoil.—Pools and ditches. Fl. May to June.

(187) Primula. PRIMROSE.

P. farinosa: leaves obovate-lanceolate, glabrous above, and usually covered underneath with white meal, which is also observable on the peduncles and calyx; peduncles much longer than the leaves, supporting a compact umbel of small, pale-lilac flowers, having a yellow eye; the corolla lobes narrow, and deeply notched.—Mountain pastures in northern England and Scotland. Fl. June, July.

P. scotica, which has broader leaves, and shorter and broader lobes to the corolla, is sometimes regarded as a variety or more northern form of *P. farinosa*. [See also p. 59.]

(188) Lysimachia. LOOSESTRIFE.

* *Stems erect.*

L. vulgaris: stem branched, 2–3 feet high, downy; leaves rather large, broadly lanceolate or ovate, in whorls of 3–4;

flowers in short, compound racemes or panicles, in the upper axils and at the summit of the branches, forming a terminal, leafy panicle, the corolla yellow, subcampanulate, deeply divided into five broad lobes; stamens connected into a cup at the base.—Wet, shady banks. Fl. July.

** *Stems trailing.*

L. nummularia : stems prostrate, 1–2 feet long, often rooting at the nodes; leaves opposite, broadly ovate or rounded, very obtuse; flowers large, yellow, solitary, on short axillary pedicels, the corolla concave, deeply divided into five ovate lobes; stamens slightly connected at the base.—Moneywort or Herb Twopence.—Banks, hedges, and moist pastures. Fl. June, July.

L. nemorum : stems slender, procumbent, ½–1 foot long; leaves opposite, broadly ovate, acute; flowers rather small, solitary, axillary, on slender pedicels longer than the leaves, the corolla rotate, bright yellow; stamens quite free.—Woods and shady places. Fl. June to August.

(189) Anagallis. Pimpernel.

A. arvensis : annual; stems procumbent, ½–1 foot long; leaves opposite, broadly ovate, sessile, entire, dotted beneath, flowers bright red, occasionally pink, or white.—Shepherd's Weather-glass.—A common cornfield weed. Fl. June to August.

A. tenella : stems slender, procumbent, and rooting, only a few inches long; leaves small, orbicular, opposite; flowers pale pink, on long, slender pedicels, the corolla narrow-campanulate, of delicate texture.—Wet mossy banks and bogs. Fl. July, August.

(190) Samolus. BROOKWEED.

S. **Valerandi**: root-leaves glabrous, obovate, spreading; flowering stems ¼-1 foot high, slightly branched, bearing a few oblong leaves, and loose racemes of small white flowers.— Maritime marshes, chiefly along the west coast. Fl. July.

(191) Statice.

S. **Limonium** : radical leaves obovate or oblong, entire, glabrous, narrowed into a long stalk; flower-stem erect, leafless, ½-1 foot or more high, repeatedly forked, so as to form a broad corymbose panicle, with a membranous bract at each division, the flowers numerous, in short, rather loose spikes at the end of the branches; calyx scarious, pale purple, with five short, broad teeth, subsequently much enlarged; petals bluish-purple. —Sea Lavender.—Maritime sands and salt marshes. Fl. July, August.

(192) Armeria. THRIFT.

A. **vulgaris** : tufted; leaves numerous, radical, all narrow-linear, entire; stems simple and leafless, 3-6 inches high, bearing a globular head of pink flowers, the petal-like border of the calyx crowned by five very short, slender teeth.—Muddy or sandy seashores. Fl. July, August.

(193) Plantago. PLANTAIN.

* *Leaves strongly ribbed longitudinally.*

P. **major** : leaves broadly ovate, with about seven prominent ribs converging at the base into the longish footstalk; spike long, slender, tapering, formed of small sessile flowers;

capsule two-celled, with several seeds.—Roadsides and waste places. Fl. June to August.

P. media: leaves ovate, nearly sessile, hoary, marked with 5–7 ribs; spike dense, cylindrical; capsule two-seeded.—Dry limestone pastures. Fl. June to August.

P. lanceolata: leaves lanceolate, slightly hairy, with 3–5 ribs; spike ovoid or oblong; capsule with two seeds.—Ribwort.—Meadows, pastures, and waste places. Fl. June, July.

** *Leaves without prominent ribs.*

P. maritima: leaves narrow-linear, fleshy, entire, or slightly toothed; spike cylindrical; ovary two-celled.—Muddy saltmarshes. Fl. August, September.

P. Coronopus: leaves linear or linear-lanceolate, or pinnatifid with linear segments, hairy; spikes cylindrical; ovary four-celled.—Star of the Earth.—Dry, stony or sandy places. Fl. June to August.

(194) Littorella. SHOREWEED.

L. lacustris: leaves tufted, narrow-linear, entire; flowers small, monœcious, the males singly or two together, on short peduncles, the females sessile among the leaves.—Margins of lakes and pools. Fl. June.

(195) Pinguicula. BUTTERWORT.

P. vulgaris: leaves spreading, ovate or oblong, somewhat succulent, covered with crystalline points which give them a wet, clammy appearance; flower-stalks 3–5 inches high, bearing one handsome, bluish-purple flower, the broad campanulate throat of the corolla attached laterally to the receptacle, and projected below into a slender spur about its own length. —Bogs. Fl. May, June.

Var. *grandiflora:* larger-flowered and very handsome, with broader lobes to the corolla.—Bogs of south-western Ireland.

(196) Utricularia. BLADDERWORT.

U. vulgaris: aquatic; root-like, floating branches, ½–1 foot long or more; leaves numerous, capillary, much divided, more or less interspersed with little green vesicles; flower-stems 6–8 inches high, bearing a few rather large yellow flowers, the corolla with a short, conical, more or less curved spur, and a broad convex palate.—Deep pools and water-channels. Fl. June, July.

U. minor differs chiefly in the small size of its parts, the floating branches being very slender, the leaves small, fine, with few forked lobes and few bladders, and the flowers half the size, pale yellow, with a very short spur.

(197) Ilex. HOLLY.

I. Aquifolium: shrub or small tree, much branched, ever-green; leaves shortly stalked, ovate, shining, entire, waved, bordered with strong prickly teeth; flowers white, in dense clusters in the axils of the leaves; berries bright red or yellow. —Hedges and woods. Fl. June.

(198) Fraxinus. ASH.

F. excelsior: tree; leaves opposite, deciduous, pinnate, with 4–8 pairs of ovate-lanceolate toothed leaflets; flowers (before the leaves) like clusters of stamens issuing from buds of the last year's shoots; clusters surrounded by a few small woolly scales, and in reality consisting of a number of pedi-cels arranged in a short raceme, each pedicel bearing a pair of sessile anthers with an ovary in the middle; capsules commonly called keys, having large oblong wings.—Woods and hedges. Fl. May.

(199) Ligustrum. Privet.

L. vulgare: shrub; leaves nearly evergreen, lanceolate or oblong, entire; flowers white, in short compact panicles at the ends of the branches; berries black, globular.—Hedges and thickets. Fl. June.

(200) Convolvulus. Bindweed.

C. arvensis: stems twining or prostrate, about two feet long; leaves stalked, ovate-sagittate, with pointed basal lobes; peduncles axillary, usually two-flowered, the corolla delicate pink, or nearly white, an inch or more in diameter, handsome. —Fields and pastures; a troublesome weed, but withal one of the most beautiful of our wild plants, and varying considerably in the colour of its flowers. Fl. June, July.

(201) Calystegia. Bindweed.

C. sepium: stems twining, climbing to the length of many feet over hedges and bushes; leaves ovate-triangular, pointed, with broad, angular basal lobes; peduncles axillary, bearing a single large pure white flower, with a pair of large, leafy bracts enclosing the calyx.—Hedges and bushy places. Fl. July, August.

C. Soldanella, found in maritime sands, has short, prostrate stems, small thick broadly rounded or kidney-shaped leaves, having broad rounded or angular lobes at the base, and large light pink flowers.

(202) Cuscuta. DODDER.

* *Corolla tube with inconspicuous appressed scales.*

C. europæa: annual leafless parasite; plant pale greenish-yellow, sometimes reddish; flowers in sessile, globular clusters; the corolla tube at first broadly cylindrical, longer than the calyx, with broad short lobes, and very minute, scarcely perceptible scales inside.—Parasitic, chiefly on herbaceous stems. Fl. August.

** *Corolla tube with prominent scales, nearly closing the orifice.*

C. Epithymum: annual leafless parasite; stems red, thread-like, much finer than in the last; flower-heads small, globular, very compact; the corolla tube cylindrical, with pointed, spreading lobes, the scales prominent, almost closing the tube.—Parasitic on small shrubs, chiefly thyme, heath, etc. Fl. July, August.

(203) Polemonium.

P. cœruleum: root leaves pinnate, forming dense tufts, and made up of 11–21 lanceolate, entire leaflets; stems erect, 1½–2 feet high, bearing a few small pinnate leaves; flowers in a showy terminal corymbose panicle, blue or white.—Jacob's Ladder or Greek Valerian.—Bushy, hilly places. Fl. July.

(204) Erythræa. CENTAURY.

E. Centaurium: annual; stems erect, from 2–12 inches high, branched in the upper part; leaves broadly ovate, forming a spreading radical tuft, the upper ones in distant pairs, ovate-oblong or narrow-linear; flowers pink, usually numerous, in a terminal, repeatedly-forked cyme or panicle, the corolla

with a slender tube, and a spreading, five-cleft limb.—Dry
pastures and sandy places. Fl. July, August. The following
are probably forms or varieties :—

E. pulchella: more branched, with numerous flowers, the
tube of the corolla not much longer than the calyx, and the
lobes of the limb narrow.—Sandy places.

E. latifolia: dwarf with rather large flowers and broad
leaves.—Sandy sea shores.

E. littoralis: much branched, usually small, with very
narrow leaves and rather large flowers, often dwindled down
to a simple stem half an inch high, with a single flower.—
Sandy sea shores.

(205) Gentiana. GENTIAN.

G. Amarella: annual; stems erect, much-branched, ¼–1
foot high, often with a purplish tinge; leaves ovate or lanceolate; flowers numerous, crowded, or sometimes forming a loose,
oblong, leafy panicle, pale purplish-blue; calyx divided to the
middle into five narrow-lanceolate, nearly equal lobes; corolla-
tube broad, fringed at the throat, the limb spreading, divided
into five ovate or oblong lobes.—Dry hilly pastures. Fl.
August, September.

G. campestris much resembles this, but is usually stouter,
more branched, and more crowded with leaves and flowers;
it is known by the parts of the flower being in fours, not in
fives, and by two of the lobes of the calyx being broadly ovate,
overlapping two other narrow ones. [See also p. 59.]

(206) Chlora.

C. perfoliata: annual; stems erect, stiff, ½–2 feet high,
pale glaucous green; leaves in a spreading tuft, elliptic-oblong,

those of the stem in distant pairs, perfoliate; flowers bright yellow, in terminal cymes, the corolla nearly rotate.—Yellow-wort.—Dry pastures and waste places, in limestone districts. Fl. July, August.

(207) Villarsia.

V. nymphæoides: aquatic; stems creeping and rooting at the base, branching and ascending to the surface of the water, bearing a single leaf at each upper branch, and a terminal floating tuft of leaves and peduncles; leaves long-stalked, deeply cordate; peduncles terminated by a rather large yellow flower.—Ponds and still waters. Fl. July, August.

(208) Datura. THORN-APPLE.

D. Stramonium: annual; stems coarse, glabrous, 1–2 feet high, with spreading, forked branches; leaves large, ovate, with irregular angular teeth; flowers solitary, on short peduncles, in the forks or at the ends of the branches; calyx loosely tubular; corolla white, fragrant; capsule roundish-ovate, very prickly, with numerous wrinkled seeds.—A roadside weed, but scarcely naturalized. Fl. July.

(209) Hyoscyamus. HENBANE.

H. niger: annual or biennial, hairy, viscid, with a nauseous smell: annual plants erect, about one foot high, the leaves sessile, the upper ones clasping the stem, ovate, irregularly pinnatifid: biennial plants producing a tuft of large, stalked, elliptic, radical leaves which perish, and in the following year a coarse branched stem 1–2 feet high; flowers in one-sided leafy spikes, rolled back at the top before flowering; calyx persistent round the fruit, strongly veined, with five stiff, broad,

U

almost prickly lobes; corolla lurid-yellow, with purplish veins.—Waste, stony places, roadsides, etc. Fl. July.

(210) Solanum.

S. Dulcamara: stem shrubby at the base, with straggling branches; leaves stalked, ovate or ovate-lanceolate, broadly cordate at the base, entire, or sometimes with an additional smaller segment on one or both sides, glabrous or downy; flowers small, purple, with yellow anthers, in loose cymes, on lateral peduncles shorter than the leaves; berries small, ovoid, red.—Bitter-sweet.—Hedges and thickets. Fl. June, July.

S. nigrum: annual; stems erect, about a foot high, with spreading branches, glabrous or nearly so; leaves stalked, ovate with coarse angular teeth; flowers small, white, in contracted cymes on short, lateral peduncles; berries small, globular, black.—A common weed of cultivation. Fl. June to September.

(211) Atropa. DEADLY NIGHTSHADE.

A. Belladonna: stems stout, erect, branching, glabrous or slightly downy, three feet high; leaves stalked, large, ovate, entire, with a smaller one usually proceeding from the same point; flowers solitary, in the forks of the stem, and in the axils of the leaves, lurid purple; berry large, globular, shining black, surrounded by the leafy calyx-lobes.—Dwale.—Waste stony places. Fl. June to August.

(212) Echium. VIPER'S BUGLOSS.

E. vulgare: biennial; stems erect, 1–2 feet high, with stiff, almost prickly hairs; leaves lanceolate, stalked, those of the stem linear-lanceolate; flowers showy, reddish purple, turning to bright blue, in numerous one-sided spikes, forming a long terminal panicle; corolla limb very oblique.—Roadsides and waste places. Fl. June, July.

(213) Lithospermum. GROMWELL.

* Nuts wrinkled.

L. arvense: annual; stems erect, branched, about a foot high, hoary with appressed hairs; leaves narrow-lanceolate or nearly linear; flowers small, white, sessile, in leafy terminal cymes; nuts conical, hard, deeply wrinkled.—Bastard Alkanet.—Cultivated and waste places. Fl. May, June.

** Nuts smooth.

L. officinale : stem erect, branched, 1–1½ foot high; leaves lanceolate; flowers small, yellowish white; nuts hard, white, very smooth and shining, without wrinkles.—Waste places, roadsides, etc. Fl. June to August.

(214) Borago. BORAGE.

B. officinalis : annual; stem erect, with spreading branches, 1–2 feet high; leaves obovate or oblong, narrowed at the base into long stalks, the upper more shortly stalked, narrower; flowers on long pedicels, in loose forked cymes, drooping, clear blue, the dark anthers very prominent.—Waste ground. Fl. June, July.

(215) Lycopsis. BUGLOSS.

L. arvensis : annual; stems procumbent at the base, branched, 1–2 feet long, covered with stiff hairs; leaves lanceolate or oblong-linear, waved, often toothed, the upper ones sessile or stem-clasping; flowers in simple or forked, terminal, one-sided spikes; corolla pale blue, the tube always curved in the middle.—A common weed of cultivation. Fl. June.

(216) Anchusa. ALKANET.

A. sempervirens: stems straggling, coarsely hairy; leaves broadly ovate; flowers in short, one-sided spikes, leafy at the base, and placed in the axils of the stem-leaves; corolla rich blue.—Waste places. Fl. May, June.

(217) Myosotis. SCORPION-GRASS.

* *Hairs of the calyx appressed.*

M. palustris: stems weak, ascending, $\frac{1}{2}$–$1\frac{1}{2}$ foot high, nearly glabrous; leaves oblong bluntish, nearly glabrous; flowers bright clear blue, with a yellow eye, rather large; calyx never divided below the middle.—Forget-me-not.—Wet ditches and sides of streams. Fl. June to August.

M. repens, which is more hairy, with narrower lobes to the calyx, reaching to about the middle; and M. cæspitosa, which has a smaller corolla, with the limb often slightly concave, are varieties sometimes considered distinct.

** *Hairs of the calyx spreading or hooked.*
   † *Pedicels equalling or exceeding the calyx.*

M. sylvatica: stems one foot high, roughly hairy; leaves oblong-lanceolate; flowers large, handsome, blue, the limb spreading flat; calyx cleft nearly to the base, with narrow segments.—Mountain pastures and shady situations. Fl. May, June.

The alpine form, with larger flowers, is by some distinguished as a species, under the name of M. alpestris.

M. arvensis: annual or biennial; stem weak, a foot high; leaves oblong, acute, hairy; flowers small, blue; the corolla with a short, concave limb; calyx deeply cleft, the narrow segments erect when in fruit.—Cultivated ground and bushy places. Fl. June to August.

†† *Peduncles shorter than the calyx.*

M. collina: annual; stems much branched, hairy, seldom six inches high; leaves in radical tufts, oblong obtuse; flowers in slender racemes, the corolla very small, bright blue, with a small, concave limb.—Dry, open places. Fl. May.

M. versicolor is a little hairy annual, something like the last, but with a more simple and erect stem, a spreading tuft of radical leaves, and small nearly sessile flowers; the corolla at first pale yellow, turning blue as it fades.

(218) **Cynoglossum.** Hound's-tongue.

C. officinale: biennial; stem stout, erect, branched, about two feet high, with rough hairs; leaves lanceolate, or the lowest ones oblong, the uppermost sessile and clasping the stem, all hoary with dense soft appressed down; racemes numerous, mostly simple, forming a terminal leafy panicle, the corolla small, dull purplish-red.—Roadsides and waste places. Fl. June, July.

(219) **Salvia.** Sage.

S. pratensis: leaves ovate, heart-shaped or oblong, coarsely toothed, wrinkled; stem 1–1½ feet high, downy, with a few narrow leaves near its base; flowers in a long, handsome, terminal, almost simple whorled spike, the corolla near thrice as long as the calyx, rich blue, with a long arched upper lip.— Dry pastures, rare. Fl. July.

S. Verbenaca : stems hairy, erect, 1–2 feet high, slightly branched; leaves stalked, ovate, coarsely toothed or lobed, wrinkled, the upper ones sessile, the bract-like floral leaves small, heart-shaped, entire; flowers in terminal hairy spikes, small, blue; the corolla seldom twice as long as the calyx.— Waste places, roadsides. Fl. June to October.

(220) Lycopus. GIPSYWORT.

L. europæus: stems erect, branching, slightly hairy, two feet high; leaves ovate-oblong, deeply toothed or pinnatifid; flowers small, numerous, in dense axillary whorls or clusters, whitish with purple dots.—Wet ditches and marshes. Fl. July, August.

(221) Nepeta. CATMINT.

N. Cataria : stems erect, branching, 2–2½ feet high, hoary with minute down; leaves ovate-cordate, coarsely toothed; flowers rather small, pale blue or nearly white, crowded in compact cymes, forming short, oblong spikes at the ends of the branches.—Hedges, roadsides and waste places. Fl. July, August. [See also p. 61.]

(222) Mentha. MINT.

* *Calyx-throat naked.*

† *Flowers in spiked whorls or terminal heads.*

M. sylvestris: stems 1–2 feet high, erect, slightly branched, and, as well as the whole plant, hoary with short down; leaves closely sessile, broadly lanceolate or narrow-ovate; flowers small, numerous, pinkish, in dense cylindrical spikes, usually several together, forming an oblong terminal panicle.—Horse Mint.—Wet pastures, sides of ditches, etc. Fl. August, September.

M. rotundifolia is a greener, and more hairy plant, with broadly ovate or orbicular leaves, and terminal cylindrical spikes of small pale pink flowers; and **M. viridis**, has glabrous green stems and leaves, the latter acutely lance-shaped, and lax cylindrical terminal flower spikes. This last is the common *Mint* of our gardens.

M. piperita: stems 1–1½ foot high, erect, glabrous, or nearly so; leaves stalked, ovate-oblong, serrated; flowers lilac, in blunt rather dense spikes, the lower whorls often distant. —Peppermint.—Wet places, rather rare. Fl. July, August.

M. aquatica: stems 1–1½ feet high, much branched, softly hairy, or sometimes nearly glabrous; leaves stalked, ovate or slightly heart-shaped; flowers in dense, terminal, globular or oblong heads, pale lilac.—Wet ditches, marshes, and edges of streams. Fl. July, August.

†† *Flowers in distant axillary whorls.*

M. sativa: stems more or less spreading or ascending; leaves stalked, ovate or elliptical serrate; flowers all in distinct axillary whorls, without any terminal head or spike, the calyx tubular, with spreading teeth.—Wet places. Fl. July, August.

M. arvensis: stems low, spreading, branched, hairy, ½–1 foot long, rarely erect; leaves stalked, ovate, toothed; flowers all in axillary whorls, mostly shorter than the leafstalks, the calyx campanulate.—Fields and moist places. Fl. July to September.

** *Calyx-throat closed with hairs.*

M. Pulegium: stems prostrate, much branched, rooting; leaves small, elliptical, entire or slightly crenate, the floral ones still smaller; flowers in dense axillary whorls.—Pennyroyal.—Wet ditches, and marshy places. Fl. August, September.

(223) **Origanum.** MARJORAM.

O. vulgare: stems erect, 1–2 feet high, hairy; leaves stalked, ovate or ovate-lanceolate, slightly toothed; flowers purplish, rarely white, in globular compact heads, forming a terminal trichotomous panicle.—Edges of woods, and hilly pastures, especially in limestone districts. Fl. July, August.

(224) **Thymus.** THYME.

T. Serpyllum: stems procumbent, slender, much branched, perennial, scarcely woody at the base, forming low dense tufts, almost covered with the purple flowers; leaves very small, ovate or oblong, fringed at the base, the floral leaves similar but smaller; flowers in short, terminal, loose, leafy spikes.—Dry hilly pastures. Fl. July, August.

(225) **Marrubium.** HOREHOUND.

M. vulgare: stem 1½ foot high, with spreading branches, thickly covered with white cottony wool; leaves stalked, orbicular, soft, much wrinkled; flowers in dense whorls or clusters in the axils of the upper leaves, small, dirty white.—Roadsides and waste places. Fl. July.

(226) **Galeopsis.** HEMP-NETTLE.

* *Calyx-teeth not longer than the tube.*

G. Ladanum: annual; stems 6–8 inches high, with spreading, almost decumbent branches, covered with short soft down, not swollen under the nodes; leaves narrow-ovate or lanceolate, coarsely toothed; flowers purple, in dense whorls in the upper axils, the upper ones forming a terminal head.—Cultivated and waste places. Fl. August, September.

** *Calyx-teeth long and almost prickly.*

G. Tetrahit: annual; stems coarse, 1–2 feet high, swollen under the nodes, with spreading branches, and stiff, spreading hairs; leaves ovate, pointed, coarsely toothed; flowers numerous, forming close whorls in the axils of the upper leaves, pale-purplish or white.—Cultivated and waste places.—Fl. July, August.

Var. *versicolor*, larger, with larger flowers, which are yellow with a purple spot on the lower lip.—Fields.

(227) **Stachys.** WOUNDWORT.

* *Stems erect.*

S. Betonica: stems 1–2 feet high, downy or hairy; leaves mostly radical, oblong, cordate at the base, coarsely crenate, the upper ones few, distant, sessile; flowers purple, in dense whorls, collected in a close terminal, oblong head or spike; the upper lip of the corolla ovate, erect, slightly concave.— Betony.—Woods and thickets. Fl. July, August. It is sometimes called *Betonica.*

S. sylvatica: stem stout, erect, branching, 2–4 feet high, coarsely hairy; leaves all stalked, cordate, ovate, crenate; flowers in distant whorls, forming long terminal spikes, dark reddish-purple, the lower lip variegated with white.—Shady banks, and edges of woods. Fl. July, August.

S. palustris: stems tall, stout, with short hairs; leaves oblong or lanceolate, slightly cordate at the base; flowers pale bluish-purple, forming shorter and more crowded spikes than in the last.—Ditches, and moist banks. Fl. August.

** *Stems decumbent.*

S. arvensis : annual; stems slender, hairy, branched, decumbent or slightly ascending; leaves small, ovate, scarcely cordate; flowers small, of a pale purple, in loose, leafy spikes. —Fields and waste places. Fl. July, August.

(228) Leonurus. Motherwort.

L. Cardiaca : stems stiff, hairy or downy, 2–4 feet high; leaves stalked, broad, palmately cut into 5–7 coarsely toothed lobes, the floral leaves narrow, 3-lobed or entire; flowers in long interrupted, terminal, leafy spikes, pink or nearly white.— Waste places, hedges, etc. Fl. July, August.

(229) Ballota. Black Horehound.

B. nigra : stems erect, hairy, branching, 2–3 feet high, with a strong, disagreeable smell; leaves stalked, ovate or cordate, coarsely toothed; flowers in dense axillary clusters, often slightly stalked, purplish, the lower lip 3-lobed, spreading.—Hedges and roadsides. Fl. July, August.

(230) Scutellaria. Skullcap.

S. galericulata : stems slightly branched, ascending, ½–1 foot high; leaves ovate-lanceolate, slightly toothed; flowers nearly sessile, opposite, in axillary pairs along the greater part of the stem, and all turned to one side, the corolla dingy blue, with a very slender tube considerably enlarged at the throat.—Wet, shady, or stony places. Fl. July, August.

(231) Prunella. Self-heal.

P. vulgaris : stem procumbent, rooting at the base, with ascending flowering branches, ¼–1 foot high; leaves stalked, ovate, nearly entire; flowers in dense spikes, violet-purple, the upper lip bending over the lower.—Pastures, banks, etc. Fl. July, August.

(232) Calamintha. CALAMINT.

* *Calyx-tube gibbous at the base beneath.*

C. Acinos : annual; stems branched, 6–8 inches high, slightly downy ; leaves small, narrow-ovate, slightly toothed ; flowers pale purple or white, in axillary whorls on short, erect pedicels.—Basil Thyme.—Fields. Fl. July, August.

** *Calyx-tube equal at the base.*

C. officinalis : stems hairy, ascending or erect, with straggling branches, 1 foot high; leaves broadly ovate, toothed ; flowers pinkish-lilac, in small loose cymes, forming terminal, one-sided, leafy panicles; calyx tubular, ribbed, half as long as the corolla, the teeth with long cilia.—Hedges, roadsides, and waste places. Fl. July to September. Mr. Bentham makes the two following plants varieties of this :—

C. Nepeta : leaves ovate, nearly entire ; flower-cymes contracted into loose whorls, the corolla half as long again as the calyx, which is shortly ciliate.—Dry open, sunny banks.

C. sylvatica : stem taller; leaves larger, broadly ovate, sharply toothed ; cymes loose, the flowers showy, the corolla fully twice as long as the calyx.—Woods in the Isle of Wight.

C. Clinopodium : stems erect or ascending, branched, softly hairy, 1–2 feet high ; leaves ovate, slightly toothed, soft, hairy ; flowers purple, in dense cymes, forming compact whorls or heads in the axils of the upper leaves, or at the ends of the branches, surrounded by subulate, hairy bracts. —Wild Basil.—Hedges and woods. Fl. July, August.

(233) Teucrium. GERMANDER.

T. Scorodonia: stems erect, hairy, 1 foot high, slightly branched, almost woody at the base; leaves ovate or lanceolate, coarsely toothed, wrinkled, downy; flowers of a pale yellow, in pairs, forming terminal and axillary one-sided racemes.— Wood Sage.—Woods and hedges. Fl. July, August.

(234) Verbena. VERVAIN.

V. officinalis: stem nearly glabrous, erect, 1–2 feet high, with long, spreading, wiry branches; leaves obovate or oblong, stalked, coarsely toothed or cut, the upper ones few, lanceolate; flowers very small, lilac, in long slender spikes. —Roadsides and waste places. Fl. July, August.

(235) Veronica. SPEEDWELL.

* *Flowers in terminal racemes.*

V. spicata: stems ascending or erect, ½–1 foot high, usually simple; leaves oblong or the lower ones ovate, downy, slightly crenate; flowers clear blue, in dense terminal spikes.—Hilly pastures, in limestone districts, rare. Fl. July, August.

V. serpyllifolia: stems shortly creeping, much branched, forming dense, leafy tufts, the flowering branches ascending, 2–6 inches high; leaves ovate, slightly crenate, glabrous; flowers very small, pale blue or white with darker streaks, sessile or shortly stalked, in terminal racemes, the bracts rather large and leaf-like.—Pastures, fields, and waste places. Fl. May, June.

V. arvensis: annual; stems hairy, 2–6 inches high, erect, simple, or sometimes diffuse and branching at the base; leaves ovate, toothed, the upper floral ones lanceolate; flowers small,

sessile, forming terminal, leafy racemes, blue or nearly white.
—Cultivated and waste places. Fl. May, June.

** *Flowers in axillary racemes.*

V. officinalis : stems much branched, creeping and rooting,
½-1 foot long ; leaves obovate or oblong, toothed, hairy ;
spikes or racemes hairy, axillary, sometimes proceeding from
the upper axils, the flowers nearly sessile, small, pale blue.—
Dry banks and heaths. Fl. June to August.

V. Anagallis : stems erect, branching, ½-2 feet high, glabrous ; leaves lanceolate, sessile or clasping, toothed ; racemes numerous, axillary, opposite, *i. e.* in the axils of both
leaves of each pair, the flowers small, pedicellate, pale blue.—
Wet ditches, and by streams and ponds. Fl. July.

V. Beccabunga : stems procumbent or floating at their
base, rooting, the flowering branches ascending, glabrous ;
leaves stalked, oblong, obtuse, slightly toothed ; flowers small,
blue, in opposite axillary racemes.—Brooklime.—Wet ditches,
and by streams and ponds. Fl. June, July.

V. scutellata : stems slender, ascending or spreading, about
6 inches high, glabrous, rarely downy ; leaves linear-lanceolate ; flowers few, in slender racemes, alternately from one
axil only of each pair of leaves, small, pale pinkish-blue.—
Marshes, ditches, and wet places. Fl. July, August.

V. Chamædrys : stems weak, ascending, about a foot long,
with two opposite lines of hairs ; leaves stalked, ovate-cordate,
crenate, hairy ; racemes longer than the leaves, with large
bright blue flowers, on longish pedicels ; capsule flat, very
broad at the top.—Woods, banks, etc. Fl. May, June.

V. montana resembles this, but is more diffuse, with looser
and more slender racemes of flowers, the stem hairy all round,
and the capsule broadest in the middle.

*** *Flowers solitary, axillary.*

V. hederæfolia: annual; stems procumbent; leaves broadly orbicular, with 5–7 coarse teeth, the middle one broad, rounded ; flowers pale blue, the sepals broadly heart-shaped. —A weed of cultivation. Fl. May, June.

V. agrestis: annual; stems hairy, branched, procumbent, ¼–1 foot long.; leaves ovate, toothed, the lowest opposite, but mostly alternate, each with a single small blue or pinkish-white flower in its axil, the sepals oblong, capsule of 2 ovoid, erect lobes.—A common weed. Fl. April to September.

V. polita is a variety with ovate sepals, and larger blue flowers ; and the allied **V. Buxbaumii**, another weed of cultivation, closely resembles this species, but is larger in all its parts, with the flowers bright blue, and the lobes of the capsule broad and divaricate.

(236) **Verbascum.** Mullein.

V. Thapsus: biennial; stem stout, erect, simple or branched, 2–4 feet high, clothed with soft woolly hairs; leaves oblong, pointed, narrowed at the base into two wings running down the stem, the lower ones often stalked ; flowers in a dense, woolly terminal spike, sometimes a foot or more long, the corolla large, yellow, three of the filaments covered with yellowish woolly hairs.—High Taper.—Roadsides and waste places. Fl. July, August.

V. nigrum: stem sparingly clothed with woolly hairs, 2–3 feet high, ending in a long, simple or slightly branched raceme; leaves crenate, nearly glabrous above, the lower ones large, cordate-oblong, on long stalks ; flowers numerous, small, yellow, with bright purple hairs to the filaments.—Banks and waysides. Fl. July, August.

(237) Sibthorpia.

S. europæa : stems very slender, creeping, rooting at the nodes; leaves small, stalked, orbicular, deeply cordate at the base, crenate, hairy; flowers minute, axillary, the corolla with the two upper lobes yellowish, the three lower broader and pink.—Cornish Moneywort.—Moist, shady places. Fl. July, August.

(238) Linaria. TOAD-FLAX.

* *Stems erect.*

L. vulgaris : stems 1–3 feet high, glaucous green; leaves crowded, linear or narrow-lanceolate; flowers large, yellow, forming a short handsome terminal raceme ; spur of the corolla long and pointed ; the projecting palate of the lower lip of a bright orange colour, completely closing the tube.— Hedges, and borders of fields. Fl. June, July.

L. minor : annual, stems much branched, 3–4 inches high, slightly glandular, downy; leaves linear, obtuse; flowers very small, on long axillary peduncles, pale purple or violet, with a short blunt spur.—A weed of cultivation. Fl. June to August.

** *Stems trailing.*

L. Cymbalaria : stems slender, glabrous, trailing, often rooting at the nodes; leaves stalked, nearly reniform, broadly 5-lobed, almost fleshy; flowers small, axillary, pale lilac, with a rather short spur, the palate yellowish.—Old walls, and stony places, naturalized. Fl. all the year.

L. spuria : annual; stems hairy, slender, branching, pro-strate, ½–1 foot long; leaves broadly ovate or orbicular; flowers solitary, in the axils of the upper smaller leaves, small yel-

lowish, with a purple upper lip, the spur slender, recurved.
—Cornfields. Fl. July, September.

L. Elatine, also a weed of cultivation, closely resembles
this, but is still more slender; leaves angular or hastate at
the base, the lowermost ovate, the peduncles more slender,
glabrous, and the spur of the corolla straight.

(239) Antirrhinum. SNAPDRAGON.

A. majus: stems branched, erect, 1–2 feet high, glabrous;
leaves narrow-lanceolate or linear, entire; flowers large, purplish-red, the corolla above an inch long.—Rocks, old walls,
and stony places. Fl. July to September.

(240) Scrophularia. FIGWORT.

S. nodosa: stems erect, 2–3 feet high, glabrous, sharply
quadrangular, the short stock emitting a number of green
knots or tubers; leaves broadly ovate or heart-shaped, acute
serrated; panicle loosely pyramidal or oblong, the corolla
lurid greenish-purple.—Moist hedges and thickets. Fl. June,
July.

S. aquatica: stem 2–5 feet high, the angles projected into
narrow wings; leaves cordate-oblong, obtuse; panicle long,
narrow, the flowers dull purple.—Ditches and sides of streams.
Fl. July, August. [See also p. 60.]

(241) Digitalis. FOXGLOVE.

D. purpurea: biennial; leaves long-stalked, ovate or ovate-
lanceolate, coarsely veined, downy; flowering stems 2–4 feet
high, terminating in a long stately raceme of purple flowers,
the corolla 1½ inch long, beautifully spotted and hairy inside.
—Dry hilly wastes, and roadsides. Fl. June to August.

(242) Pedicularis. LOUSEWORT.

P. palustris: annual; stems glabrous, erect, or in dry situations decumbent at the base, much branched, one foot, or in water two feet high; leaves pinnate, with short ovate deeply cut segments, the floral ones twice pinnate; flowers almost sessile in the axils, deep purple-red, the calyx broad, with two broad, short, irregularly cut or jagged lobes, and the upper lip of the corolla with two minute teeth below its middle.—Red Ratile.—Wet meadows and watery ditches. Fl. June, July.

P. sylvatica: stems prostrate or spreading, branching, seldom above six inches long; leaves pinnate, with deeply cut, small segments; flowers sessile in the upper axils, pink-red, the calyx broadly oblong, with five unequal teeth, the upper lip of the corolla with one minute tooth on each side, under the point.—Moist heathy pastures and meadows. Fl. June, July.

(243) Bartsia.

B. Odontites: annual; stems erect, branching, a foot high, downy; leaves lanceolate, toothed; flowers purplish-red, in numerous one-sided spikes, the upper lip of the corolla longer than the lower one.—Fields and waste places. Fl. July, August.

(244) Euphrasia. EYEBRIGHT.

E. officinalis: annual; stem 2–8 inches high, usually much branched, glabrous or slightly downy; leaves small, sessile, opposite, ovate, deeply toothed, the teeth of the upper ones finely pointed; flowers in loose, terminal, leafy spikes, the corolla white or reddish streaked with purple, and having a

X

yellow spot in the throat.—Heaths and pastures. Fl. July
to September.

(245) Rhinanthus. RATTLE.

R. Crista-galli: annual; stems erect, glabrous or slightly
hairy, ½–1 foot high, simple or slightly branched; leaves opposite, lanceolate, and coarsely-toothed, the floral ones broader,
shorter, and more cut at the base; flowers in a loose leafy
spike, yellow, often with a purple spot on the upper, or on
both lips.—Meadows and pastures. Fl. June.

(246) Melampyrum. COW-WHEAT.

M. pratense: stem erect or ascending, glabrous or nearly
so, ½–1 foot high, with very spreading, opposite branches;
leaves lanceolate, the floral ones distant from each other,
short, and often toothed at the base; flowers yellow, in distant axillary pairs, all turned one way.—Woods and bushy
places. Fl. July, August.

(247) Orobanche. BROOMRAPE.

O. major: plant at first pale-yellow, soon becoming a dingy
purplish-brown, stem simple, stout, erect, 1–2 feet high, with
lanceolate scales; flowers closely sessile, forming a dense
spike half the length of the stem, the corolla-tube nearly as
broad as long, curved, with a very oblique limb, the upper lip
entire or shortly two-lobed, the lower one three-lobed; upper
part of the style and stamens usually covered with short glandular hairs, which are wanting in the lower parts.—Parasitical
on the roots of the *Broom*, more rarely on those of the *Furze*.
Fl. June, July.

Several other species are found, though less frequently, as—

O. elatior, a tall yellowish species, with the stamens hairy ; *O. rubra*, red-brown ; *O. caryophyllacea*, light or dark-brown ; and *O. minor*, light yellowish-brown, the three latter being seldom over a foot high, with a short flower-spike.

(248) Salicornia. GLASSWORT.

S. herbacea : annual ; stems erect, glabrous, succulent, jointed, leafless, six inches high, with few erect branches, or when luxuriant branching from every joint : often procumbent and rooting, extending to a foot or more, and in favourable situations hardening and acquiring the appearance of under-shrubs, but probably not lasting beyond the second year ; flowers immersed in the upper joints forming terminal, succulent, cylindrical spikes, each joint having six flowers, three in a triangle on each side, the perianth succulent, flat, nearly closed at the top, the stamens protruding through the minutely three- or four-toothed orifice ; style included, divided into 2–3 stigmas ; nut enclosed in the unchanged, succulent perianth.—Salt-marshes and muddy sea-shores. Fl. August, September.

(249) Suæda.

S. maritima : annual or biennial ; stems erect or procumbent, much branched, seldom a foot high ; leaves acute, semi-cylindrical, succulent ; flowers small, green, sessile, solitary, or 2–3 together in the axils of the leaves.—Salt-marshes and maritime sands. Fl. July to September.

(250) Salsola. SALTWORT.

S. Kali : annual ; stems procumbent, glabrous, much branched, ½–1 foot long ; leaves awl-shaped, ending in a stout prickle, the uppermost nearly triangular; flowers sessile in the upper axils, the inconspicuous perianth regular, five-cleft.—Sea-coasts. Fl. July, August.

(251) **Beta.** Beet.

B. maritima : stems erect or spreading, branched, two feet high; leaves broad, triangular-ovate, thick, green, the upper ones small, narrow; flowers green, single or clustered, in long, loose, terminal spikes, often branching into a leafy panicle; the ripe perianth forming a hard, angular, often prickly mass, enclosing a single horizontal seed.—Sea-shore. Fl July to September.

(252) **Chenopodium.** Goosefoot.

  * *Perianth covering the seed-like fruit.*

   † *Leaves undivided.*

C. Vulvaria : annual; stems procumbent, branched; leaves small, ovate, entire, covered with granular mealiness, and remarkable for a strong stale-fish smell when rubbed; flower-clusters in short axillary and terminal racemes.—Waste, rubbishy places. Fl. August, September.

C. polyspermum : annual; stems procumbent or spreading, much branched, sometimes erect; leaves green, ovate; flower-clusters small, in short axillary spikes, the upper ones forming a narrow terminal panicle.—Waste places. Fl. July to September.

  †† *Leaves angular-lobed.*

C. album : annual; stems erect, 1–2 feet high, mealy; leaves ovate or rhomboidal, sinuately toothed or angular, the upper ones narrow, entire; flower-clusters in compound, branched, nearly leafless racemes.—Waste places. Fl. July, August.

C. urbicum: annual; stems erect, slightly branched, 1–2 feet high, not mealy; leaves long-stalked, triangular, sinuately-toothed; flower-clusters in erect, nearly leafless, compound spikes.—Waste places. Fl. August.

C. murale: annual; stems erect or decumbent, much branched, a foot high; leaves rhomboid, ovate, coarsely-toothed; flowers chiefly axillary, in branched, spreading, leafless cymes.—Waste places. Fl. August, September.

** *Perianth not covering the seed-like fruit.*

C. rubrum: annual; stems erect, a foot high; leaves rhomboid, irregularly sinuate-toothed; flowers in erect, compound, dense leafy spikes, mostly with only 2–3 segments to the perianth, the seed erect, not horizontal.—Waste places, especially near the sea. Fl. August and September.

C. Bonus-Henricus: stems a foot high, scarcely branched; leaves broadly triangular, sinuate or slightly toothed; flowers numerous, in clustered spikes, forming a narrow terminal panicle.—Good King Henry.—Waste ground near villages and dwellings. Fl. May to August.

(253) Atriplex. ORACHE.

A. patula: annual; stems erect or prostrate, more or less mealy-white; leaves stalked, usually hastate, sometimes opposite, the upper ones narrow, entire; flowers clustered in slender spikes, forming narrow, leafy, terminal panicles; segments of the fruiting perianth ovate or rhomboidal, pointed, often toothed at the edge and warted or muricate on the back. —Waste places and sea-coasts. Fl. June to September.

Mr. Bentham places as varieties the four following plants often distinguished as species:—

A. deltoidea, erect or spreading; lower leaves broadly triangular or hastate, often coarsely and irregularly-toothed.—Waste ground and sea-shores.

A. erecta: stem erect; leaves lanceolate, the lower ones broader and hastate.—Cultivated ground.

A. angustifolia : stem spreading or decumbent ; leaves mostly lanceolate, or the upper ones linear.—Waste places.

A. littoralis : stems erect; leaves still narrower than in the last, often toothed.—Sea-shores.

A. rosea : annual ; stems procumbent, the whole plant frosted with a white scaly meal ; leaves broadly triangular or rhomboidal, coarsely-toothed ; fruiting perianth mealy-white, thick, rhomboidal or orbicular, often warted, the segments united to above the middle.—Sea-coasts.—Fl. July to September.

(254) **Ceratophyllum.** Hornwort.

C. demersum: aquatic ; stems floating, submersed ; leaves whorled, twice or thrice forked, with subulate segments, usually slightly-toothed on the edge ; flowers small, sessile, axillary, without any real perianth, the males consisting of 12–20 sessile, oblong anthers, the females of a small ovary with a simple style.—Pools, slow streams, and shallow lakes. Fl. June, July.

(255) **Callitriche.** Water Starwort.

C. aquatica : aquatic ; stems slender, glabrous, floating in water, or creeping and rooting in wet mud ; leaves obovate or oblong, or the lower submerged ones narrow-linear, and obtuse or notched at the top : the upper ones obovate, and spreading in little tufts on the surface of the water, or all submerged and linear ; flowers minute, the males consisting of a single stamen with a conspicuous filament, the females of a sessile or stalked ovary, with two erect or recurved styles.—Shallow waters or wet mud. Fl. May to September.

(256) Urtica. NETTLE.

U. urens : annual, with stiff, stinging hairs; stems erect, branching, ½–1 foot high, glabrous ; leaves ovate, deeply toothed ; flowers in small, loose, almost sessile, axillary clusters.—Cultivated and waste places. Fl. June to October.

U. dioica : stems erect, 2–3 feet high, with copious stinging bristles ; leaves cordate-ovate, coarsely-toothed ; flowers diœcious, both males and females clustered in axillary, branched, spreading spikes.—Hedges and waste places. Fl. July, August.

(257) Parietaria. PELLITORY.

P. officinalis : stems branching, diffuse or procumbent, ½–1 foot long, downy with short soft hairs ; leaves stalked, ovate or oblong, entire ; flowers in sessile clusters.—Old walls, and waste, stony places. Fl. July to September.

(258) Humulus. HOP.

H. Lupulus : stems twining to a considerable height over bushes and small trees ; leaves opposite, stalked, broadly heart-shaped, deeply 3–5-lobed, sharply-toothed, rough but not stinging ; flowers diœcious : the males in loose panicles in the upper axils, small, yellowish-green ; the females in shortly stalked, axillary, ovoid or globular spikes or heads, conspicuous for their broad, closely-packed bracts, each with two sessile flowers in its axil, the perianth a concave scale enclosing the ovary ; scales of the spike becoming enlarged after flowering, concealing the seed-like fruits.—Hedges, thickets, and open woods. Fl. July.

(259) Rumex. Dock.

* *Leaves never hastate at the base, sometimes obtusely auricled.*

† *Inner perianth-segments entire.*

‡ *Segments cordate-ovate.*

R. crispus : stem 2–3 feet high, with few branches; leaves long narrow, waved or crisped; whorls of flowers crowded in a long narrow panicle, one of the segments of the fruiting perianth bearing an oblong coloured tubercle or grain.— Roadsides and waste places. Fl. June to August.

‡‡ *Segments ovate, not cordate.*

R. Hydrolapathum : stem 3–5 feet high, slightly branched; leaves long, lanceolate, pointed; panicle long, dense, leafy at the base, the branches scarcely spreading; inner perianth-segments ovate, entire or scarcely toothed, with a large oblong tubercle on all three.—Edges of streams, pools, and ditches. Fl. July, August.

R. conglomeratus : stem 2–3 feet high; leaves cordate-oblong, pointed; panicle with spreading branches and distant whorls; inner perianth-segments narrow-ovate, equal, each with an oblong tubercle.—Waste places. Fl. June to August.

R. sanguineus : stem two feet high, branched : leaves lanceolate, cordate at the base; panicle leafy at the base, with stiff, slender, spreading branches, the whorls of flowers distinct; inner perianth-segments narrow, entire or scarcely-toothed, one with a large tubercle.—Roadsides and waste places. Fl. June to August.

†† *Inner perianth-segments toothed.*

R. obtusifolius : stem 2–3 feet high, slightly branched; leaves cordate-oblong, obtuse; whorls of flowers distant; inner segments of the perianth bordered below the middle by a few small teeth, usually ending in a narrow entire point.—Waste ground. Fl. July to September.

** *Leaves hastate*, i. e. *with acute basal auricles.*

R. Acetosa: stems scarcely branched, 1–2 feet high ; leaves oblong, sagittate at the base, very acid ; flowers diœcious, in long, terminal, leafless panicles, usually turning red, the inner segments of the fruiting perianth enlarged, orbicular, entire, almost petal-like.—Sorrel.—Meadows and pastures. Fl. May, June.

R. Acetosella: stems slender, ¼–1 foot high, often turning red; leaves narrow-lanceolate, sagittate, the lobes of the base usually spreading ; flowers small, diœcious, in slender terminal panicles ; inner perianth-segments broadly ovate or orbicular, scarcely enlarged.—Sheep Sorrel.—Dry open pastures. Fl. May to July.

(260) Oxyria.

O. reniformis: leaves glabrous, cordate-orbicular or kidney-shaped, acid; stem slender and almost leafless, 6–10 inches high, terminating in a simple or slightly-branched raceme; flowers small.—High mountain ranges. Fl. July, August.

(261) Polygonum.

* *Stems prostrate wiry branched, or simple and floating.*

P. aviculare: annual; stems often 1–2 feet long ; stipules white, scarious, becoming ragged at the edges; leaves small, narrow-oblong ; flowers small, shortly stalked, clustered in the axils of the leaves.—Knotgrass.—Cultivated and waste places. Fl. May to September.

P. maritimum is distinguished by its thicker stems, larger

thicker and more glaucous leaves, larger scarious stipules, and larger flowers.

P. amphibium: stems usually floating in water, rooting at the lower nodes; leaves oblong or lanceolate, spreading on the surface; spikes terminal, solitary, supported on short peduncles above the water, dense, cylindrical, rose-red.—Ponds and watery ditches. Fl. June to September. On dry ground it becomes a prostrate weed.

** *Stems twining.*

P. Convolvulus: annual; stems glabrous; stipules short; leaves stalked, heart-shaped or broadly sagittate, pointed; flowers in little loose clusters, the lower axillary, the upper forming loose irregular, terminal racemes.—Cultivated and waste places. Fl. July to September.

P. dumetorum is more luxuriant, and the three angles of the fruiting perianth are more or less expanded into white scarious wings.

*** *Stems erect or ascending.*

P. Bistorta: leaves long-stalked, ovate, subcordate; stems simple, erect, 1–2 feet high, terminating in a dense, oblong, or cylindrical spike of pretty pink flowers.—Bistort or Snakeweed.—Moist pastures and meadows. Fl. June.

P. Persicaria: annual; stems erect or spreading, branched, glabrous, 1–2 feet high; leaves lanceolate, often marked in the centre with a dark spot; stipules more or less fringed at the top with short fine bristles; spikes terminal, rather numerous, oblong or cylindrical, dense, the flowers reddish, or sometimes green.—Roadsides and waste places. Fl. June to September.

P. lapathifolium, which closely resembles this, is distinguished by the pedicels and perianths being dotted with small prominent glands, the plant being usually pale-green, and the stipules seldom fringed.—Waste damp places.

P. **Hydropiper** : annual; stems 1-2 feet high, often decumbent at the base; stipules fringed; leaves lanceolate, wavy; flowers in slender, drooping spikes, the clusters of flowers almost all distinct, the perianths dotted with small glands.—Water-pepper.—Wet ditches and edges of streams. Fl. August, September.

(262) **Thesium.**

T. **linophyllum** : stems glabrous, procumbent or ascending, 6–8 inches long; leaves linear-lanceolate; flowers small, in a terminal, leafy raceme, greenish-white.—Chalky pastures, parasitical. Fl. June, July.

(263) **Euphorbia.** SPURGE.

* *Glands of the involucre rounded on the outer edge.*

E. **Helioscopia** : annual; stems erect, ½-1 foot high, simple, or with a few branches ascending from the base; stem-leaves obovate or broadly oblong, the floral ones broadly obovate or orbicular, very obtuse, minutely-toothed; umbel of five rays, each ray once or twice forked at the end, flowers and floral leaves crowded into broad leafy heads; glands of the involucre entire and rounded.—Waste and cultivated ground. Fl. June to September.

** *Glands of the involucre crescent-shaped, with the horns turned outwards.*

† *Floral leaves all distinct.*

E. **Peplus** : annual; stems erect, glabrous, ½-1 foot high, branching from the base; stem-leaves obovate, entire, the floral-leaves broadly ovate or cordate; umbel of 2-3 repeatedly forked rays ; flower-heads small ; glands of the involucre crescent-shaped, with long points ; seeds pitted.—Cultivated and waste places. Fl. July, August.

E. exigua: annual; stems slender, glabrous, erect or ascending, 3–8 inches high ; stem-leaves numerous, small, narrow, the floral ones usually lanceolate ; umbels of 3–5 rays, sometimes contracted into terminal heads, more frequently elongated and forked ; glands of the involucre crescent-shaped, with fine points ; seeds slightly wrinkled.—Cornfields. Fl. June to August.

E. Lathyris: annual or biennial ; stem tall, stout, three feet high, smooth, glaucous ; stem-leaves narrow-oblong, the upper broader, all opposite ; umbels of 3–4 long rays, once or twice forked, with large ovate-lanceolate floral-leaves ; glands of the involucre crescent-shaped, with short blunt points.— Naturalized about cottage gardens. Fl. June, July.

†† *Floral leaves connate.*

E. amygdaloides: stems erect, reddish, almost woody, 1–2 feet high ; leaves rather crowded towards the middle of the stem, lanceolate or narrow-oblong, the upper more distant and shorter ; umbel of five long rays, not much divided, with a few axillary peduncles below it ; floral leaves of each pair always connected into one large orbicular one, pale yellowish-green ; glands of the involucre crescent-shaped, with rather long points.—Woods and thickets. Fl. March, April. [See also p. 16.]

(264) **Mercurialis.** MERCURY.

M. annua: annual; stems erect, glabrous, ½–1 foot high, with opposite branches ; leaves ovate or oblong, coarsely-toothed, of a thin texture ; male flowers clustered along slen-

der peduncles nearly as long as the leaves ; females 2–3 together in the axils of the leaves, usually on separate plants.—
Cultivated and waste places. Fl. August, September

(265) Aristolochia. BIRTHWORT.

A. Clematitis : stems erect, simple, 1½ foot high, smooth ;
leaves broadly cordate ; flowers axillary, clustered, yellow, the
tube slender with an oblique mouth.—Stony, rubbishy places,
naturalized. Fl. July, August.

(266) Paris. HERB PARIS.

P. quadrifolia : stems ¾–1 foot high, with a whorl of four
broadly-ovate or obovate leaves at top; perianth yellowish-
green, stalked, the four outer segments narrow-lanceolate, the
four inner ones linear and rather more yellow; berry bluish-
black.—Woods and shady places. Fl. May, June.

(267) Tamus. BLACK BRYONY.

T. communis : stems slender, smooth, twining to a considerable length over hedges and bushes ; leaves shining, heart-
shaped, with a tapering point; flowers small, yellowish-green,
the males in slender often-branched racemes, the females in
short close racemes; berries scarlet.—Hedges, open woods,
and bushy places. Fl. June.

(268) Acorus. SWEET FLAG.

A. Calamus : aquatic, leaves highly aromatic, linear, erect,
2–3 feet long; flowering-stem simple, erect, the long linear
leaf-like spathe forming a flattened continuation ; spike sessile,
appearing lateral, cylindrical, very dense, yellowish-green.—
Sweet Sedge.—Edges of lakes and streams. Fl. June.

(269) Typha. BULL-RUSH.

T. latifolia : aquatic ; stems erect, reed-like, 3–6 feet high ; leaves long, erect, linear, sheathing at the base, flat above ; flowers in a dense continuous spike, often more than a foot long, the upper male portion yellow, the lower part dark brown.—Cat's-tail or Reed-mace.—Margins of ponds, lakes, and ditches. Fl. June, July.

T. angustifolia, which is not quite so common, differs chiefly in the interruption in the spike between the male and the female flowers, for a space varying from a few lines to an inch in length, and in having narrower stiffer leaves, and more slender spikes.

(270) Sparganium. BUR-REED.

S. ramosum : aquatic ; leaves long, linear, sheathing at the base ; stems erect, two feet high or more, with a few branches at the summit, each bearing several smaller male heads, and below them a few larger female heads, which are glabrous with prominent stigmas.—Margins of ponds, lakes, and streams. Fl. July.

S. simplex is rather smaller, with narrower leaves, the flower-heads much fewer towards the top of the simple stem ; and **S. natans** has weaker stems, long narrow floating leaves, and very few flower-heads.

(271) Lemna. DUCKWEED.

L. trisulca : aquatic ; fronds floating, thin, oblong, minutely-toothed at the top, and ending in a little stalk at the base, with young ones growing at right angles from opposite sides, each with a single root beneath.—Ponds and still waters. Fl. June, but rarely.

L. minor: aquatic; fronds floating, small, broadly ovate or orbicular, thickish, cohering three or four together, with one root under each, not stalked.—Ponds and still waters. Fl. June, July, not unfrequent.

L. gibba: aquatic; fronds floating, obovate, larger and much thicker than the last, flat above, spongy and almost hemispherical beneath, with a single root to each.—Ponds and still waters, rather rare. Fl. June, July, rarely.

L. polyrrhiza: aquatic; fronds larger, floating, broadly ovate or orbicular, thickish, each with a cluster of roots beneath.—Still waters. Fl. not observed.

(272) Potamogeton. PONDWEED.

* *Upper leaves floating on the surface of the water.*

P. natans: aquatic; leaves opaque, ovate or oblong, usually rounded, sometimes cordate or tapering at the base, marked by several longitudinal nerves, the submerged ones thinner and narrower, all stalked or reduced to a mere stalk; spike dense and cylindrical.—Stagnant or running waters. Fl. June, July.

P. heterophyllus resembles this, but is smaller, and has the lower submerged leaves sessile or nearly so, linear, and 1–3 nerved; while **P. lucens**, which is larger, has the lower leaves lance-shaped, with 5–7 longitudinal nerves.

** *Leaves all submerged.*
† *Leaves alternate.*

P. perfoliatus: aquatic; leaves broadly ovate, thin, obtuse, completely clasping the stem, the auricles united on the opposite side, spike shortish.—Rivers and ponds. Fl. July.

P. crispus: aquatic; leaves thin, narrow-oblong or broadly

linear, obtuse, waved and sinuated on their edges; spikes
small.—Ponds, streams and ditches. Fl. June.

P. pusillus: aquatic; stems thread-like; leaves narrow-
linear, 1–3 inches long, not dilated at the base into a scarious
sheath; spike short, close.—Pools and still waters. Fl. June,
July.

P. pectinatus: aquatic; stems thread-like; leaves narrow
linear, 2–3 inches long, dilated at the base into a rather long
sheath, scarious at the edge; spike slender, interrupted.—
Pools and still waters. Fl. June, July.

†† *Leaves opposite.*

P. densus: leaves short, ranged in two rows on opposite
sides of the stem, submerged, thin, broadly lanceolate, folded
and clasping the stem at their base; spikes 2–4-flowered, becoming reflexed.—Shallow pools and ditches. Fl. June, July.

(273) Zannichellia. HORNED PONDWEED.

Z. palustris: aquatic; stems slender, branched, floating;
leaves finely linear, mostly opposite, with a small, sheathing,
membranous stipule; flowers axillary, sessile or nearly so.—
Ponds or lagoons of fresh or brackish water. Fl. June to
August.

(274) Stratiotes. WATER SOLDIER.

S. aloides: aquatic; leaves tufted at the bottom of the
water, sessile, long, narrow, succulent, bordered by small
pointed teeth; flower-bracts with delicate diœcious peduncles,
bearing a spathe of two bracts, which enclose several stalked
male flowers with twelve or more stamens, and one solitary
sessile female. The plant rises to the surface to flower and

then sinks again.—Lakes and ditches; common in the fens of the eastern counties. Fl. July.

(275) Hydrocharis. FROGBIT.

H. Morsus-ranæ: aquatic; stems floating, runner-like, with tufts of thick, entire, stalked, orbicular-cordate leaves; flowers diœcious: the males on short peduncles, bearing 2–3 flowers on long pedicels, enclosed at the base in a spathe of two thin bracts, each flower with the outer segments pale-green, the inner ones large white, and having 3–12 stamens; the females resembling the males, but with the spathe sessile among the leaves, and the flowers with six styles having 2-cleft stigmas.—Ditches and ponds. Fl. July, August.

(276) Anacharis.

A. Alsinastrum: aquatic, dark-green, much-branched, entirely floating under water; leaves numerous, opposite or in whorls of 3–4, sessile, linear-oblong, transparent; female flowers, the only ones known in this country, sessile in the upper axils, the slender perianth-tube elongated, so as to reach the surface of the water, where it terminates in 3–6 small spreading segments.—Ponds, canals, and slow streams, in which it spreads so rapidly as to choke up the channels; recently introduced from North America. Fl. July to September.

(277) Iris. FLAG.

I. Pseudacorus: aquatic; stem two feet high; leaves sword-shaped, the lower ones longer, stiff and erect, glaucous-green; flowers 2–3, large, erect, bright yellow, the outer perianth-segments spreading, broadly ovate, clawed, the inner oblong,

erect.—Yellow Flag.—Wet meadows and watercourses.   Fl.
June and July.

I. fœtidissima : leaves narrow, sword-shaped, deep green,
having a peculiar smell when bruised; flowers smaller, several together, dull livid purple, rarely pale-yellowish white, the
outer perianth-segments narrow-ovate.—Gladdon, or Roast-
beef-plant.—Woods and shady places.—Fl. May to July. The
smell of this plant is sometimes compared to that of roast beef,
and Hooker and Arnott have a curious remark, that in Devon-
shire it is so frequent that one can hardly avoid walking on it
when herborizing, and being annoyed by the smell.

(278) Gladiolus.   Cornflag.

G. communis : stem 1½–2 feet high ; leaves linear-lanceolate ; flowers red, all turned to one side, sessile, the perianth
about 1½ inches long, the expanded part of the segments oblong-lanceolate, the uppermost broader and rather longer than
the others.—New Forest and Isle of Wight.   Fl. June.   The
English plant is sometimes considered to be G. *illyricus.*

(279) Listera.   Twayblade.

L. ovata : stem 1–1½ foot high ; leaves two, broadly ovate ;
flowers small, green, in long slender racemes ; lip with two
linear parallel lobes.—Moist pastures and woods.   Fl. June.

L. cordata : stem 4–6 inches high, slender ; leaves two,
small, broad or cordate at the base ; flowers very small,
greenish, in a short raceme ; lip linear, with four lobes.—
Mountain heaths.   Fl. July.

(280) Epipactis.   Helleborine.

E. latifolia : stem 2–3 feet high ; leaves strongly ribbed,

ovate, stem-clasping, the upper ones narrower, lower bracts often longer than the flowers; flowers pendulous, in a long, one-sided raceme, greenish with a purple lip.—Woods and shady places. Fl. July, August.

E. palustris : stem 1-1½ feet high; leaves lanceolate; bracts all shorter than the flowers; racemes loose, the flowers slightly drooping, pale greenish-purple, the lip white, streaked with pink.—Moist places. Fl. July, August.

(281) Cephalanthera.

C. grandiflora : stem 1-1½ foot high; leaves prominently veined, broadly ovate, the upper ones broadly lanceolate; flowers large, yellowish white, in a loose leafy spike, the bracts all longer than the ovary, and the lower ones leaf-like and longer than the flowers.—Woods and thickets. Fl. June.

C. ensifolia has narrower leaves and white flowers; and C. rubra has narrower leaves and red flowers. Both are rare.

(282) Orchis.

* *Spur conical, nearly as long as the ovary.*

O. latifolia : tubers palmate; stems one foot high; leaves large, oblong, blunt, not always spotted; flowers purplish-rose, dotted and marked with purple; lip 3-cleft; stalks of the pollen-masses each having a distinct gland.—Marshes and moist meadows. Fl. June, July.

** *Spur slender, longer than the ovary.*

O. pyramidalis : tubers entire; stem 1-1½ foot high; leaves lanceolate; spike dense, ovoid, oblong, 2-4 inches long; flowers rosy or purplish-red; lip broad, 3-lobed; stalks of the pollen-masses connected by a common gland.—Pastures in limestone districts. Fl. July.

[See also p. 69.]

(283) Gymnadenia.

G. conopsea : tubers palmate ; stem 1–2 feet high ; leaves linear or narrow-lanceolate ; spike oblong or cylindrical ; flowers light rosy-purple, small, sweet-scented, with a long slender spur.—Heaths and pastures. Fl. June, July.

(284) Habenaria.

H. bifolia : stem 1–1½ foot high ; leaves two, rather large, broadly ovate or oblong ; flowers pure white or with a slight greenish tinge, large and sweet-scented, in a loose spike, 3–8 inches long ; lip linear, entire, the spur slender, twice as long. —Moist pastures and meadows. Fl. June, July.

H. albida : stem 6–8 inches high ; leaves few, oblong ; spike dense, cylindrical, 1–2 inches long, with numerous small, yellowish-white, sweet-scented flowers ; lip with three deep acute lobes, the middle one largest.—Mountain pastures. Fl. June, July.

H. viridis : stem 6–8 inches high ; leaves few, ovate or oblong ; flowers yellowish-green, in a close spike ; lip linear, with three teeth, the middle one smallest.—Frog Orchis.— Dry hilly pastures. Fl. June, July.

(285) Aceras.

A. anthropophora : stem 8–9 inches high ; leaves ovate, oblong, or nearly lanceolate ; spike slender, 2–4 inches long, the flowers small, dull yellowish-green, the lip narrow-linear, twice as long as the sepals, and fancifully compared to a hanging man, two lateral lobes representing the arms, and the middle one, which is longer and 2-cleft, the body and legs. —Man Orchis.—Dry chalky pastures. Fl. June.

(286) Asparagus.

A. officinalis: stems erect, much-branched, 1–2 feet high, elegantly feathered by the numerous clusters of fine subulate leaves; flowers small, greenish-white, hanging, 2–3 together in the axils of the principal branches; berries small, red, globular.—Maritime sands. Fl. June, July. The garden *Asparagus* is a form improved by cultivation.

(287) Simethis.

S. bicolor: leaves radical, long, linear, grass-like; stem leafless, one foot high, with a loose terminal panicle of erect flowers; perianth spreading, of six oblong segments, white, purplish on the outside.—Sandy heaths in Kerry and Dorsetshire. Fl. May.

(288) Narthecium. Bog Asphodel.

N. ossifragum: stem stiff, erect, ½–1 foot high; leaves linear, vertically flattened, sheathing at their base in two opposite ranks; flowers bright yellow, in a stiff terminal raceme. —Bogs and wet moors. Fl. June, July.

(289) Tofieldia. Scottish Asphodel.

T. palustris: leaves small, sword-shaped; flower-stem 6–8 inches high, terminated by a globular or ovoid spike of small yellowish-white flowers.—Mountain bogs. Fl. July.

(290) Allium. GARLIC.

* Stems leafy.

A. oleraceum : stems 1–2 feet high ; leaves narrow-linear, nearly flat, rather thick, their sheathing bases covering the stem a considerable way up ; spathe-valves with long, linear points ; flowers pale-brown, in loose umbels, intermixed with bulbs.—Cultivated and waste places. Fl. July, August.

A. vineale : stem two feet high, leafy below ; leaves terete, hollow ; flowers pale rose-colour with green keels, in globose umbels, having bulbs intermixed with the flowers.—Crow Garlic.—Cultivated and waste places. Fl. July

** Stems leafless.

A. ursinum : leaves thin, flat, spreading, ovate-lanceolate ; flower-stem scarcely a foot high, bearing a loose umbel of about a dozen white flowers, the perianth-segments lanceolate, spreading.—Ramsons.—Woods and shady places. Fl. May, June.

A. triquetrum : leaves broadly linear, flat but folded and keeled ; flower-stem 6–8 inches high, bearing a loose, slightly drooping umbel of rather large white flowers, the perianth-segments oblong, not spreading.—Hedges in Guernsey. Fl. May, June.

(291) Butomus. FLOWERING RUSH.

B. umbellatus : aquatic ; leaves long, erect, triangular ; flower-stem leafless, 2–4 feet high, rush-like, bearing a large umbel of showy, rose-coloured flowers, the perianth an inch across, of six ovate, spreading segments.—Watery ditches and sides of rivers.—Fl. June, July.

(292) Alisma. WATER PLANTAIN.

A. Plantago: leaves radical, ovate or lanceolate; flower-stem 1–3 feet high, with whorled unequal branches, forming a loose pyramidal panicle; flowers rather small, very pale rose-colour, on long whorled pedicels.—Watery ditches, ponds, and edges of streams. Fl. July, August.

A. ranunculoides leaves narrow-lanceolate, sometimes reduced to a linear leafstalk; flower-stems simple, with a single terminal umbel, rarely a second whorl below it; flowers larger than in the last, sometimes nearly an inch diameter; pale purple.—Wet ditches and bogs. Fl. June, July.

(293) Actinocarpus. STAR-FRUIT.

A. Damasonium: aquatic, annual; leaves glabrous, radical, long-stalked, ovate or oblong; flower-stems erect, 3–9 inches high, bearing a terminal umbel and a few whorls of flowers lower down, the inner segments of the perianth delicate, white, with a yellow spot at the base; carpels six, radiating horizontally like a star.—Watery ditches and pools. Fl. June, July.

(294) Sagittaria. ARROWHEAD.

S. sagittifolia: leaves radical, on long stalks, the blade long, sagittate; flower-stem leafless, erect, longer than the leaves, bearing several distant whorls of rather large white flowers.—Ditches and shallow streams. Fl. July, August.

(295) Triglochin. ARROW-GRASS.

T. palustre: leaves slender, rather succulent, dilated and sheathing at the base; flower-stems ½–1 foot high, bearing a slender spike of very small, yellowish-green flowers; ripe fruit linear.—Wet meadows and marshes. Fl. June, July.

T. maritimum is very near this, but is usually rather stouter, with more succulent leaves, and the ripe fruit is ovoid or oblong.—Salt marshes.

(296) Juncus. Rush.

* *Leaves none ; barren and fertile stems subulate, with sheathing scales at the base.*

† *Panicle lateral.*

J. communis: stems leafless, densely tufted, cylindrical, 2–4 feet high, erect but soft and pliable, sheathed at the base by a few brown scales; some barren resembling leaves, others bearing towards the top a panicle of flowers; the flowers brown in close dense clusters (*J. conglomeratus*), or paler in loose open panicles (*J. effusus*); perianth-segments very pointed; capsule about as long, very obtuse or even notched.—Wet situations. Fl. July.

J. glaucus: stems seldom two feet high, thin, hard, stiff, often glaucous; panicle erect, loose, branched; perianth-segments lanceolate-subulate; capsule shining brown, rounded or almost pointed.—Wet places. Fl. July, August.

†† *Panicle terminal.*

J. maritimus: stems 2–3 feet high, tufted, rigid, and sharp-pointed; flowers numerous, in loose, irregularly compound panicles; perianth-segments lanceolate, acute; capsule elliptical, rather shorter.—Maritime sands. Fl. July, August.

J. acutus: stems 3–6 feet high, very rigid, prickly-pointed; flowers numerous, in compound panicles; capsule roundish-ovate, considerably longer than the perianth-segments.—Maritime sands. Fl. July, August.

** *Stems all leafy.*

† *Leaves rounded or subcompressed, jointed internally.*

J. articulatus: stems erect, 1–3 feet high; leaves sheathing the stem below, cylindrical upwards, hollow, but divided inside by cross partitions of pith, which give them the appearance of being jointed; flowers in little clusters, arranged in a compound terminal panicle; perianth-segments all pointed, or the inner ones obtuse; capsule more or less pointed.—Boggy places. Fl. June to August.

J. obtusiflorus differs only from the common larger erect form of *J. articulatus* in having all the segments of the perianth obtuse or nearly so, and about as long as the very pointed capsule.—Marshes, rather rare.

†† *Leaves grooved above, not jointed internally.*

J. compressus: stems 1–1½ foot high, erect, rather slender, with a few nearly radical leaves shorter than the stem, and 1–2 higher up, all narrow and grooved; flowers scarcely clustered, in a loose terminal panicle, shining brown; perianth-segments obtuse; capsule roundly-ovate, about as long.— Wet marshy places, especially near the sea. Fl. June to August.

J. bufonius: annual; stems numerous, forming dense tufts, 1–8 inches high, branching and flowering almost from the base; leaves chiefly radical, short, slender, angular, grooved; flowers solitary along the branches, the lower bracts leaf-like; perianth-segments narrow, pointed, pale-green, with scarious edges; capsule oblong, shorter.—Wet places. Fl. July, August.

*** *Leaves all radical, numerous.*

J. squarrosus: leaves all radical or nearly so, short, numerous, very narrow, grooved, stiff, spreading; flower-stem nearly a foot high, rigid, with a terminal panicle; flowers usually distinct; perianth-segments rather broad, glossy brown, with broad scarious edges.—Moors and heaths. Fl. June, July.

(297) Luzula. WOOD-RUSH.

L. sylvatica: stems 1½–2 feet or more; leaves linear-lanceolate, hairy; flowers in large, loose, compound panicles, the perianth-segments broad, bristle-pointed, equalling the capsule.—Woods. Fl. May, June.

L. spicata: stems 3–12 inches high; leaves short, hairy; flowers small, in dense clusters, all sessile, forming an ovoid or oblong terminal spike, more or less drooping.—Lofty mountains. Fl. July. [See also p. 73.]

(298) Carex. SEDGE.

* *Spikelets several, the terminal one composed of male and female flowers intermixed.*

C. arenaria: creeping; stems 1–1½ foot high, leafy at the base; spikelets large, ovoid, simple, sessile, crowded 8–10 together in a terminal spike.—Maritime sands. Fl. June, July.

** *Spikelets several, one or more terminal ones wholly male.*
† *Styles bifid or two-cleft.*

C. cæspitosa: stem 1–3 feet high, densely tufted; leaves narrow; spikelets 3–6, each ½–1½ inch long, the terminal one and the upper portion or whole of the next male, the remainder female, the lowest usually shortly stalked, with one or two of the outer bracts leafy; glumes dark-brown or black, often with a green midrib.—Pastures, meadows, and marshes. Fl. May to July. Mr. Bentham includes as varieties:—

C. rigida: a dwarf alpine form, six inches high, with short, flat, and rigid leaves.—Wet mountainous places.

C. stricta: stems about two feet high, glaucous with narrow leaves, rather long spikelets, the fruits arranged in 8–9 rows.—Marshes.

†† *Styles trifid, or three-cleft.*

C. **distans**: stems tufted, slender, 1–2 feet high; leaves flat, rather narrow; spikelets few and far apart, the terminal one male, the others female, oblong-cylindrical, ½–1 inch long, stalked, the stalks enclosed in the long sheaths of the leafy bracts; glumes brown.—Marshes and wet moors. Fl. June, July.

C. **capillaris**: stems slender, densely tufted, 3–9 inches high, longer than the leaves; terminal spikelets male, small; female spikelets 2–3, much lower down, on long threadlike peduncles, pale coloured, loose-flowered; glumes very scarious on the edges.—Alpine meadows. Fl. June.

C. **limosa**: stem slender, one foot high; leaves narrow; terminal male spikelet ½–1 inch long; females 1 or 2, on slender stalks, drooping, rather loose; bracts leafy; glumes rather dark-brown, ovate, the upper ones pointed.—Bogs and mountain marshes. Fl. June. [See also pp. 73–76.]

(299) Schœnus. Bog-rush.

S. **nigricans**: stems stiff, rush-like, about a foot high; leaves short, stiff, almost radical; spikelets several, dark, shining brown, almost black, closely sessile, in compact terminal heads, with an involucre of 2–3 broad brown bracts, one of which has a stiff, erect, leaf-like point.—Bogs and marshes, chiefly near the sea. Fl. June.

(300) Blysmus.

B. **compressus**: stems 6–8 inches high; leaves flat, keeled; spike terminal, about an inch long, consisting of 10–12 oblong 6–8 flowered spikelets, closely sessile on opposite sides of the axis, the broad, brown, glume-like outer bract shorter than the mature spikelet.—Bogs and marshes. Fl. June, July.

B. rufus: stems ½–1 foot high, stiff, slender, with few very narrow leaves near the base; spike terminal, consisting of about six sessile 2–4 flowered spikelets, dark, shining brown, almost black, nearly concealed by the outer bract, which is dark brown, and shining.—Marshy places.   Fl. July.

(301) Eleocharis.

E. acicularis: plant small, slender, tufted, the fine subulate stems scarcely two inches high, with short sheaths at their base, most of them bearing a single terminal oblong spikelet, of a dark brown colour, the outer bract similar to the glumes; flowers usually 6–8 in the spikelet.—Wet sandy places.   Fl. July, August.

E. palustris: stems numerous, erect, densely tufted, ½–1 foot high, leafless; spikelets solitary and terminal, oblong; glumes numerous, closely imbricated, brown, with scarious edges, green on the midrib, the outer bract only differing from the glumes in being rather larger.—Edges of pools and watery ditches.   Fl. June.

E. multicaulis is very much like this, but smaller, the stems more slender, many of them barren and leaf-like, and the spikelets rather smaller.—Marshy places.

(302) Scirpus.   CLUB-RUSH.

* *Hypogynous bristles none.*

S. fluitans: aquatic; stems long, slender, branching, either floating in water, or forming soft densely matted masses on its margin; leaves linear-subulate; spikelets solitary, terminal, oblong, greenish, the outer bract without any leafy point.—Pools and still waters.   Fl. June, July.

S. setaceus: stems slender, 2–3 inches high, forming dense tufts, with 1–2 short, subulate leaves on each stem, sheathing it at the base ; spikelets ovoid, solitary, or 2–3 together in a little cluster, appearing lateral, the subulate point of the outer bract forming a continuation of the stem ; glumes broad, short, dark brown, with a green midrib.—Muddy places, margins of pools, etc. Fl. July.

** *Hypogynous bristles* 4–6.

S. cæspitosus : stem ½–1 foot high, densely tufted ; spikelets solitary and terminal, ovoid, brown, the outer bract like the glumes but larger, with an almost leafy tip, about the length of the spikelet; flowers usually 6–8 in the spikelet.—Marshes and bogs. Fl. June to August.

*** *Hypogynous bristles* 6.
† *Stem triangular ; panicle leafy.*

S. maritimus : stems sharply triangular, 2–5 feet high ; leaves long, flat, pointed, often far exceeding the stem; spikelets rich brown, ovoid or lanceolate, sometimes only 2–3 in a close sessile cluster, more frequently 8–10 in a compound cluster, the outer ones stalked ; glumes notched, with a fine point.—Salt marshes. Fl. July.

S. sylvaticus : stems triangular, 2–3 feet high ; leaves long, grass-like; spikelets ovoid, dark shining green, very numerous, in clusters of 2–3 together, forming a terminal, much branched, compound umbel or panicle, with an involucre of 2–3 linear leaves ; glumes keeled and pointed.—Moist woods, and banks of rivers. Fl. July.

†† *Stem triangular ; panicle leafless.*

S. triqueter : stems acutely triangular, 2–3 feet high, leafless ; spikelets ovoid 8–10 or more, the central ones sessile, the others stalked, forming a compound lateral cluster or umbel, the stiff, triangular outer bract continuing the stem for an inch or more; glumes brown, broad, usually notched or fringed at the top, with a minute point.—Marshes and edges of pools. Fl. August.

††† *Stem terete.*

S. lacustris: stems stout, erect, 2–6 or 8 feet high, cylindrical at the base, tapering upwards, sometimes obtusely triangular near the top, with a single short leaf near the base; spikelets ovoid or oblong, rather numerous, in a compound lateral umbel or cluster, the outer bract continuing the stem; glumes numerous, broad, brown, fringed at the edge, notched at the top, with a little point in the notch.—Bull-rush.—Margins of lakes and ponds. Fl. June, July.

(303) Eriophorum. COTTON-SEDGE.

E. vaginatum: stems tufted, one foot high or more, with one or two linear, almost subulate leaves; upper sheaths inflated, without any or only a very short blade; spikelet solitary, terminal, ovoid, deep olive-green; hypogynous bristles numerous, forming at length dense cottony tufts.—Bogs and wet moors. Fl. May.

E. polystachyum: leaves few, mostly radical, triangular, or channelled; stems 1–1½ feet high, with a terminal umbel of 2–3 to 8–10 or more spikelets, the inner ones sessile, the outer ones more or less stalked, often drooping; spikelets ovoid or oblong; hypogynous bristles very numerous, forming dense cottony tufts, often attaining 1–1½ inch in length.—Bogs and wet moors. Fl. May, June.

E. gracile is a small form, with very slender leaves, and few almost erect spikelets; and E. latifolium, a tallish slender plant, with the leaves nearly flat. Both are rare.

(304) Rhynchospora. BEAK-SEDGE.

R. alba: stems 6–9 inches high, slender, forming dense tufts; leaves chiefly radical, short, subulate, the floral ones scarcely exceeding the flowers; spikelets nearly white, in a small, loose terminal cluster, often with one or two smaller clusters on slender peduncles in the axils of the next leaves.— Turfy bogs. Fl. July.

(305) Milium. MILLET-GRASS.

M. effusum: stem tall, slender, 4–5 feet high; leaves short, flat; panicle long, loose, slender, spreading.—Moist woods. Fl. June, July.

(306) Digraphis.

D. arundinacea: stems reed-like, 2–3 feet high; leaves rather broad, long; panicle upright, with short spreading branches; spikelets numerous.—River-banks and marshes. Fl. July. The Ribbon-grass of gardens is a variety with variegated leaves.

(307) Phleum. CAT'S-TAIL-GRASS.

P. pratense: stem 1–3 feet high; leaves rough on the edges; spike or spike-like panicle cylindrical, very compact, 1–4 inches long; outer glumes truncate at top, with broad scarious edges.—Timothy-grass.—Meadows and pastures. Fl. June to October.

P. arenarium: annual; stem erect, 6–8 inches high; leaves short; spike about an inch long, dense, nearly cylindrical, tapering at the base; spikelets about $1\frac{1}{2}$ line long; outer glumes lanceolate, tapering into a short point.—Maritime sands. Fl. June.

(308) **Alopecurus.** Fox-tail-grass.

* *Outer glumes glabrous.*

A. agrestis : annual ; stem 1–2 feet high, erect, slightly decumbent at the base ; leaves short, with long, scarcely loose sheaths ; spike 2–3 inches long, thin, tapering ; two outer glumes united to about the middle.—Waste places, roadsides, etc. Fl. May to September.

** *Outer glumes hairy on the keel.*

A. pratensis : stems erect, scarcely decumbent at the base, 1–2 feet high ; upper sheaths loose ; spike 2–3 inches long, dense, obtuse, hairy ; outer glumes free or scarcely united at the base.—Meadows and pastures. Fl. May, June.

A. geniculatus : stem procumbent at the base ; upper sheaths loose ; spike 1–2 inches long, cylindrical ; outer glumes united at the base.—Moist meadows, marshy places. Fl. June, July. Sometimes the stems thicken at the base into a kind of bulb, and it is then called **A. bulbosus.**

(309) **Lagurus.** Hare's-tail-grass.

L. ovatus : annual ; stems erect, ¼–1 foot high ; leaves hoary with soft down, their sheaths rather swollen ; spikelets 1-flowered, crowded in an ovoid or oblong, softly hairy head, ¼–1 inch long.—Maritime sands in the Channel Islands. Fl. June.

(310) **Polypogon.** Beard-grass.

P. monspeliensis : annual ; stems procumbent at the base, rarely erect, 1–1½ feet high ; leaves flat, rather flaccid ; panicle contracted into a slightly branched spike, 2–3 inches long, yellowish shining green, and thickly bearded with numerous straight smooth awns, which are 3–4 times as long as the spikelets.—Fields, near the sea, rare. Fl. July, August.

(311) **Agrostis.** BENT-GRASS.

A. alba : stem procumbent at the base, 1–2 feet high; leaves flat; panicle slender, spreading when in flower, contracted after flowering; outer pales thin, awnless or rarely with a minute awn from its base.—Pastures and waste places. Fl. July.

A. vulgaris : smaller and dwarfer than the last; panicle spreading both before and after flowering.—Pastures and waste places. Fl. July.

A. canina: similar to the foregoing, but the panicle less spreading, and the outer pales bear on their back below the middle a fine awn, which slightly protrudes beyond the glumes. —Moist heaths. Fl. July.

(312) **Apera.** WIND-GRASS.

A. Spica-venti : annual; stems 1–2 feet high, slender, elegant; leaves narrow; panicle long, spreading, with slender, hair-like branches, and little shining spikelets, the outer pales with a hair-like awn, three or four times their own length.— Fields and sandy pastures. Fl. June, July.

A. interrupta has the spikelets shorter and more crowded, in a narrow panicle, with nearly erect branches.—Sandy fields.

(313) **Ammophila.** SEA-REED, or MARRAM.

A. arenaria: stems stiff, erect, 2–3 feet high; leaves narrow, erect, glaucous, rolled inwards on their edges; panicle contracted into a close, narrow-cylindrical, tapering spike, 5–6 inches long.—Maritime sands. Fl. July.

(314) Calamagrostis. SMALL-REED.

C. Epigeios : stems 3–4 feet high, erect, firm ; leaves long, narrow, somewhat glaucous; panicle branched, not spreading, from a few inches to near a foot long, often purplish; outer glumes very narrow-lanceolate, pointed, almost subulate.—Moist woods and thickets. Fl. July.

C. lanceolata is usually more slender, with a much looser panicle, more frequently assuming a shining purple colour.

(315) Aira. HAIR-GRASS.

A. cæspitosa : stems 2–4 feet high ; leaves stiff, flat, rough above, forming dense tufts; panicle ½–1 foot long, very elegant, with spreading, slender, almost capillary branches, the spikelets silvery-grey or purplish; outer pales scarcely projecting from the glumes, with a fine hair-like awn inserted near the base, not so long as the glume.—Moist, shady places. Fl. June, July.

A. flexuosa : stems rather slender, 1–1½ foot high ; leaves narrow, rolled inwards on the edges, almost subulate ; panicle spreading, 2–3 inches long; spikelets very shining, the fine hair-like awns protruding beyond the glumes.—Heaths and hilly pastures. Fl. July.

A. caryophyllea : annual ; stems slender, graceful, six inches high; leaves short, fine; panicle loose, spreading, with long, capillary branches, the awns shortly protruding from the glumes.—Sandy and hilly pastures. Fl. June, July. [See also p. 77.]

(316) Avena. OAT.

A. fatua : annual ; stems erect, glabrous, 2–3 feet high ; panicle loose, of large spikelets, hanging from filiform pedicels of unequal length, arranged in alternate bunches along the main axis; awn twice as long as the spikelet, twisted at the base, abruptly bent about the middle.—Haver.—A common weed of cultivation. Fl. June, July.

A. pratensis: stems erect, 1–1½ foot high; leaves scabrous; panicle slightly compound or reduced to a simple raceme, the spikelets erect, 3–4-flowered, glabrous, shining; awn twice the length of the glumes.—Meadows and limestone pastures. Fl. July.

(317) Trisetum.

T. flavescens: stems erect, 1–2 feet high; panicle oblong, 3–5 inches long, with slender, somewhat spreading branches, the spikelets erect, shining, often yellowish; awn twisted, but short, very fine and hair-like.—Dry meadows and pastures. Fl. July.

(318) Arrhenatherum. FALSE OAT, OR OAT-GRASS.

A. avenaceum: stems erect, 2–3 feet high; leaves few, flaccid; panicle narrow, loose, 6–8 inches long, spreading only whilst in flower; outer pales with a fine bent awn on the middle of the back about twice its own length.—Meadows, hedges, and thickets. Fl. June, July.

(319) Holcus. SOFT-GRASS.

H. lanatus: roots fibrous; stems 1–2 feet high, clothed, like the leaves, with short down; panicle 2–3 inches long, whitish, or sometimes reddish; outer glumes obtuse, concealing the awn of the pales.—Meadows, pastures, and waste places. Fl. June, July.

H. mollis: similar to the last, but creeping; outer glumes

tapering to a fine point; awn of the pales usually projecting beyond the glume.—Pastures and hedges. Fl. July.

(320) Spartina.

S. stricta: stems stiff, erect, 1–2 feet high; leaves short, erect, flat, excepting at the top, the edges always rolled inwards when dry; panicle 3–4 inches long, consisting of 2–4 erect, spike-like branches, with the spikelets arranged in two rows, erect, turning to one side; glumes downy.—Muddy salt-marshes. Fl. August.

(321) Lepturus.

L. incurvatus: annual; stems decumbent, branched at the base, the flowering-stems curved upwards or erect, 3–9 inches high; leaves short, fine, the uppermost close under the flowers; spike 2–4 inches long, usually curved, the spikelets imbedded in the axis, which breaks off readily at every notch.—Sea Hard-grass.—Salt-marshes and maritime sands. Fl. August.

(322) Nardus. Mat-grass.

N. stricta: stems densely tufted, erect, wiry, ½–1 foot high; leaves fine, stiff, bristle-like; spike erect, slender, simple, one-sided, often assuming a purplish hue.—Moors, heaths, and hilly pastures. Fl. July.

(323) Elymus. Lyme-grass.

E. arenarius: stems 2–4 feet high; leaves stiff, glaucous, rolled inwards on the edges, ending in a hard point; spike close, 3–4 inches to 8–9 inches long, the spikelets in rather distant pairs; glumes lanceolate.—Maritime sands. Fl. July.

E. geniculatus is a related plant, with the spike much elongated and abruptly bent down, the spikelets distant, and the glumes enlarged, awl-shaped.—Salt marshes, rare.

(324) Hordeum. BARLEY.

H. pratense : stems erect or decumbent, two feet high; leaves glabrous and rather narrow; spike $1\frac{1}{2}$–2 inches long, close, cylindrical; flowers of the central spikelet only perfect, the outer glumes of all the spikelets awn-like from the base.— Moist meadows and pastures. Fl. June.

H. murinum: annual; stems decumbent at the base, 1–2 feet long; leaves often hairy; spike dense, cylindrical, 3–4 inches long, thickly beset with the long rough awns; outer glumes of the three spikelets all awn-like, but those of the central spikelet somewhat broader at the base and ciliate.— Mouse Barley.—Waste places, roadsides, etc. Fl. June to August.

(325) Triticum. WHEAT-GRASS.

T. repens : stems stiff, ascending or erect, 1–3 feet high, sometimes glaucous; spikelets placed on alternate sides of a spike, 2–6 inches long; glumes 5–7 ribbed, acute, or terminating in an awn.—Couch-grass.—Fields and waste places. Fl. July.

T. junceum, a plant with glaucous herbage and very blunt glumes to the spikelets, is common on sandy sea-shores.

T. caninum also closely resembles *T. repens*, but the stems are tufted, more leafy, and less glaucous; the glumes rather thinner, with five prominent ribs, and terminating in a rather long awn.—Woods and banks. Fl. July.

(326) Lolium. RYE-GRASS.

L. perenne : stems erect or slightly decumbent, 1–$1\frac{1}{4}$ foot

high, leafy below; spike ¼–1 foot long, distichous, the spikelets at a considerable distance from each other, longer than
the glumes.—Meadows, pastures, and waste places. Fl. June.

L. temulentum: annual; stems erect, three feet high;
spike long, the outer glume of the spikelets usually as long as
the spikelet itself; pales sometimes with an awn longer than
themselves.—Darnel.—Fields and waste places. Fl. July.

(327) Brachypodium. FALSE BROME-GRASS.

B. sylvaticum: stems rather slender, erect, 2–3 feet high;
leaves flat, rather long; spike simple, drooping; spikelets
usually 6–7, each an inch or more in length, nearly cylindrical,
becoming flattened when in fruit; outer pales ending in an
awn, usually as long as or longer than the glume itself.—
Woods, hedges, and thickets. Fl. July.

B. pinnatum has the spikelets more erect, the flowering
glumes smaller, and more open, and the awn very much
shorter.—Dry limestone wastes. Fl. July.

(328) Bromus. BROME-GRASS.

* *Outer pales narrow-lanceolate.*

B. erectus: stems erect, two feet high; leaves narrow;
panicle 3–5 inches long, compact, the branches erect; spikelets not numerous; awn straight, scarcely half as long as the
pale.—Fields and waste places. Fl. July.

B. asper: annual or biennial; stems 3–6 feet high; leaves
long, flat, with long hairs on their sheaths; panicle loose, with
long, drooping branches, bearing a few loose spikelets above
an inch long; awn straight, shorter than the pale.—Hedges
and thickets. Fl. July, August.

B. sterilis: annual or biennial; stems erect, 1–2 feet high;

leaves softly downy; panicle six inches long or more, with numerous drooping branches, the spikelets linear-lanceolate; awns much longer than the pales.—Waste places, waysides, etc. Fl. June, **July.**

** *Outer pales oblong, turgid.*

B. arvensis: annual or biennial; stems erect, 1-3 feet high, softly downy or glabrous; panicle small, slender, elongated or compact, nearly erect, more frequently more or less drooping; pales always short, oblong or ovoid, turgid, closely packed; awn slender, usually about the length of the glumes. —Cultivated and waste places. Fl. June, July.

The following closely allied plants are included in *B. arvensis* by Mr. Bentham :—

B. secalinus: a tall cornfield variety, with a loose, more or less drooping panicle, the flowers not so closely imbricated, becoming quite distinct and spreading when in fruit.

B. mollis: common in open waste places, with a more erect panicle, either short and compact or long and slender, the whole plant softly downy.

B. racemosus: like the last, but much more glabrous.

(329) **Festuca.**

* *Awns none, or very short.*

F. ovina: stems densely tufted, $\frac{1}{2}$-$1\frac{1}{4}$ foot high; leaves very narrow, almost cylindrical; panicle rather compact, $1\frac{1}{2}$-4 inches long; glumes narrow, glabrous, or downy, almost always bearing a fine point or awn.—Hilly pastures. Fl. June.

F. rubra, found in light sandy or loose stony places, resembles this, but has the rootstock more or less creeping.

F. pratensis: stems 2-3 feet high; leaves flat; panicle

narrow, erect, subsecund, slightly branched, 5–6 inches long.
—Meadows and moist pastures. Fl. June, July.

Var. *loliacea :* spikelets almost sessile, in a simple spike.—
Meadows.

Var. *elatior :* taller, reed-like, with broader leaves, the panicle more branched and spreading.—Banks of rivers and wet
places.

     ** *Awns as long as or longer than the glumes.*

     F. Myurus: annual; stems tufted, one foot high; leaves
narrow, convolute; panicle slender, one-sided, 2–6 inches long,
contracted, sometimes reduced to a simple spike, the branches
always short and erect; glumes narrow, the outer ones very
unequal, the pales ending in an awn at least as long as themselves.—Sandy places, roadsides, etc.   Fl. June, July.

(330) Dactylis.   Cock's-foot-grass.

     D. glomerata: stems coarse, 1–2 feet high; leaves flaccid,
rough on the edges; clusters of spikelets dense and ovoid, collected into a close spike of about an inch, or in a broken spike
of several inches long, or on the branches of a short, more or
less spreading panicle; each spikelet much flattened, ovate;
glumes lanceolate, ciliated on the back.—Meadows, pastures,
and waste ground.   Fl. June, July.

(331) Cynosurus.   Dog's-tail-grass.

     C. cristatus: stems slender, wiry, erect, 1–2 feet high;
leaves short, narrow; flowering spike semicylindrical, oblong
or nearly linear, 1–3 inches long, the clusters regular, all
turned to one side.—Dry, hilly pastures, and downs.   Fl.
August.

(332) Briza. QUAKE-GRASS

B. media: stems erect, rather stiff, 1–1½ foot high; leaves flat, narrow, few; panicle 2–4 inches long, loose, spreading, the spikelets hanging from the long, slender branches, at first orbicular, then ovate, variegated with green and purple.—Meadows and pastures. Fl. June.

B. minor: annual; stems erect, ¼–1 foot high; panicle spreading, much-branched, slender, the spikelets numerous, bluntly triangular.—Fields and waste places. Fl. June.

(333) Poa. MEADOW-GRASS.

* *Pales rounded on the back.*

P. maritima: stems decumbent or erect, a foot high; leaves short, narrow, usually convolute; panicle erect, rather stiff, 3–4 inches long, the branches erect, or the lower ones spreading; spikelets not numerous, shortly stalked, all turned to one side of the branches.—Maritime sands. Fl. June, July.

P. distans is very near this, but the leaves are flatter, the stems taller and more slender, and the panicle more spreading, with long, slender branches.

P. procumbens is a dwarfer plant, with decumbent stems, 6–8 inches long, flat leaves, and a more compact, branched, one-sided panicle.

P. rigida: annual; stems tufted, six inches high, stiff, erect or slightly decumbent at the base; panicle lanceolate, one-sided, about two inches long, rather crowded, the branches slightly spreading.—Waste, dry, or stony places. Fl. June.

** *Pales keeled.*

P. nemoralis: stems 1–2 feet high, erectish, but weak and slender; leaves narrow; panicle contracted or spreading, with slender branches.—Woods and shady places. Fl. June, July.

P. alpina: stems tufted, ½–1 foot high; leaves short, broad, mostly radical; panicle close ovoid, about two inches long, spreading; spikelets crowded, ovate, 3–5-flowered.—Alpine pastures. Fl. June, July.

(334) Glyceria.

G. aquatica: aquatic; stems stout, reed-like, 4–6 feet high; leaves flat, very rough on the edges; panicle much branched, spreading, nearly a foot long, the spikelets numerous.—Wet ditches and shallow waters. Fl. July.

G. fluitans: aquatic; stems 2–3 feet high, rather weak, creeping at the base; leaves often floating; panicle erect and slender, a foot long, the branches few, usually erect; spikelets ½–1 inch long.—Flote-grass.—Wet ditches and stagnant waters. Fl. June to September.

(335) Catabrosa. WATER HAIR-GRASS.

C. aquatica: aquatic; stems glabrous, procumbent, creeping at the base or floating, often 2–3 feet long; flowering branches erect; leaves short, flat, flaccid; panicle 4–6 inches long, with unequal, slender, spreading branches, in rather distant whorls.—Shallow pools and ditches. Fl. June, July.

(336) Melica. MELIC-GRASS.

M. uniflora: stems 1–2 feet high, slender, elegant; leaves long and narrow; panicle reduced to an almost simple raceme with 3–4 spikelets, or having a few long, slender distant branches, each bearing several spikelets, the spikelets brownish, erect, containing one flower.—Woods and shady places. Fl. May, June.

M. nutans: stems 1–2 feet high, slender, erect; leaves erect, flat; panicle one-sided, 2–3 inches long; spikelets 10–15 drooping, two-flowered; outer glumes brown or purple with scarious edges.—Woods and shady rocky places. Fl. June.

(337) **Arundo.** REED.

A. Phragmites: aquatic; stems stout, 5–6 feet high or more, leafy all the way up; leaves broad; panicle from a few inches to a foot long, with numerous branches, more or less drooping, purplish-brown, the spikelets very numerous, narrow; pales surrounded by silky hairs which lengthen as the seed ripens, giving the panicle a beautiful silvery appearance.— Wet ditches, marshes, and shallow waters. Fl. July.

There are many other wild flowers to be occasionally found in the summer season, but those we have noticed are the most likely to be met with, and a knowledge of these will furnish a key to the rest.

AUTUMN FLOWERS AND FRUITS.

"It was an eve of Autumn's holiest mood.
The corn-fields, bathed in Cynthia's silver light,
Stood ready for the reaper's gathering hand;
And all the winds slept soundly. Nature seemed
In silent contemplation to adore
Its Maker. Now and then the aged leaf
Fell from its fellows, rustling to the ground;
And, as it fell, bade man think on his end.
On vale and lake, on wood and mountain high,
With pensive wing outspread, sat heavenly Thought,
Conversing with itself. Vesper looked forth
From out her western hermitage, and smiled;
And up the east, unclouded, rode the moon
With all her stars, gazing on earth intense,
As if she saw some wonder working there."

Pollok.

ILLUSTRATIONS.

At this season many summer flowers will be lingering on,
some almost exhausted with their summer's bloom, and others
blossoming anew at intervals with almost pristine vigour.
These we must not stay to gather. Our object must be rather to pluck a few fresh illustrations from the Flora of the
waning year; and we shall take them in the order which
has been adopted in treating of the flowers of the foregoing
seasons.

The Grass of Parnassus* is by some botanists classed amongst the Thalamiflores, and referred to the Hypericaceous family already adverted to, but by others it is included with that of the Droseras, or removed to that of the perigynous Saxifrages. It is a very pretty dwarf perennial herb, producing a tuft of stalked smooth heart-shaped leaves, and slender erect flower scapes, six to ten inches high, bearing a single sessile leaf below the middle, and terminated by a rather large white spreading flower. This flower has a calyx of five ovate segments, a corolla of five distinct obovate petals; five stamens, which are perigynous, and alternating with them five fringed glands or nectaries, which may be taken to represent groups of imperfect stamens; these glands are short and thick, placed opposite the petals, and margined with ten or twelve white filaments, each bearing a small globular yellow gland. There is a one-celled four-valved ovary, tipped by four sessile stigmas, and growing into a roundish capsule. The plant is found not unfrequently in boggy and moist heathy situations, flowering in the wane of summer.

Belonging to the Calyciflores is the Dwarf Furze,† a humble and often prostrate plant found decorating heaths and sandy wastes during the latter part of the floral season, and belonging to the family of Leguminous plants. It differs from the Common Furze (*Ulex europæus*) chiefly in its smaller size. This latter is, however, rather a spring or early summer-flowering plant. The present smaller species is, like the larger, a thorny shrub, but, instead of growing into an erect branching bush, its branches are procumbent, spreading along upon the surface of the heathy waste or stony bank on which it grows. The main branches are thickly clothed with short

---

* *Parnassia palustris*—Plate 22 C.

† *Ulex nanus*—Plate 22 D.

intricate branchlets, thorny at the points and branched at the base, slender, smooth, and striated. The leaves are, for the most part, reduced to thorns very much resembling the branchlets, so that the stems seem to be formed of an intricate mass of sharp thorns. The flowers spring from the axils of the smaller thorns, which branch out from the primary ones, and are scarcely so long as the latter: indeed

> " Every flower has a troop of swords
> Drawn to defend it."

The flowers consist of a calyx, coloured yellow like the corolla, and divided nearly to the base into two concave segments, which are entire or minutely-toothed at the tips ; a papilionaceous corolla, also yellow, the petals scarcely separating even when the flower is fully blown, but consisting of the usual standard keel and wings; ten stamens, united by their filaments into a complete sheath around the ovary, which becomes a turgid, few-seeded pod, scarcely longer than the calyx.

Nearly related to the Umbelliferous family, and like it one of the epigynous Calyciflores, is the common Ivy,[*] a woody evergreen climber, belonging to the Araliaceous family. The earlier-formed stems of this very beautiful plant climb up against trees or walls or rocks, clinging as they go by means of small root-like protuberances, and spreading out the leaves right and left flat against the body to which they adhere. In this way its stems cover a large space. The dark green glossy leaves borne on this part of the plant are angular and three- or five-lobed, this being the form to which the term ivy-leaved is applied. When the plants have grown to a considerable height, or to what may be considered mature age, they throw out bushy tufted branches, like those of other shrubs,

---

[*] *Hedera Helix*—Plate 22 F.

and on these, which bear the flowers, the leaves are ovate.
Sometimes the Ivy may be seen climbing flat against the surface of a wall till it reaches the top, and then developing a large
branched head, as if planted on the wall top.

> " Emong the rest, the clamb'ring yvie grew,
>   Knitting his wanton arms with grasping hold,
>   Lest that the poplar happely should rew
>   Her brother's strokes, whose boughs she doth enfold
>   With her lythe twigs, till they the top survew,
>   And paint with pallid green her buds of gold."
>
> *Spenser.*

The flowering branches bear a short raceme or panicle of
several nearly globular umbels of yellowish-green flowers.
These consist of a very slight entire calyx-border about half
way up the ovary, five short broad petals, five erect stamens,
and the styles cohering into a single mass. The cells of the
ovary are from five to ten in number, and the berry, which is
smooth and black, contains from two to five seeds. The Ivy
is made the emblem of friendship, from the closeness of its
adherence to the tree on which it has once fixed itself. It has
also been called " the critic's ivy." The plant was dedicated
by the ancients to Bacchus, whose statues were often found
crowned with a wreath of its leaves. The priests of the Greeks,
it is said, presented a wreath of Ivy to newly-married persons,
as a symbol of the closeness of the tie which ought to bind
them together.

The Devil's-bit,* an example of the Dipsacaceous family,
belongs to the same series of epigynous Monopetals as the
family of Composites. It is found abundantly in meadows
and pastures, and in heathy places, and is a perennial herb,
with a short thick root-stock, ending abruptly as if it had been

* *Scabiosa succisa*—Plate 23 A.

bitten off, whence the vulgar name. The leaves mostly spring
from the root, and are stalked, ovate or oblong, entire, nearly
or quite free from hairs ; those of the stems are few, opposite,
oblong, and occasionally slightly toothed. The stems grow
from one to two feet high, and produce from one to three or
five heads of flowers, a pair of flowering branches springing
from one or both of the upper pairs of leaves, according to the
vigour of the plant. The flower-heads are surrounded by an
involucre of two or three rows of lanceolate bracts, the outer
of which are as long as the flowers, the inner ones passing
gradually into the scales of the receptacle. Unlike those of
the allied Teasels, these bracts are not prickly. The receptacle bears a globular head of florets, between which small
pointed scales are placed. The florets are nearly equal, the
outer series being scarcely larger than the others ; they consist of an ovary crowned by the little cup-shaped calycine
border with four bristle-shaped teeth, a four-lobed tubular corolla, four stamens inserted in the corolla tube, and having
free anthers, and a long simple style. Each of the florets is
inserted in an involucre, which is tubular and angular, bordered by very small green teeth, and completely enclosing the
ovary and fruit. It flowers towards the latter part of summer,
and during autumn. This genus is represented in gardens by
the Sweet Scabious or Blackamoor's Beauty, an annual of not
unfrequent occurrence, especially in cottage gardens.

The Ericaceous family has been already referred to, but is
very well illustrated by a favourite autumn-blooming shrub,
the Strawberry Tree or Common Arbutus.* This shrub,
which is frequent in hilly districts in the south of Europe, is
found abundantly about the Lakes of Killarney, but is even
better known as a very ornamental garden shrub, an evergreen

* *Arbutus Unedo*—Plate 23 B.

always pleasing in habit and foliage, and in the autumnal season becoming decorated with the greenish-white pitcher-shaped flowers of the present season, and the large red strawberry-like fruits produced by the preceding year's flowers, so that, the fruit taking a year to arrive at maturity, the shrub has perfect flowers and ripe fruit all at one time. The Arbutus forms a tall evergreen shrub or bushy tree, sometimes of considerable size, furnished with shortly-stalked, ovate or oblong-lanceolate, toothed leaves, shining on the upper surface. The flowers grow in terminal drooping panicles, consisting of an inferior calyx of five small sepals, an ovoid or pitcher-shaped, waxy-looking corolla, of a greenish-white often tinged with pink, ten enclosed stamens, and a five-celled ovary, which becomes a globular berry, red and granulated on the surface, and thus having, when seen at a distance, considerable resemblance to a strawberry, whence the name of Strawberry Tree. The flowers have been compared by Miss Twamley to pearls :—

"Small bell-shaped flowers, each of an orient pearl
Most delicately modelled, and just tinged
With faintest yellow, as if, lit within,
There hung a fairy torch in each lamp-flower."

A beautiful autumnal flower is the Calathian Violet, or Marsh Gentian,* a type of the Gentianaceous family. It is a perennial herb, growing in moist heaths and pastures, with an upright stem, six inches to a foot high, bearing opposite oblong-lanceolate leaves on the lower part of the stem, and linear ones above, all blunt at the point and rather thick in texture. The flowers are nearly sessile, and grow in pairs from the axils of the upper leaves. The calyx is tubular, with five narrow lobes ; and the corolla has a narrowly-bell-shaped tube an inch and a half long, without hairs in the throat, and a five-lobed

* *Gentiana Pneumonanthe*—Plate 23 C.

2 A

limb, with short, broad, spreading lobes, of a deep blue within, marked with a broad greenish band down the middle of each segment, and yellowish towards the base. There are five stamens affixed to the corolla tube, two stigmas, and a one-celled ovary. This genus contains some of the most beautiful of dwarf herbaceous plants known, some of which are met with in gardens. One of them, *G. acaulis*, the Gentianella of gardens, a native of the South of Europe, is not uncommon, and a good deal resembles one of our native species, *G. verna*.

Our autumnal flora furnishes us with another illustration of the Labiate family, in the Pennyroyal,* one of the Mint genus, and a familiar medicinal herb. It is a prostrate, powerfully-scented plant, with slender stems rooting at the joints as they slowly spread over the surface, ascending at the points, and furnished with small ovate-obtuse, slightly-crenated leaves, which become still smaller towards the top, where they serve as bracts to the axillary verticillasters or half-whorls of flowers. The flowers are small, with a tubular calyx, divided into five regular lobes, the throat closed with hairs, and a labiate corolla, tubular below and somewhat bell-shaped above, with a four-lobed limb, the upper lobe broader and sometimes slightly notched; within the corolla are four equal stamens, a single style cleft at top into two stigmatic lobes, and rising up from the centre of the four-lobed ovary, which becomes developed into four smooth nuts. To the same genus belong the various Mints, several of which are familiar as culinary or medicinal herbs.

Among the Monochlamyds occur a group of plants of weedy character, called the Chenopodiaceous family. An example of it is seen in the Many-seeded Goosefoot,† an annual weed found

* *Mentha Pulegium*—Plate 23 D.
† *Chenopodium polyspermum*—Plate 23 E.

in the southern and central parts of England. It is a spreading, much-branched plant, variable in size, with entire ovate-elliptic leaves, and clusters of minute flowers in short axillary spikes. As usual in the family, the perianth consists of five thin green sepals, within which are five stamens, and an ovary crowned by two styles. This ovary grows into a lenticular fruit, enclosed within, but not covered by, the perianth. Though an inconspicuous and weedy race, so far as our native species are concerned, the Chenopodiaceous plants are not without their beauty and utility; for amongst them are comprised the Spinach and Beet of our culinary gardens, and the Mangel-Wurzel of our fields; while others yield abundance of soda, and some are famous anthelmintics.

Of autumnal Monocotyledons, we find the Meadow Saffron,* which illustrates the family of Melanthaceous plants. It has considerable first-sight resemblance to the autumn-flowering Crocuses (*Crocus sativus* and *C. nudiflorus*), but is at once distinguished by the flowers having six stamens instead of three, and by other peculiarities. The plant is one of those with that kind of solid bulb-stem called a corm, which is large and ovate. At the flowering period there are no leaves, the corm ending in a sheath of brown scales which enclose the base of the flowers. These flowers have a funnel-shaped six-parted perianth, like that of *Crocus*, with a long slender tube running down in a stalk-like form, and enclosing in its base the ovary, which is underground; the colour is a bright light purple. There are six stamens inserted in the throat of the tube of the perianth, and three long filiform styles running up from the ovary the full length of the long tube, and terminating in somewhat clavate stigmas. The capsule is three-celled, containing numerous seeds. The leaves are produced in spring along with the

* *Colchicum autumnale*—Plate 24 A.

capsules, which are elevated above ground by the lengthening of the stalks. These leaves are broadly lance-shaped, eight or ten inches long, not unlike those of the tulip, except that they are smooth dark green instead of glaucous. The corms and the seeds of this plant, which is found abundantly in moist rich pastures in various parts of England, are very powerful medicinal agents, extensively employed in modern practice.

The little family of Restiaceous plants is represented in our flora by a small aquatic tufted plant, found in the Hebrides and in Ireland, and called the Jointed Eriocaulon.* The plant has a slender rootstock, creeping in the mud under water, and forming tufts of linear, soft, pellucid, beautifully-cellular leaves, one to three inches long, from among which rises the peduncle, two inches to a foot high, bearing a head of numerous compact small flowers, the central of which are chiefly males, and the outer ones chiefly females, all intermixed with small bracts, of which the outer ones are rather larger, forming an involucre round the head. The perianth is very delicate, of four segments, with a minute black gland on the two inner ones; the male flowers contain four stamens, and the females a two-lobed ovary, with two long subulate stigmas.

A more detailed summary of the flowers of the pleasant autumn season, when vegetation becomes languid with its summer's efforts, will be given in the following pages; meanwhile we may ask with the poet—

> " Who loves not Autumn's joyous round,
> Where corn and wine and oil abound ?
> Yet who would choose, however gay,
> A year of unreserved decay ?"

* *Eriocaulon septangulare*—Plate 24 B.

SUMMARY OF AUTUMN FLOWERS.

[I.—GROUPS AND ORDERS.]

EXOGENOUS PLANTS or DICOTYLEDONS.

Leaves with netted veins. *Flowers* usually quinary—the parts in fives, or quaternary—the parts in fours. *Embryos* with two (rarely more) cotyledons; hence dicotyledonous.

Thalamiflores: Polypetalous dichlamydeous plants, with the petals distinct (separable) from the calyx, and the stamens hypogynous; Orders numbered 1 to 7.

* *Ovary apocarpous.*

1. **Ranunculaceous plants**—herbs or climbing shrubs; stamens indefinite, inserted on the receptacle.

** *Ovary syncarpous.*
   † *Placentas parietal* (i.e. *seeds attached to sides of carpels*).
   ‡ *Stamens six, united in two sets.*

2. **Fumariaceous plants**—herbs; flowers very irregular, with two sepals and four petals.

   ‡‡ *Stamens tetradynamous, distinct.*

3. **Cruciferous plants**—herbs; flowers regular, of four sepals and four petals, arranged crosswise.

   †† *Placentas axile* (i.e. *seeds attached to axis of carpels*).
   ‡ *Flowers regular.*
     § *Sepals overlapping at the edge (imbricate).*
     ‖ *Ovary one-celled; stamens distinct.*

4. **Caryophyllaceous plants**—herbs; leaves opposite, undivided,

without stipules; flowers symmetrical, with 4–5 sepals and petals, and definite stamens.

|||| *Ovary many-celled.*
   (a) *Stamens monadelphous.*

5. **Geraniaceous** plants—herbs; flowers symmetrical, with five sepals and petals; carpels fixed around a persistent central axis.

   (b) *Stamens polyadelphous.*

6. **Hypericaceous plants**—shrubs; leaves opposite, often dotted; sepals and petals five, stamens indefinite.

   §§ *Sepals parallel at the edge (valvate).*

7. **Malvaceous** plants—herbs or shrubs; sepals and petals five, surrounded by an involucre of three or more bracts; stamens united by their filaments into a column around the pistil.

Calyciflores: Polypetalous dichlamydeous plants, with the petals usually distinct, and the stamens perigynous or epigynous; Orders 8 to 12.

   * *Ovary superior, the calyx distinct from the carpels; (stamens perigynous).*

8. **Leguminous** plants—shrubs or herbs; flowers very irregular, papilionaceous; stamens ten, all or nine of them united; ovary a legume or pod.

9. **Crassulaceous** plants—herbs, with succulent leaves; flowers regular, isomerous; carpels as many as the petals free.

   ** *Ovary superior or half inferior, the calyx adhering more or less to the carpels; (stamens perigynous).*

10. **Saxifragaceous plants**—herbs; flowers regular; stamens 5–10, separate.

*** *Ovary inferior; (stamens epigynous).*

11. **Umbelliferous plants**—herbs; fruit dry, of two carpels, separating from the axis.

12. **Araliaceous plants**—shrubs; fruit succulent, of more than two carpels, not separating.

Monopetals: Dichlamydeous plants, with the petals united (from the base more or less upwards) into a single piece; Orders 13 to 18.

  * *Stamens epigynous; (ovary inferior).*
  † *Stamens attached to the corolla, separate.*

13. **Dipsacaceous plants**—herbs; flowers in compact heads or spikes; calyx surrounded by an involucre; stamens equalling in number the divisions of the corolla, the anthers free.

  †† *Stamens attached to the corolla, cohering by their anthers.*

14. **Composite plants**—herbs; florets in heads; stamens equalling in number the divisions of the corolla, the anthers united into a ring around the style.

  ** *Stamens hypogynous.*

15. **Ericaceous plants**—shrubs; stamens as many or twice as many as the corolla lobes; anthers opening by two pores.

  ** *Stamens perigynous.*
  † *Seeds attached to a free central placenta.*

16. **Primulaceous plants**—herbs; stamens attached to the tube of the corolla, isomerous, opposite its lobes.

  †† *Seeds attached to a parietal placenta.*
  ‡ *Corolla regular.*

17. **Gentianaceous plants**—herbs; stamens attached to the tube of the corolla, isomerous, alternating with its lobes.

‡‡ *Corolla more or less irregular.*

18. **Labiate plants**—herbs; corolla two-lipped; stamens didynamous; ovary four-lobed, the lobes one-seeded.

Monochlamyds: Perianth single, consisting of a calyx only.

19. **Chenopodiaceous plants**—herbs; perianth small, inferior;
carpels solitary, one-seeded.

ENDOGENOUS PLANTS or MONOCOTYLEDONS.

Leaves with parallel veins. *Flowers* usually ternary—the
parts in threes. *Embryos* with one cotyledon, hence Monocotyledonous. This group includes Orders 20 to 24.

* *Flowers perfect, with a petal-like whorled perianth.*
† *Ovary inferior.*

20. **Orchidaceous plants**—herbs; flowers six-leaved, irregular;
stamens and style united into a central column.

†† *Ovary superior.*

21. **Liliaceous plants**—herbs; flowers six-leaved, the segments
sometimes combined, regular; anthers turned inwards; style
and ovaries consolidated.

22. **Melanthaceous plants**—herbs; flowers six-leaved, regular,
long-tubed; anthers turned outwards; styles distinct;
ovaries subterraneous in the flowering stage, separable.

** *Flowers glumaceous, or formed of imbricated colourless
scales, not truly whorled.*
† *Ovary 2-3-celled.*

23. **Eriocaulaceous plants**—aquatic herbs, with spongy cellular
leaves, and unisexual flowers in small crowded scaly heads.

†† *Ovary one-celled.*

24. **Graminaceous plants**—herbs, with grassy sheathing leaves,
the leaves split on the side opposite the blade.

[II.—GENERA OR FAMILIES.]

1. Ranunculaceous Plants. RANUNCULACEÆ.

(1) **Ranunculus**—calyx of five, rarely three, sepals; petals five, the nectariferous pore naked, or covered by a scale; carpels collected into globular or elliptical heads.

2. Fumariaceous Plants. FUMARIACEÆ.

(2) **Fumaria**—calyx of two sepals; petals four, the upper one spurred at the base; fruit indehiscent, one-seeded.

3. Cruciferous Plants. CRUCIFERÆ.

* *Fruit a silique or pod.*

(3) **Diplotaxis**—calyx spreading; pod compressed; seeds oval-oblong in two rows.

** *Fruit a silicule or pouch.*

(4) **Senebiera**—pouch kidney-shaped, 2-celled, entire or notched so as to be almost two-lobed; cells one-seeded.

4. Caryophyllaceous Plants. CARYOPHYLLACEÆ.

* *Calyx with 2–4 scales or bracts at its base.*

(5) **Dianthus**—calyx tubular, with 2–4 opposite imbricated basal scales; petals five, clawed, crenate or jagged; stamens ten; stigmas two; capsule one-celled; seeds peltate.

** *Calyx without basal scales.*

(6) **Saponaria**—calyx tubular, naked at the base; petals clawed, two-cleft; stamens ten; stigmas two; capsule one-celled; seeds globose or reniform.

5. Geraniaceous Plants. GERANIACEÆ.

(7) **Geranium**—stamens ten; fruit beaked, separating into five one-seeded capsules, each with a long awn, glabrous internally, and ultimately recurved.

(8) **Erodium**—stamens five fertile; fruit beaked, separating into five one-seeded capsules, each with a long awn, bearded internally, and becoming spirally twisted.

6. Hypericaceous Plants. HYPERICACEÆ.

(9) **Hypericum**—calyx five-parted or five-sepaled; styles three; capsule more or less perfectly three-celled.

7. Malvaceous Plants. MALVACEÆ.

(10) **Malva**—calyx surrounded by an involucre of three leaves; styles numerous; carpels numerous, one-seeded, arranged in a circle.

(11) **Althæa**—calyx surrounded by an involucre having from six to nine divisions; fruit one-seeded, collected into a five-lobed head.

8. Leguminous Plants. LEGUMINOSÆ.

* *Stamens monadelphous.*

(12) **Ulex**—calyx coloured, with two bracts, two-lipped; pod oval-oblong, turgid, few-seeded.

(13) **Ononis**—calyx campanulate, five-cleft, with linear segments; standard large, streaked; keel pointed; pod turgid, sessile, few-seeded.

** *Stamens diadelphous.*

(14) **Melilotus**—calyx tubular, five-toothed; corolla deciduous; keel simple; wings shorter than the standard; pod longer than the calyx, one or few-seeded, indehiscent.

(15) **Trifolium**—calyx tubular, five-cleft ; corolla persistent ; keel shorter than both wings and standard ; pod small, indehiscent, shorter than the-calyx, 1–2 rarely 3–4-seeded.

9. Crassulaceous Plants. CRASSULACEÆ.

(16) **Sedum**—sepals five, cohering at the base, often foliaceous ; petals five, spreading ; stamens ten.

10. Saxifragaceous Plants. SAXIFRAGACEÆ.

(17) **Parnassia**—calyx deeply five-cleft ; petals five ; stamens five, accompanied by five scales, each having a tuft of globular-headed filaments.

11. Umbelliferous Plants. UMBELLIFERÆ.

* *Albumen solid.*

† *Fruit ovate or elliptical, rounded or slightly compressed dorsally.*

‡ *Vittæ single between the ribs.*

(18) **Fœniculum**—fruit nearly taper ; carpels with five prominent obtusely-keeled ridges, of which the lateral form a margin, and are rather broader than the others.

‡‡ *Vittæ two or more between the ribs.*

(19) **Silaus**—fruit nearly taper ; carpels with five sharp winged equal ridges, of which the lateral form a margin ; interstices with many vittæ.

†† *Fruit much compressed dorsally.*

(20) **Angelica**—fruit two-winged on each side ; carpels with three dorsal raised filiform ridges, and two marginal ones dilated into a wing twice as broad as the rest ; interstices with single vittæ.

(21) **Archangelica**—fruit two-winged on each side; carpels with thick dorsal ridges, and two marginal ones dilated into a wing twice as broad as the rest, without vittæ; seed with numerous vittæ.

(22) **Peucedanum**—fruit surrounded by a flat dilated margin; carpels with equidistant ridges, the three dorsal filiform, the two lateral more obsolete, contiguous to the dilated margin; interstices with single vittæ.

** *Albumen furrowed or involute at the margin.*

(23) **Physospermum**—fruit contracted at the side, double; carpels roundish or reniform-globose, with five fine slender filiform equal ridges, of which the lateral are placed within the margin; interstices with single vittæ.

12. Araliaceous Plants. ARALIACEÆ.

(24) **Hedera**—calyx-limb of five teeth; petals, stamens, and styles 5–10; berry five-celled, five-seeded, crowned with the calyx.

13. Dipsacaceous Plants. DIPSACACEÆ.

(25) **Scabiosa**—inner calyx of five bristles, outer membranous and plaited; receptacle scaly; fruit nearly cylindrical, with eight excavations.

14. Composite Plants. COMPOSITÆ.

* *Florets all tubular, or those of the disk tubular with the outer ones ligulate, forming a ray; style not swollen below its branches* (Corymbiferæ).

  † *Pappus none, or consisting of a mere border.*

  ‡ *Receptacle scaly; (heads radiant).*

(26) **Anthemis**—involucre hemispherical, the scales nearly equal,

scarious at the margin; receptacle conical, paleaceous; pappus a membrane.

‡‡ *Receptacle without scales; (heads discoid).*

(27) **Artemisia**—involucre ovate or round; florets all tubular, those of the disk hermaphrodite, five-toothed, of the ray slender, less numerous, entire, female; receptacle naked or hairy; pappus none.

(28) **Tanacetum**—involucre hemispherical; florets all tubular, those of the disk hermaphrodite, five-lobed, of the ray female, three-lobed; receptacle naked; pappus membranous, entire.

†† *Pappus pilose.*

‡ *Anthers without bristles at the base.*

(29) **Senecio**—involucre with bracteoles at the base, the scales scorched at the apex; flowers flosculous or radiant; receptacle naked; pappus soft, hairy.

(30) **Aster**—involucre imbricated, the scales linear, acute; flowers radiant, of the ray female, in a single row, oblong; receptacle naked; pappus hairy.

(31) **Chrysocoma**—involucre imbricated, hemispherical or ovate, the scales linear; florets all hermaphrodite, tubular; receptacle excavated; pappus hairy, ciliated.

‡‡ *Anthers with two bristles at their base.*

(32) **Inula**—involucre imbricated; flowers radiant or tubular; receptacle naked; pappus hairy, simple.

(33) **Pulicaria**—involucre imbricated; flowers radiant; receptacle naked; pappus double, the outer membranous.

** *Florets all tubular; style swollen below its branches* (Cynarocephaleæ).

† *Pappus pilose in many rows, equal, long, the hairs united into a ring at the base.*

(34) **Onopordum**—involucre imbricated, the scales pungent;

receptacle excavated like honey-comb; fruit compressed, four-cornered, furrowed transversely.

†† *Pappus pilose in many rows, unequal, the second row longest, equal to or shorter than the fruit.*

(35) **Centaurea**—involucre imbricated, the scales leafy, scarious, or spiny in various ways; receptacle paleaceous, the scales jagged; fruit inserted obliquely at the base.

*** *Florets all ligulate; style not swollen* (Cichoraceæ).
† *Pappus sessile, in one row.*

(36) **Crepis**—involucre double, the outer row short; receptacle naked; fruit terete, not ribbed, narrowed upwards or beaked; pappus of simple hairs, which are usually white.

(37) **Hieracium**—involucre imbricated; receptacle naked, or with a few short hairs; fruit terete, ribbed, truncate and margined above; pappus of stiff tawny hairs.

†† *Pappus sessile, in many rows.*

(38) **Sonchus**—involucre oblong, imbricated, ovate at the base; receptacle naked; pappus short, hairy; fruit compressed, striated longitudinally.

15. Ericaceous Plants. ERICACEÆ.

(39) **Arbutus**—calyx five-parted; corolla globose or ovate-campanulate, with a small contracted border; stamens ten, with flattened filaments; anthers furnished with two reflexed awns; berry globose, granular.

16. Primulaceous Plants. PRIMULACEÆ.

(40) **Cyclamen**—calyx campanulate, five-cleft; corolla with the tube ovate, the limb five-parted and reflexed; stamens inserted in the base of the tube; fruit globose, coriaceous, or rather fleshy, many-seeded.

17. Gentianaceous Plants. GENTIANACEÆ.

(41) **Gentiana**—calyx 4–5-cleft; corolla funnel- or salver-shaped, with a 4–5-cleft limb.

18. Labiate Plants. LABIATÆ.

(42) **Mentha**—calyx equal, five-toothed ; corolla with a short tube, four-cleft, nearly regular ; stamens four, equal, erect.

(43) **Lamium**—calyx bell-shaped, five-toothed ; corolla-tube enlarged at the throat, the upper lip erect, arched or concave, the lower spreading with a broad middle lobe, and two minute lateral ones ; stamens four, the two lower longest.

19. Chenopodiaceous Plants. CHENOPODIACEÆ.

(44) **Chenopodium**—perianth 3–5-parted, persistent, unaltered ; stamens five, springing from the receptacle ; stigmas 2–3.

20. Orchidaceous Plants. ORCHIDACEÆ.

(45) **Spiranthes**—sepals coloured, and with the petals converging, parallel with the lip ; lip shovel-shaped, unguiculate, with two fleshy projections at the base ; column club-shaped, with two teeth at the apex.

21. Liliaceous Plants. LILIACEÆ.

(46) **Scilla**—perianth-segments spreading deciduous ; filaments filiform, smooth, inserted into the base of the perianth.

22. Melanthaceous Plants. MELANTHACEÆ.

(47) **Colchicum**—perianth with a long tube, and a campanulate six-parted limb ; stamens inserted in the orifice of the tube ; capsules three, inflated, erect, united at the base, many-seeded.

23. Eriocaulaceous Plants. ERIOCAULACEÆ.

(48) **Eriocaulon**—barren flowers central, the perianth 4–6-cleft, its inner segments united nearly to the summit; stamens 4–6: fertile flowers in the circumference, the perigone deeply four-parted; stigmas 2–3.

24. Graminaceous Plants. GRAMINACEÆ.

(49) **Leersia**—panicle loose; spikelets one-flowered, consisting of only two keeled glumes.

(50) **Cynodon**—panicle digitate; spikelets one-sided, one-flowered, attached to a flat rachis; glumes two, keeled, nearly equal; pales two, keeled, the upper enwrapped by the lower.

(51) **Poa**—panicle loose; spikelets two- or many-flowered, ovate; glumes two, rather unequal, usually keeled; pales two, nearly equal, awnless, scarious at the top.

(52) **Molinia**—panicle contracted, elongate; spikelets two- or many-flowered, lanceolate; glumes two, acute, unequal; pales two, nearly equal, awnless.

[III.—SPECIES AND VARIETIES.]

(1) **Ranunculus.** CROWFOOT.

* *Leaves smooth.*

R. sceleratus: annual; stems 1–2 feet high; lower leaves broad, stalked, tripartite, the segments blunt crenate, upper leaves trifid, inciso-dentate; flowers very small, pale yellow. —By ditches and ponds. Fl. June to September.

** *Leaves hairy.*

R. hirsutus: annual; stems 4–18 inches high; radical leaves with three stalked, trifid, and cut leaflets, peduncles furrowed; flowers pale yellow.—Damp wastes and meadows. Fl. June to October.

(2) Fumaria. FUMITORY.

F. officinalis: annual; stem much-branched; leaves much cut; flowers rose-coloured, sepals ovate-lanceolate, acute, toothed, narrower and two-thirds shorter than the corolla; fruit globose, truncate, slightly emarginate.—A common weed. Fl. June to September.

F. capreolata: annual; stem climbing by means of the twisting petioles; leaves much cut; flowers cream-coloured, tipped with red, sepals ovate, acute, toothed, as broad as the corolla and half its length; fruit globose, emarginate.—Cultivated ground. Fl. June to September.

(3) Diplotaxis.

D. tenuifolia: stem 1–1½ foot high, shrubby below, branched, glabrous, leafy; leaves glaucous, linear-lanceolate, very acute, sinuate-dentate or pinnatifid, the segments linear, remotely dentate; flowers large, yellow; plant fetid.—Old walls and wastes. Fl. June to October.

(4) Senebiera.

S. didyma: annual; stem spreading, prostrate, a foot or more in length; leaves pinnatifid; flowers small, white, in long, lax, slender clusters; pouch notched, of two wrinkled lobes; style extremely short.—Waste ground near the sea. Fl. July to September.

(5) Dianthus. PINK.

D. prolifer: annual; stem erect, wiry, 1–1½ foot high;

2 B

flowers capitate, small, rose-pink, the bracts ovate-obtuse, membranous, overtopping the calyx.—Gravelly pastures. Fl. July to September.

(6) **Saponaria.** SOAPWORT.

S. officinalis : stem 1–2 feet high, stout, erect, leafy ; leaves elliptic-lanceolate, ribbed ; flowers fasciculate, corymbose, flesh-coloured or pale pink, large and handsome, the petals retuse, crowned.—Roadsides and hedges, mostly near villages. Fl. August, September.

(7) **Geranium.** CRANE'S-BILL.

G. Robertianum : annual or biennial ; stems ½–1 foot high, bright red ; leaves ternate, the segments divided ; flowers small, reddish-purple, with hairy sepals.—Hedge-banks. Fl. June to October.

(8) **Erodium.**

E. cicutarium : annual ; stem hairy, tufted ; leaves pinnate, the leaflets deeply pinnatifid ; flowers small, purple or pinkish, in an umbel.—Waste places. Fl. June to September.

(9) **Hypericum.**

* *Stamens in five bundles.*

H. Androsæmum : stem two feet high, shrubby, compressed, two-edged ; leaves cordate-ovate, with a strong aromatic smell when rubbed ; flowers large, in terminal cymes, yellow, the sepals unequal, subcordate-ovate, the petals oval, obtuse.—. Tutsan.—Hedges and thickets. Fl. July to September.

** *Stamens in three bundles.*

H. perforatum : stem erect, 1–2 feet high, two-edged ; leaves elliptic-oblong or linear-oblong, with pellucid dots; flowers small, yellow; sepals erect, lanceolate, acute, denticulate near the apex; petals obliquely-oblong.—St. John's-wort.— Thickets and hedges. Fl. July to September.

(10) Malva. MALLOW.

M. sylvestris: stem erect, 2–4 feet high; leaves kidney-shaped, with seven deep crenate lobes; stipules lanceolate; flowers large, purple, on axillary aggregated peduncles, the petals much longer than the hairy calyx; fruit-stalks erect, the fruit glabrous.—Waste places. Fl. June to September.

M. rotundifolia: annual; stem decumbent; leaves roundish-heartshaped, with five shallow, acutely-crenate lobes; stipules ovate-acute; flowers small, purple; fruit-stalks decurved, the fruit pubescent.—Waste places. Fl. June to September.

(11) Althea.

A. officinalis: stem 2–3 feet high, covered thoughout with soft, velvety pubescence; leaves soft on both sides, cordate or ovate, 3–5-lobed; peduncles axillary, many-flowered, shorter than the leaves; flowers pale rose.—Marsh Mallow.—Marshes near the sea. Fl. August, September.

(12) Ulex. FURZE.

U. nanus: stem procumbent, with the primary spines short, spreading, branched at their base only; flowers smaller and of a deeper golden yellow than the common Furze, the calyx nearly glabrous, with scarcely perceptible bracts.—Dry heaths. Fl. August to October.

(13) Ononis. REST-HARROW.

O. arvensis: stem procumbent, uniformly hairy, usually

without spines, rooting at the base ; leaves trifoliate, the leaflets broadly-oblong; flowers axillary, solitary, stalked, pink, the standard streaked with darker lines; pods ovate, erect, shorter than the calyx.—Barren sandy places. Fl. June to September.

O. antiquorum : stem erect or ascending, bifariously hairy, usually spinous ; leaves trifoliate, the leaflets oblong ; flowers axillary, solitary, stalked, pink ; pods ovate, erect, longer than the calyx.—Barren places and pastures. Fl. June to September.

(14) **Melilotus.** MELILOT.

M. officinalis : annual or biennial; stem erect, 2–3 feet high; leaflets serrate, truncate, narrowly ovate; stipules setaceous, entire; flowers in lax lateral racemes, yellow; corolla twice as long as the calyx, the wings keel and standard equal; pods ovate-acute, hairy.—Waste places. Fl. June to September.

M. alba : stem erect, 3–4 feet high ; leaflets obovate, the upper ones oblong, serrate, obtuse ; stipules awl-shaped, entire ; flowers white, in lax racemes; corolla twice as long as the calyx, the wings and keel equal, shorter than the standard ; pods ovate-obtuse, glabrous.—Sandy and gravelly places. Fl. July to September.

(15) **Trifolium.** CLOVER.

* *Flowers stalked, white.*

T. repens : stems creeping; leaflets obovate or obcordate, often with a dark spot at their base ; flowers in roundish heads, white, the standard striated ; peduncles axillary, longer than the leaves, the flowers at length deflexed; calyx glabrous.— Dutch or White Clover.—Meadows and pastures. Fl. May to September.

** *Flowers sessile.*

† *Calyx not inflated.*

T. pratense: stem erect, 1–2 feet high; leaflets oval, emarginate, the upper ones entire, apiculate; flowers purplish, sometimes white, the heads ovate, dense, sessile; calyx 10-nerved, hairy.—Purple Clover.—Mountainous pastures, fields.   Fl. May to September.

T. arvense: annual; stem erect, or in a maritime form, procumbent, finely hairy; leaflets linear-oblong; flowers small, whitish, almost concealed by the very hairy calyx, the heads nearly cylindrical, stalked; calyx 10-nerved, hairy.—Hare's-foot Trefoil.—Sandy fields.   Fl. July to September.

†† *Calyx with the upper lip inflated after flowering.*

T. fragiferum: stem creeping; leaflets obovate, emarginate, minutely serrate; flowers purplish-red, the heads globose, large, remarkable when in fruit for their curious calyces; peduncles axillary longer than the leaves; calyx of the fruit membranous reticulated downy.—Damp pastures.   Fl. July to September.

(16) Sedum.

S. Telephium: stem erect, 1–2 feet high; leaves large and broad, oval-oblong, dentate, smooth, rounded at the base, and sessile; flowers purple, handsome, in dense corymbs.—Orpine, or Live-long.—Hedge-banks on a gravelly soil.   Fl. July to September.

(17) Parnassia.

P. palustris: leaves mostly radical, cordate, stalked; the stem-leaves amplexicaul; petals white, veined, with a short claw; filaments of the petaloid scales 9–13; glands of the scales yellow.—Wet, boggy places.   Fl. August to October.

(18) Fœniculum. FENNEL.

F. vulgare : stem terete below, 3-4 feet high, completely filled with pith, branching; leaves decomposite, the segments elongate, capillary; umbels of many rays, concave, large; flowers yellow; whole herb aromatic.—Rocky banks near the sea. Fl. July to September.

(19) Silaus.

S. pratensis : stem angular, 1-2 feet high; leaves mostly radical, those of the stem decreasing upwards, 3-4 times pinnate, the leaflets lanceolate, entire or bifid, the terminal tripartite; flowers pale yellow, with an involucre of 1-2 leaves. —Damp meadows and pastures. Fl. June to September.

(20) Angelica.

A. sylvestris : stem 2-3 feet high, slightly downy above, purplish; leaflets equal, ovate-lanceolate or ovate, often subcordate at the base, inciso-serrate not decurrent, lateral ones rather unequal at the base; flowers pinkish-white, with an involucre of about three leaves.—Wet places. Fl. July to September.

(21) Archangelica.

A. officinalis : stem 3-5 feet high, the foliage, stalks, and flowers bright green; leaves 2-3 feet wide, the leaflets ovate-lanceolate, all sessile, partly decurrent, the terminal one trifid; petioles much dilated at the base.—Watery places. Fl. July to September.

(22) Peucedanum.

P. officinale : stem terete, striated, 2-3 feet high; leaves five times tripartite, the leaflets linear-acute, flaccid, very long, narrow; flowers yellow, in large umbels of twenty or more rays; general involucre three-leaved, deciduous; pedicels much longer than the fruit.—Hog's Fennel or Sulphur-weed. —Salt marshes, rare. Fl. July to September.

(23) Physospermum.

P. cornubiense: stem 1–3 feet high, erect, round, striated, minutely scabrous, bearing a few small ternate leaves, with linear-lanceolate, entire segments, the uppermost represented by a barren lanceolate-acute sheath; radical leaves triternate, the leaflets wedge-shaped, cut, or deeply three-lobed, with acute segments; flowers white, in terminal umbels.—Cornwall and Devon. Fl. August, September.

(24) Hedera. Ivy.

H. Helix: shrubby, climbing by means of rootlike fibres; leaves coriaceous, ovate or cordate, and five-lobed with angular lobes, those of the flowering branches ovate-oblong, acute, entire; umbels simple, erect, collected in panicles.— Rocks, old walls, and hedges. Fl. October, November.

(25) Scabiosa.

S. succisa: stems 1–2 feet high; leaves mostly from the root, stalked, ovate-oblong, entire, those of the stem with a few teeth; flowers deep blue, in roundish heads.—Devil's Bit. —Pastures and heaths. Fl. August to October.

(26) Anthemis.

* *Scales of the receptacle awl-shaped, acute.*

A. Cotula: annual; stem 1–2 feet high, branched, angular, furrowed; leaves bipinnatifid nearly glabrous, the lobes linear, acute, mostly entire; heads solitary, on long terminal stalks, the disk florets yellow, those of the ray white; involucral scales obtuse, with white membranous margins; fruit terete, tubercular-striated; whole plant fetid and acrid.—Fields and waste places. Fl. July to September.

** *Scales of the receptacle thin, membranous, obtuse.*

A. nobilis: stem procumbent, one foot long, much branched; leaves bipinnate, the leaflets linear-subulate, slightly downy and fleshy, acute; heads solitary, terminal, the florets of the disk yellow, of the ray white; fruit subtrigonous, smooth; whole plant pleasantly aromatic. — Chamomile. — Gravelly places. Fl. July to September.

(27) **Artemisia.**

* *Receptacle hairy.*

A. Absinthium: stem bushy, 1–2 feet high; leaves silky, in many deep lanceolate-obtuse segments; heads drooping, hemispherical, in erect aggregate leafy panicles; floral-leaves simple; florets dull yellow, the outer row female.—Wormwood.—Waste ground. Fl. July to September.

** *Receptacle naked.*

A. vulgaris: stem 2–3 feet high, erect, leafy; leaves woolly and white beneath, pinnatifid, with lanceolate-acuminate cut and serrated segments; heads ovate, the clusters leafy, nearly simple, erect; flowers few, reddish or brownish yellow.—Mugwort.—Waste ground. Fl. July to September.

A. maritima: stem decumbent or ascending, woolly, much branched; leaves downy, pinnatifid, with linear, obtuse segments; racemes unilateral, the heads drooping, oblong; florets few, all perfect; reddish-yellow.—Salt marshes. Fl. August, September.

(28) Tanacetum. TANSY.

T. vulgare: stem 2–3 feet high; leaves bipinnatifid, the leaflets serrated; heads in a terminal corymb; florets golden yellow; whole plant powerfully scented.—Way-sides. Fl. August, September.

(29) Senecio.

* *Florets all tubular, no ray.*

S. vulgaris: annual; stem 6–12 inches high, branching; leaves half-clasping, pinnatifid, the segments distant, oblong, obtuse, and together with the rachis and auricles acutely and unequally toothed, the lower ones narrowed into a stalk; heads small, in clustered racemes; florets yellow.—Groundsel. —Common everywhere. Flowers the whole year.

** *Florets of the ray small, and rolled back.*

S. sylvaticus: annual; stem 1–2 feet high, erect, more or less branched, hairy; leaves deeply pinnatifid, downy, the segments oblong, unequally toothed; heads corymbose; florets yellow, those of the ray sometimes wanting.—Dry and gravelly hills. Fl. July to September.

*** *Florets of the ray spreading, conspicuous.*

S. Jacobæa: stem 2–3 feet high, smooth, striated, branched, leafy; leaves glabrous, the radical ones oblong-obovate, attenuated below, lyrate-pinnatifid, stalked, those of the stem sessile, bipinnatifid, with spreading, oblong, deeply and irregularly-toothed and cut segments, the lowermost of them much divided, clasping; florets yellow, the heads collected in erectly-branched corymbs.—Ragwort.—Waste ground. Fl. July to September.

(30) Aster.

A. Tripolium: stem 1–2 feet high, erect, hollow, leafy, many-flowered, smooth corymbose; leaves linear-lanceolate, fleshy, smooth; flower-heads large, with a yellow disk, and bright blue rays, the rays often wanting.—Muddy salt marshes. Fl. August, September.

(31) Chrysocoma.

C. Linosyris: stem smooth, ½–1 foot high; leaves numerous, narrow-linear, entire; flower-heads yellow, in compact terminal corymbs. — Rocky cliffs, on the south and west coasts. Fl. August, September.

(32) Inula.

I. Helenium: stem 3–4 feet high, round, furrowed, solid, leafy, branched above; leaves unequally dentate, downy beneath, cordate-ovate, acute, clasping, the root-leaves stalked elliptic-oblong; heads few together or solitary, terminal, very large, the florets bright yellow.—Elecampane.—Moist pastures. Fl. July to September.

I. Conyza: stem 1–2 feet high, leafy; leaves ovate-lanceolate, downy, denticulate, the lower ones narrowed into a footstalk; heads corymbose, the florets yellow, those of the circumference between tubular and ligulate, with a short ligule on the inner side.—Plowman's Spikenard.—Calcareous soils. Fl. July to September.

(33) Pulicaria.

P. vulgaris: annual; stem 6–12 inches high, leafy, much branched, downy; leaves lanceolate, wavy, narrow at the

base and somewhat clasping; heads small, lateral and terminal, hemispherical, with very short rays; florets yellow.—
Moist sandy heaths. Fl. August, September.

(34) Onopordum.

O. Acanthium: biennial; stem erect, many-headed, 4-5
feet high, woolly, with broad spinous wings, branched; leaves
elliptic-oblong, woolly on both sides, sinuate, spiny, decurrent; involucre nearly globose, large, somewhat cottony, the
scales fringed with minute spinous teeth; florets purple.—
Cotton Thistle.—Waste ground. Fl. August, September.

(35) Centaurea.

* *Involucral scales with a scarious, jagged, or pectinated, not
decurrent appendage.*

C. nigra: stem 1-2 feet high; leaves scabrous, lanceolate,
sinuate-dentate, the upper entire or nearly so, stem-clasping;
heads usually not radiant; florets all fertile, purple or rarely
white; involucre globular, nearly black externally, the appendages of the outermost scales smaller and narrower than the
others, those of the innermost row roundish, dark brown,
membranous, jagged, all contracted just below the appendage.
Knapweed.—Pastures and roadsides. Fl. June to September.

** *Involucral scales lanceolate, their upper half with a sca-
rious fringed decurrent margin.*

C. Scabiosa: stem 2-3 feet high, rough, furrowed; leaves
hispid, pinnatifid, roughish, the segments lobed, with callous
points, the lobes of the upper ones entire; heads on long,
naked stalks, solitary; involucres rather woolly, the scales
pale, blunt, with dark acute membranous pectinated decurrent appendages; florets purple, the outer row radiant.—

Great Knapweed.—Fields and hedges. Fl. July to September.

*** *Involucral scales horny at the end, with palmate spines.*

C. solstitialis : annual ; stem 1–2 feet high, branched, spreading, winged with the decurrent bases of the linear-lanceolate entire hoary leaves, the root-leaves lyrate ; heads terminal, solitary ; involucral scales woolly, palmately-spinous, the central spine of the intermediate scales very long, needle-shaped ; florets yellow.—Yellow Star-thistle.—Cornfields and waste places near the sea. Fl. July to September.

(36) Crepis.

C. virens : annual ; stem 1–3 feet high, subcorymbose ; leaves very variable, lanceolate, remotely-dentate, runcinate or pinnatifid, the uppermost linear-arrowshaped, clasping, with flat margins ; outer involucral scales adpressed-linear, inner ones glabrous within ; florets yellow.—Common in dry pastures and wastes. Fl. June to September.

C. paludosa : stem two feet high, leafy, simple, angular, subcorymbose ; leaves large, ovate-oblong, taper-pointed, runcinate-dentate, narrowed into a footstalk, glabrous, the upper ones ovate-lanceolate, cordate, clasping, acute, entire or dentate ; involucral scales lanceolate, much attenuated, glandular-pilose, the outer ones short ; florets yellow.—Damp woods and shady places. Fl. July to September.

(37) Hieracium. HAWK-WEED.

H. vulgatum : stem erect, leafy, 1–3 feet high, with several leaves, the radical ones few, persistent, all oblong-lanceolate, narrowed into a footstalk, usually coarsely-toothed, the teeth all pointing upwards ; heads panicled ; involucre pubescent ;

florets yellow.—A very variable plant.—Woods, banks, and walls. Fl. July to September.

H. boreale : stem erect, scabrous, 1–2 feet, hispid below, leafy, slightly-branched and corymbose above; leaves ovate-lanceolate or lanceolate, sinuate-dentate or nearly entire, the lower ones narrowed into a dilated petiole, the upper sessile with a rounded base, the uppermost small; radical leaves evanescent; corymb irregular; involucral scales with a slightly hispid keel; florets yellow.—Thickets. Fl. August, September.

(38) Sonchus. Sow-thistle.

S. arvensis : stem 3–4 feet high, simple, leafy; leaves long, acute, lanceolate, runcinate, finely-toothed, cordate at the base, the uppermost entire; heads corymbose, large; florets yellow; involucre and peduncles glandular-hairy; root creeping.—Cornfields. Fl. August, September.

S. palustris : stem 4–6 feet high, simple, leafy; leaves linear-lanceolate, the lower runcinate, the upper simple, all arrow-shaped, spinosely-ciliated, with acute auricles; heads corymbose; florets lemon-coloured; involucre and peduncles glandular-hairy; root without scions.—Marshes, rare. Fl. July to September.

(39) Arbutus.

A. Unedo : an evergreen tree, with rough bark; leaves elliptic-lanceolate, serrated, coriaceous, glabrous; panicle terminal, nodding, pedicels glabrous, the flowers whitish, pendulous; fruit globular, granulated, red, resembling a strawberry, but dry and flavourless.—Strawberry Tree.—Lakes of Killarney. Fl. September, October.

(40) Cyclamen. SOW-BREAD.

C. europæum : tuberous; leaves long-stalked, heart-shaped, with shallow angular lobes, blotched with white above, purplish beneath; flowers pale rose-coloured or whitish, with two-horned spots at the mouth.—Woods in Sussex and Kent, rare. Fl. August, September.

(41) Gentiana. GENTIAN.

*   * Throat of corolla fringed with long hairs.*

G. Amarella : annual, very variable in size and in the number of the flowers; stem 3–12 inches high, erect, square, much-branched; leaves sessile, ovate-lanceolate, the radical ones obovate; flowers pale purple, the corolla salver-shaped, five-cleft; calyx-segments five, nearly equal, linear-lanceolate. —Dry limestone pastures. Fl. August, September.

G. campestris : annual; stem 3–10 inches high; leaves elliptic-oblong; flowers blue, salver-shaped, four-cleft, the tube of the corolla slightly thicker upwards; calyx-segments four, the two outer ones very large, ovate.—Dry limestone pastures. Fl. August, September.

*   ** Throat of corolla naked, not fringed.*

G. Pneumonanthe : stem 4–10 inches high, leafy, simple, erect or ascending; leaves linear, obtuse; flowers very large, mostly solitary, slightly stalked, bell-shaped, five-cleft, deep blue, with five greenish bands down the middle of each segment.—Calathian Violet.—Moist turfy heaths. Fl. August, September.

(42) Mentha. MINT.

M. Pulegium : stem prostrate; leaves stalked, elliptical, obtuse, slightly crenate, all similar; flowers in distant, globose, many-flowered semi-whorls, small, lilac. A small, creeping, powerfully-scented medicinal herb. — Penny-royal. — Wet places. Fl. August, September.

⸱ (43) **Lamium.**

L. amplexicaule: annual; stem dwarf, decumbent; leaves roundish-cordate, obtuse, inciso-crenate, the lower ones stalked, the upper sessile, clasping; flowers in 1–3 compact semi-whorls, the corolla purple, its tube slender, much longer than the softly hairy calyx; lateral lobes of the lower lip of the corolla toothless.—Henbit Nettle.—Waste and cultivated ground. Fl. May to September.

L. purpureum: annual, with low spreading stems; leaves cordate-obtuse, crenate-serrate, stalked, the upper ones crowded, shortly-stalked, ovate; flowers pale purple, the lip spotted with red, the lateral lobes of the lower lip of the corolla with two teeth.—Red Dead-nettle.—Waste and cultivated ground. Fl. May to September.

(44) **Chenopodium.** GOOSEFOOT.

C. polyspermum: annual; stem erect or procumbent; leaves ovate-elliptical, sessile, acute or obtuse; flowers in axillary leafless cymose or spicate racemes, small, inconspicuous.—Damp waste places. Fl. August, September.

(45) **Spiranthes.** LADY'S TRESSES.

S. autumnalis: tuberous; stem 4–6 inches high; radical leaves ovate-oblong, those of the stem bract-like; spike spiral, dense; flowers greenish-white, the column and its lid acute with an obtuse-ovate membranous process between them on each side.—Dry chalky and gravelly places. Fl. August, September.

(46) Scilla. SQUILL.

S. autumnalis: bulbous; leaves linear, numerous; flowers purplish-blue, with a green line down the back, in perfection before the leaves appear, forming a slightly corymbose raceme on a scape 4–6 inches high; peduncles ascending, without bracts.—Dry pastures in the south and west. Fl. August or September.

(47) Colchicum.

C. autumnale: tuberous; leaves flat, lanceolate, erect, a foot long, and often an inch broad, dark green, smooth; flowers several, bright purple, rising from the root with very long tubes, the germen at first remaining under ground, appearing in the spring with the leaves.—Meadow Saffron.— Meadows. Fl. September, October.

(48) Eriocaulon.

E. septangulare: aquatic: stem varying in height according to the depth of the water, usually with six or eight, rarely seven or ten angles, each corresponding with a bundle of vessels surrounding a central bundle; scapes striated, longer than the cellular compressed subulate glabrous leaves; flowers four-cleft, hairy at the extremities, each flower with a broad, blunt, black scale in front, shorter and broader than the flower.—Pipewort. —Peaty lakes and pools.

(49) Leersia.

L. oryzoides: stem 1–2 feet high, the leaves and sheaths very rough; panicle loose, with wavy branches, usually more or less included in the leaf-sheaths; spikelets numerous, all turning in one direction, half-oval, the keel ciliate.—Marshy ditches in the south. Fl. August, September.

(50) Cynodon.

C. Dactylon: roots creeping; stem with long branched scions, flowering stem 4–6 inches high, terminating in a single cluster of 3–5 digitate, spreading, many-flowered slender spikes; spikelets purplish; leaves on the barren shoots flat, spreading.—Sandy shores of Devon and Cornwall. Fl. August, September.

(51) Poa.

P. annua: annual; stem ascending or procumbent; leaves flaccid, often wavy, broad; panicle spreading erect with a triangular outline; the branches patent or divaricated; spikelets ovate-oblong.—Very common. Fl. March to November.

(52) Molinia.

M. cœrulea: stem 1–2 feet high, the upper part naked; leaves long, linear, attenuated; panicle erect, elongate, narrow, the spikelets 1–3-flowered. The var. *depauperata* has the few spikelets one-flowered.—Wet heaths. Fl. July to September.

WINTER FLOWERS AND FRUITS.

———◆———

"The seasons came and went, and went and came,
To teach men gratitude; and, as they passed,
Gave warning of the lapse of time, that else
Had stolen unheeded by. The gentle flowers
Retired, and stooping o'er the wilderness,
Talked of humility and peace and love."

Pollok.

ILLUSTRATIONS.

OF real genuine winter flowers we have none, and in their
stead we introduce the Holly and the Mistletoe, the emblems
of the closing year. True that here and there stray flowers
of various forms may meet the eye, but they are the chance
productions of favoured spots or of favoured seasons, some
remnants of the past, some mayhap, like the Primrose, harbingers of the coming year, and not one of them to be de-
pended on. We must be content with the coloured berries of
the Holly and the Mistletoe.

The Holly* belongs to the regular-flowered Monopetals,
and affords an illustration of the Aquifoliaceous family. It
is a small bushy evergreen tree or large shrub, of erect habit,
furnished with abundant leaves, which are of a shining green,
thick, stalked, ovate, and much waved at the margin, where

* *Ilex Aquifolium*—Plate 24 C.

they are provided with strong coarse very prickly teeth. The leaves of the Holly however are not always prickly, especially on the upper part of the tree, which fact has led to the poetic fancy of Southey, who writes—

> " Below, a circling fence, its leaves are seen
>   Wrinkled and keen ;
> No grazing cattle through their prickly round
>   Can reach to wound ;
> But as they grow where nothing is to fear,
> Smooth and unarm'd the pointless leaves appear."

The flowers grow in dense clusters in the axils of the leaves ; they are white, with a calyx of four, or rarely of five small teeth ; a regular monopetalous corolla, with a short tube, and a limb deeply divided into four or five segments ; as many stamens as there are lobes of calyx or corolla ; and a four-celled ovary crowned by four minute sessile stigmas. The ovary becomes a bright red (sometimes yellow) berry, or rather a small drupe enclosing four nuts, containing each a single seed. These coral-coloured berries of the Holly give the plant a brilliant appearance at a time when there is little of brilliancy left to us in the vegetable world :

> " Summer trees are pretty—very,
>   And I love them all ;
> But this Holly's glistening berry
>   None of them excel.
> While the fir can warm the landscape,
>   And the ivy clothes the wall,
> There are sunny days in winter
>   After all !"

The Mistletoe* is one of the Calyciflores, and a member of the Loranthaceous family. It is a singular plant, a real parasite, growing on, and deriving its nourishment from a

* *Viscum album*—Plate 24 D.

great variety of trees, and occurring very commonly in the
south and west of England. The stems, which become woody
when old, consist of repeatedly-forked succulent branches,
which grow into dense tufts two or three feet in diameter, of
a yellowish-green colour, attached by a thickened base to the
branches of the foster trees. The leaves are narrow oblong
or ovate, lanceolate, obtuse, one pair growing at the end of
each branchlet. The flowers are sessile in the axils of some of
these pairs of leaves, the male and female ones produced on separate plants. The males, which grow three or five together
in a somewhat cup-shaped fleshy bract, are without evident
calyx, but have a corolla of four ovate fleshy petals, united at
the base, the anthers being sessile in the centre of the petals,
and opening by several pores. The females are solitary, or
rarely two or three together, in a cup-shaped bract, with an
obscure entire superior calycine margin, four minute somewhat triangular petals, and a stigma sessile on the ovary.
The fruit is a white semi-transparent berry, enclosing a single
seed, which is surrounded by a mass of very glutinous pulp.

The Mistletoe is ever associated in our ideas with the festival of Christmas, and with the well-known custom of decorating churches and houses with evergreens at that season—
a custom which has been in existence ever since Christianity
has been planted amongst us, and appears to have been
derived from a similar practice of the Pagans. The plant is
sometimes—but now, at least, very rarely—found upon the
Oak. The Druids held it in high veneration when it was
seen growing on that tree; and at certain seasons, especially,
it is said, at Yuletide or Christmas, they were accustomed to
gather it with great solemnity. "When the end of the year
approached, they marched with great solemnity to gather the
Mistletoe, in order to present it to God, inviting all the

world to assist at the ceremony, in these words—'The new year is at hand, gather the Mistletoe!' Their sacrifices being ready, the priest ascended the Oak, and with a golden hook cut the Mistletoe, which was received in a white garment, spread for the purpose. Two white bulls that had never been yoked were then brought and offered to the Deity, with prayers that He would prosper those to whom He had given so precious a boon." The new year's day of the Druids did not however exactly correspond with ours, for according to Toland it was the 10th of March; and this idolatrous veneration was confined to the Mistletoe of the Oak, perhaps on account of its rarity. Now, however,

> "Past is the time when, bending low,
> Druids revered thee, Mistletoe!
> Error's broad shades are chased away
> By Revelation's brilliant ray,
> And superstition can no more
> Bid us an humble plant adore.
> Yet who, in hour of Christmas mirth,
> Can place thee o'er the social hearth,
> With ivy and with holly gay,
> Or twine thee with the fragrant bay,
> Nor lift with joy his heart above,
> Nor hymn the notes of praise and love?"

The Mistletoe has in later times been rather associated with mirth and glee, than with religious superstition—" Forth to the woods did merry men go, to gather in the Mistletoe;" and then, they "opened wide the baron's hall to vassal, tenant, serf, and all."

The popular regard in which it came to be held appears —so says Mr. Lees—"to have arisen from a superstition extending back as far as Druidical times, when the young bride wore a branch of Mistletoe suspended from her neck, which was supposed (as it was considered a remedy against barrenness) to ensure an offspring as numerous as the spotless berries produced by the plant itself. So that formerly it seems to have been the exact converse of the dreaded Willow; for while those that had lost their loves were conducted to that hopeless barren tree, or, at least, recommended to sojourn beneath its shade, those damsels who were not in such an unfortunate predicament were either merrily or strategetically escorted to the Mistletoe, whose berries being pure white, of course could not fail to intimate the bridal wreath and white satin ribbon. Archdeacon Nares, who has written very learnedly on this subject, and seems to be a great friend to the mystic rites of the Mistletoe, deprecates any unreasonable resistance on the part of ladies taken to or caught under the sacred plant; as he states that a non-performance of the usual ceremonial brings in its train all the evils of old-maidenism "—whatever they may be. It appears that in the berries of the plant alone resided its privilege; one of these was to be plucked at every salute, and various authorities insist that when the last berry was plucked from the bush its potential and venerated character ceased.

The habit of the Mistletoe is very peculiar. Unlike the Ivy, which like it grows upon trees, but derives support only, and not nourishment, from that to which it has attached itself, the Mistletoe is a true parasite, sucking the vitals of its prop. It is doubtless planted by those members of the feathered tribe, the mistle-thrush especially, which feed upon its berries, the viscid juice causing the seeds to adhere to the under side of the branches when brought in contact with them by the birds in the process of cleaning their bills. In this position the seeds germinate, and the young plants insinuate their root-point into the sap-wood of the tree, forming there, instead of roots, in each case, an ever-active sucker, by means of which they appropriate to themselves the juices of the tree on which they grow.

It is a rather singular coincidence that the two plants so intimately associated in our Christmas festivities should be also associated by their products. Thus the viscid substance called birdlime is obtained, by maceration and trituration, not only from the bark of the young shoots of the Holly, but also from the glutinous berries of the Mistletoe.

Though wild flowers are wanting in the cheerless winter, and we have been content to record the Holly and Mistletoe as illustrations of the hiemal flora (albeit at that season it is their fruits and not their flowers which are their attractive features), the field botanist is not to conclude that the winter is to him a period barren of interest, for at that season the Cryptogamic tribes abound in full perfection. There are the Fungi, with their various forms and brilliant colours, as Mr. Berkeley's ' Outlines of British Fungology ' bears evidence, —its glowing pictures and its stores of information being enough to tempt every reader to become a fungologist. There are the Lichens, those time-stains of grey and yellow, which give a venerable air to wall and tower, or paint the rock with tints of orange or sienna. And then there are the Mosses, green and beautiful plant miniatures, abounding everywhere, on moors and rocks, in woods and fields, on walls, by streams, in bogs, wherever the soil or atmosphere is moist, so that no mean portion of the earth's clothing is furnished by them. These, and more than these, of the cryptogamic stores, which Nature spreads before her votaries even in the bleak and barren winter, furnish material for a life-time study, and are full of interest for those who deign to study them. Our

task, however, is accomplished, and we can only point out
these as fresh fields and pastures new, to those who can brace
their nerves for a pleasant winter's ramble.

> " These as they change, Almighty Father, these
> Are but the varied God.   The rolling year
> Is full of thee.   Forth in the pleasant Spring
> Thy beauty walks, thy tenderness and love
> Wide flush the fields ; the softening air is balm ;
> Echo the mountains round ; the forest smiles ;
> And every sense and every heart is joy.
> Then comes thy glory in the Summer months,
> With light and heat refulgent.   Then thy sun
> Shoots full perfection through the swelling year ;
> And oft thy voice in dreadful thunder speaks,
> And oft at dawn, deep noon, or falling eve,
> By brooks and groves in hollow-whispering gales.
> Thy bounty shines in Autumn unconfined,
> And spreads a common feast for all that lives.
> In Winter awful thou !   With clouds and storms
> Around thee thrown, tempest o'er tempest rolled,
> **Majestic darkness !**   On the whirlwind's wing
> Riding sublime, thou bid'st the world adore,
> And humblest nature with thy northern blast.
>
> So roll your incense, herbs, and fruits, and flowers,
> In mingled clouds to Him, whose sun exalts,
> Whose breath perfumes you, and whose pencil paints."
>   *Thomson.*

GLOSSARY.

Abortive, defective, barren.

Abrupt, appearing as if suddenly terminated or broken off.

Accrescent, persistent and increasing in size.

Accrete, having contiguous parts or organs naturally grafted together.

Acerose, linear and sharp-pointed, as in the leaves of the fir-tribe.

Achene, a dry, hard, single-seeded indehiscent fruit, with the pericarp inferior, and consequently invested by the calyx, as in the seeds of compound flowers.

Achlamydeous, without any distinct perianth, as in the willows.

Acicular, of slender form, like a needle.

Acotyledonous, wanting cotyledons.

Aculeate, sharply pointed or prickly.

Acuminate, ending in a long, taper point.

Acute, when the extremities present an angle less than a right angle.

Adherent, having parts originally or normally distinct, united or grafted together.

Adnate, attached throughout the long length, as in the case of anthers when their lobes are attached throughout their whole length to the filament, or of stipules when they adhere to the peduncles, etc.

Adpressed, having one part lying close to another throughout its length, as hairs to the surface of a leaf.

Aggregated, having similar but distinct parts crowded together.

Albumen, a substance of a farinaceous, oily, or horny consistency, found in many seeds, surrounding the embryo wholly or in part.

Alternate, having the parts or organs so placed, that the one is not directly before or over-against the other.

Amplexicaul, having the peduncle, leaf, or stipule dilated at the base, and extended partially round the stem, so as to clasp it.

Angular, having a determinate number of angles.

Annual, applied to those plants which produce seed and die in the same year in which they germinate.

Anther, that portion of the stamen which contains the pollen, most

frequently formed of two distinct
cells, and generally attached towards the summit of a filament.

Antheriferous, bearing or supporting
the anthers.

Apetalous, applied to flowers which
are destitute of true corolla.

Apex, the opposite extremity of any
organ to that by which it is attached, which is considered its
base.

Apiculate, terminating in a sharp
but short point.

Apocarpous, having the carpels quite
free from adhesion, as in the
Buttercup.

Appendage, a part superadded to
another, as the leaves to the stem.

Appressed, closely applied to some
other part throughout its whole
length, as the pubescence on some
leaves and branches.

Aquatic, living or growing in water.

Aril, *Arillus*, an expansion of the
placenta, rising around certain
seeds in the form of an integu-
ment, generally more or less fleshy,
as in the genus Euonymus.

Aristate, awned.

Articulation, a point where a discon-
tinuity of tissue naturally takes
place, without the appearance of
its having been torn asunder, as
where leaves fall from the stem.

Articulated, furnished with articulations.

Artificial, a term applied to some
characteristic by which a plant
may be distinguished or separated from others, without reference to those other circumstances
by which its affinities are established.

Ascending, starting horizontally or
rising obliquely from the base,
curving upwards, and ultimately
attaining a vertical position, as in
many stems.

Attenuated, gradually diminished in
breadth towards either extremity.

Auricle, a prolonged appendage at
the base of some leaves.

Auriculate, having auricles or ear-
like appendages.

Awn, a stiff bristle-like appendage,
such as may be met with on the
glumes and pales of grasses.

Awned, furnished with an awn.

Axil, the upper angle formed by the
attachment of a leaf or branch to
its support.

Axile, of or belonging to the axis.

Axillary, situated in an axil.

Axis, an imaginary line forming a
centre round which an organ is
developed.

Baccate, bearing berries.

Bark, the external coating of the
stems and roots of flowering
plants.

Basal, attached to another part at
that extremity by which it is
joined to its support.

Beak, a long slender termination.

Beaked, terminated by a beak, as
the pod of the Radish.

Bearded, bearing tufts of hair-like
pubescence.

Bell-shaped, having a tubular and
inflated form, resembling a bell,
as the corolla of many Campa-
nulas.

Berry, a succulent seed-vessel, with
the seeds embedded in pulp.

Biennial, applied to those plants
which produce only leaves during

the first year of their growth, and in the second bear seed, and then die.

Bifid, divided about halfway to the base into two parts.

Bilobed, divided into two lobes, as the anthers of most flowers.

Bipartite, deeply divided into two parts, the incision extending beyond the middle.

Bipinnate, having the leaflets on the secondary petioles of a doubly compound leaf arranged in a pinnate manner, the secondary petioles themselves being similarly disposed.

Bipinnatifid, having the divisions in a pinnatifid leaf themselves divided in a pinnatifid manner.

Bisaccate, having two bags or sacs.

Biternate, having the leaflets of a doubly compound leaf arranged in a ternate manner.

Bloom, a whitish, waxy secretion produced on the surface of some fruits, as in the plum.

Blunt, terminating in a rounded manner, without tapering.

Bony, hard and brittle, and of close texture, as the stones of plums, etc.

Bracts, the leaves more or less modified in form, which are seated on the peduncles and pedicels: frequently reduced to mere scales; sometimes highly coloured.

Bracteate, having bracts.

Bracteoles, small bracts seated on the flower-stalk.

Bristle, any short or stiff hair.

Bristle-pointed, terminating gradually in a fine point.

Bush, a low shrub, densely branched from the surface of the ground.

Cæspitose, densely crowded in turf-like patches.

Calyciflores, a group of dicotyledons, in which the stamens adhere to the calyx, whether they are perigynous or epigynous.

Calycine, of the nature or appearance of a calyx.

Calyx, the outermost whorl of the perianth, composed of the sepals, either free or cohering; in those flowers which have only a single floral whorl, it is generally considered a calyx rather than a corolla.

Campanulate, bell-shaped.

Capillary, hair-like, both as to size and form.

Capitate, applied to those slender organs, such as the style, in which the summit is swollen out, somewhat like the head of a pin.

Capitule, an inflorescence consisting of numerous flowers, sessile or nearly so, collected into a dense mass at the summit of a peduncle.

Capsule, a dry dehiscent seed-vessel, with one or more cells, and many seeds.

Carpel, one of the parts, whether combined or distinct, which compose the innermost of the four sets of floral whorls, into which a complete flower is separable.

Cartilaginous, tough and hard, like the skin of an apple pip.

Catkin, a form of spiked inflorescence, in which the flowers are unisexual and closely crowded, and the place of each perianth is supplied by a bract.

Cell, one of the hollow divisions of the seed-vessel, or anther.

Chaffy, furnished with chaff.

Channeled, hollowed out and gutter-like.

Ciliate, furnished with cilia, or stiffish hairs, so as to form a fringe on the margin of an organ.

Clavate, slender at the base and gradually thickening towards the apex, so as to become club-shaped.

Claw, the narrowed base of a petal.

Clustered, similar parts collected in a close, compact manner.

Cocci, the closed cells of a fruit which separate from each other when ripe, as in the Spurge.

Column, the solid body formed by the union of the filaments and style, as in Orchids.

Commissure, the face by which the two parts of the fruit of the Umbellifers cohere.

Compact, closely agglomerated or pressed together.

Complete, where no essential part is wanting, as where a flower is furnished with both stamens and pistils.

Cone, a dense aggregation of scale-like carpels, **arranged round an axis, as in the fir tribe.**

Compressed, flattened lengthwise, as in the pod of a pea, or in some stems.

Conical, approaching the form of a cone.

Connate, when the bases of two opposite leaves are united round the stem, so that this appears to pass through them.

Connective, a portion of the stamen, distinct from the filament, which connects the cells of the anthers.

Connivent, converging.

Continuous, where there is no break in the arrangement of parts; the opposite of interrupted.

Contorted, when an organ is folded or twisted back upon itself.

Convolute, rolled up in a longitudinal direction, so that one edge overlaps the other, as the spathe of an Arum.

Cordate, resembling the heart in a pack of cards.

Coriaceous, leathery.

Corky, resembling cork in texture.

Corm, the swollen, succulent, bulb-like mass which composes the stem of certain monocotyledons, as in the Crocus; frequently termed a solid-bulb.

Corolla, the floral whorl next in succession within the calyx, composed of petals, which are free or more or less united together.

Corolline, of or belonging to a corolla.

Coronet, certain appendages, free or united, seated on the inner surface of the perianth, as the cup in Narcissus, or the scales in Silene.

Corymb, an inflorescence of which the branches originate at different **parts along the main axis, and elevate all the flowers to about** the same height.

Corymbose, approaching the form assumed by the corymb.

Cottony, bearing long soft entangled and interlaced hairs.

Cotyledon, a seed-lobe, or seed-leaf as it is often called; some embryos possess only one (monocotyledons), others have two (dicotyledons), or more.

Crenate, having a series of rounded marginal prominences.

Crescent-shaped, approaching the figure of a crescent, as the glands on the involucre of Euphorbia.

Crest, an irregular, elevated, notched ridge, resembling the crest of a helmet.

Crisped, curled.

Cryptogam, a term applied to the lower tribes of plants, which are not furnished with true flowers.

Cryptogamic, having the characteristics of the cryptogams, that is, not bearing true flowers.

Culm, the stem of a grass.

Cuneate, wedge-shaped.

Cupuliferous, bearing cupules or involucres composed of bracts adhering by their bases, and forming a sort of cup in which the fruit is seated, as in the oak or nut.

Cuspidate, gradually tapering into a sharp stiff point.

Cylindrical, approaching closely to the form of a cylinder; as the stems of grasses.

Cyme, an inflorescence in which numerous peduncles are given off in all directions from the summit of a branch, and these bear a terminal flower and secondary pedicels from below it, as in the common elder.

Deciduous, applied to trees or shrubs the leaves of which are shed annually, so that the branches become bare; the opposite of evergreen; applied also to other organs which fall off after their functions have been performed.

Decompound, Decomposite, subdivided to a considerable extent.

Decumbent, reclining upon the surface of the earth, but with a tendency to rise again towards the extremity.

Decurrent, prolonged below the point of insertion, as if running downwards, as in the leaves of most thistles.

Dehiscent, bursting by a regular line of suture.

Dentate, toothed.

Denticulate, furnished with small teeth.

Diadelphous, having the stamens united into two distinct sets or bundles.

Dichotomous, forked or subdivided once or repeatedly, with the branches in pairs.

Dichlamydeous, having both calyx and corolla.

Diclinous, applied to those plants of the same species which bear unisexual flowers on distinct plants.

Dicotyledons, plants whose seeds have two or more cotyledons or seed-lobes.

Dicotyledonous, possessing two cotyledons.

Didynamous, having four stamens, of which two are longer than the other two.

Diffuse, spreading widely in an irregular horizontal manner.

Digitate, applied to simple leaves, where the lobes are very narrow and cut nearly to the base of the limb; and to compound leaves, where the leaflets are all placed at the very extremity of the petioles.

Diœcious, bearing unisexual flowers of the same species on distinct individual plants.

Discoid, disk-like, round, and somewhat thickened, with the margins also rounded.

Disk, a fleshy expansion between the stamens and pistil, occurring in some flowers, and considered to result from the abortion of an inner whorl of stamens ; also the central portion occupied by the flowers in a capitule.

Dissected, having the segments very numerous and deeply cut, as in some leaves.

Dissepiment, the vertical partitions in the interior of an ovary, dividing it wholly or partially into two or more cells.

Distichous, longitudinally arranged in two rows, on opposite sides of a common axis.

Distinct, wholly unconnected with adjoining parts or organs.

Divaricate, branching off and spreading irregularly at an obtuse angle.

Dorsal, attached to the back of any organ.

Downy, covered with short weak close hairs.

Drupaceous, possessing the character of a drupe.

Drupe, an indehiscent, superior, one-celled fruit, fleshy externally and bony within, containing one or two seeds, as the plum, peach, etc.

Elliptical, approaching the form of an ellipse, that is an oval rounded at the ends, or an oblong widened in its smaller diameter.

Emarginate, slightly notched at the summit.

Embryo, the rudiment of a plant contained in the seed.

Endogen, the same as a monocotyledon.

Endogenous, possessing the internal structure of monocotyledons, that is with the newest cellular and vascular tissue of the stem produced within the older, and not collected in concentric rings.

Entire, without any traces of division, incision, or separation.

Epigynous, having the outer whorls of the flower adherent to the ovary, so that their upper portions alone are free, and appear to be seated on it, as in the Umbellifers.

Equal, where one part is of the same general form, disposition, and size, as some other part with which it is compared ; synonymous with regular.

Epipetalous, growing on the petals.

Erect, when any part or organ stands perpendicularly, or very nearly so, to the surface to which its base is attached.

Evergreen, bearing green leaves all the year round.

Exogen, the same as a dicotyledon.

Exogenous, the peculiar structure of dicotyledonous stems, wherein the successive deposits of newly organized wood are exterior to the old ones, and form concentric layers.

Falcate, plane and curved, with the edges parallel, like a sickle.

Fasciculate, where several similar parts originate at the same spot, and are collected, as it were, into a bundle.

Filament, the stalk which in many stamens supports the anther.

Filiform, cylindrical and slender, like a thread.

Flaccid, bending without elasticity.

Fleshy, firm but of succulent texture.

Floret, one of the little flowers in a head, as in Compositæ.

Flosculous, when the corolla of a floret is tubular.

Flower, the apparatus destined for the production of seed, and necessarily including one or other, or both, of the sexual organs.

Foliaceous, of the nature of a leaf.

Follicle, a one-valved inflated pericarp, opening by a suture along one of the sides to which the seeds are attached.

Forked, separating into two distinct branches, more or less apart.

Foveolate, impressed with small holes or depressions.

Fringed, clothed with hair-like appendages or cilia on the margin.

Fugacious, soon falling off or perishing, as the calyx of poppies.

Funnel-shaped, tubular, small below and widening upwards.

Furcate, forked.

Furrowed, marked by depressed lines.

Germen, the same as *Ovary*; the base of the pistil, containing the ovules.

Germination, the act by which seeds begin to grow.

Gibbous, convex as though swollen.

Glabrous, wholly destitute of pubescence.

Gland, a secreting organ of cellular tissue, sometimes sunk or sometimes elevated.

Glaucous, dull green with a peculiar whitish-blue lustre.

Globular, nearly spherical.

Glume, the outermost husks of the floral envelopes of grasses.

Glumaceous, resembling the dry scale-like glumes of grasses.

Granulated, covered with, or composed of, small tubercles resembling grains.

Granule, a small grain; also a small wort-like appendage on the calyx of certain species of dock.

Gynandrous, having the stamens and style coherent into a common body.

Hair, capillary expansions of cellular tissue, which coat the surface of various parts of many plants.

Hastate, shaped like the head of a halbert, the base diverging on each side into an acute lobe.

Herbaceous, not woody; any portion of a plant which is more particularly green and succulent.

Hermaphrodite, where both stamens and pistil occur in the same flower.

Hiemal, belonging to winter.

Hirsute, hairy.

Hispid, hairy, the hairs long and rigid.

Hoary, greyish-white.

Horizontal, when a plane surface lies perpendicularly to the axis of the body which supports it, as most leaves.

Horny, of a hard close texture, resembling horn.

Hypocrateriform, that form (of corolla) in which the tube is long and cylindrical, and the limb flat, spreading at right angles to it.

Hypogynous, seated below the base of the ovary, but not attached to the calyx.

Imbricate, overlapping.

Immersed, when one part or organ is completely imbedded in another.

Impari-pinnate, unequally pinnate, that is, pinnate with an odd terminal leaflet.

Incised, cut.

Incurved, gradually bending from without inwards.

Indefinite, where the number of any particular description of organ is uncertain.

Indehiscent, without dehiscence, or regular line of suture.

Inferior, placed below another organ ; used especially to express the connection of the tube of a calyx with the ovary.

Inflated, bladdery.

Inflorescence, the general arrangement or disposition of the flowers.

Interrupted, having symmetry or regularity of outline or composition partially destroyed.

Involucel, a partial involucre.

Involucre, a whorl of bracts, free or united, seated on the peduncle, either near to or at some distance below the flower or flowers.

Involute, rolled inwards at the edge, as some leaves.

Irregular, wanting symmetry ; also unequal.

Isomerous, having an equal number of parts.

Joints, certain parts where the uniformity of the tissue is altered, and where it may readily be ruptured or falls asunder in decay.

Keel, a projecting ridge, rising along the middle of a flat or curved surface ; also the two lowermost and more or less combined petals of a papilionaceous corolla.

Keeled, furnished with a keel.

Knotted, swollen at intervals into knobs, somewhat resembling a knotted cord.

Labellum, the lip of a flower.

Labiate, a tubular calyx or corolla with the limb divided into two unequal portions, or lobes placed above and below, like the lips of a mouth.

Lanceolate, narrow and tapering at each end, like the head of a spear.

Lateral, fixed on or near the side of any organ.

Lax, loose.

Leaf, an appendage to the stem, composed of cellular tissue with fibres of vascular tissue intermixed.

Leaflet, each separate portion or subordinate expansion in the limb of a compound leaf.

Legume, the one-celled and two-valved seed-vessel of Leguminosæ.

Lenticular, of the form of a double-convex lens.

Leprous, covered with flat scurfy scales.

Ligule, a membranous appendage at the summit of the sheathing petiole, in grasses.

Ligulate, strap-shaped.

Limb, the superior flat expanded part of a petal or leaf.

Linear, having the margins parallel, and the length considerably longer than the breadth.

Lip, a term applied to each of the two large divisions of an unequally parted monopetalous corolla; also applied to one of the segments of an irregular perianth, when of a shape remarkably different from the rest, as in Orchids.

Lobe, a rounded projecting part of some organ.

Lobed, divided into lobes.

Loose, having the separate parts arranged at some distance from each other upon a common axis.

Lurid, of a dingy brown; grey with orange.

Lyrate, having several pairs of small lobes near the base, with deep sinuses between them.

Marginal, placed upon, or attached to, the edge of any thing.

Mealy, covered with a scurfy powder.

Membranaceous, Membranous, thin, and more or less transparent.

Mericarp, one of the carpels in the fruit of Umbellifers.

Midrib, the principal nerve or vein, which runs from the base to the apex of a leaf.

Milk, an opaque white juice, found in many plants.

Monadelphous, having the filaments of the stamens all united in one set or bundle.

Monochlamydeous, having only one whorl to the perianth.

Monocotyledons, plants whose seeds have only one cotyledon.

Monocotyledonous, possessing but one cotyledon.

Monopetals, a group of dicotyledons, in which the corolla is monopetalous.

Monœcious, bearing two kinds of unisexual flowers on the same individual plant.

Monopetalous, having the corolla in one piece which is formed by the union of several petals.

Mucronate, abruptly pointed by a sharp spinous process.

Multifid, having deep and numerous subdivisions or laciniations.

Muricated, rough, with short hard tubercular excrescences.

Nectariferous, possessing a nectary; also secreting nectar, that is, a sweetish exudation, secreted by glands in different parts of plants.

Nectary, any supplementary organ in the flower, whether glandular or not, which cannot readily be referred to the parts forming the floral whorls; also certain parts of the whorls themselves, of anomalous character, whether secreting nectar or not.

Nerves, the fibrous bundles extending through the leaves, often ramified, like veins or nerves in the animal structure.

Nerveless, without nerves.

Netted, resembling network.

Nodding, having the summit so much curved that the apex is directed perpendicularly downwards.

Node, one of the parts of the stem from whence a leaf springs, especially when a little thickened or swollen.

Nodose, knotted.

Nut, a hard indehiscent pericarp.

Ob, in composition signifies that the point of attachment is at the opposite extremity to where it occurs in the form defined by the simple word; thus *ob-clavate* is the inverse of clavate, the attachment being at the thicker end; *ob-cordate*, is the inverse of cordate, the attachment being at the narrow end; *ob-ovate*, is the inverse of ovate, the attachment being at the narrow end.

Oblique, when the parts of an organ are so divided that the sides are more or less unequal.

Oblong, of a bluntly elliptical shape, where the major and minor axes bear a proportion to each other of about four to one.

Obtuse, blunt.

Ochrea, a tubular membranous stipule, through which the stem seems to pass.

Ochreate, furnished with ochrea.

Opposite, applied to similar parts or organs when so arranged in pairs, that one of them is immediately on the opposite side of some interposed body, or of the axis about which they are disposed; thus—*oppositiflorus* is where the peduncles are opposite; *oppositifolius*, where the leaves are opposite.

Orbicular, perfectly or very nearly circular.

Ovary, the lowermost portion of the pistil, containing the ovules, and ultimately becoming the fruit.

Ovate, of the form of an egg, if applied to a solid body; or of the figure presented by a longitudinal section of an egg, if applied to a superficial area.

Ovoid, a solid with an ovate figure.

Ovule, the rudimentary state of a seed.

Palate, the inferior surface of the throat in ringent and personate corollas.

Pales, the membranaceous bracts forming the perianth of grasses; also the chaff-like scaly bracts on the receptacle of some Composites.

Palmate, having the arrangement of parts on any organ such as to imitate the form of an open hand.

Palmatifid, having the subdivisions of a simple organ (usually a leaf) arranged palmately.

Panicle, an inflorescence where the rachis either subdivides into several branches, or is furnished with distinct branching peduncles.

Panicled, having the flowers arranged in a panicle.

Papilionaceous, an irregular corolla composed of five petals, the upper of which forms the standard; two others, placed laterally, the wings; and two (opposite the standard, and more or less cohering) form the keel.

Papillose, covered with small elongated cellular protuberances.

Pappus, the peculiar limb to the calyx of the florets of Composite flowers, and frequently hairy or downy as in thistles.

Parallel, where the axes of two parts lie parallel to each other.

Parasite, a plant which grows on some other plant, and obtains nourishment directly from its juices.

Parietal, attached to the inner surface of the pericarp.

Partite, divided nearly to the base.

Patent, spreading.

Pectinate, pinnatifid with the segments parallel, narrow, and close, like the teeth in a comb.

Pedate, having the parts arranged in a palmate manner, with the addition of further subdivisions in the lateral portions.

Pedicel, the partial stalk or immediate support of a flower in an inflorescence composed of flowers arranged upon a main peduncle.

Pedicellate, furnished with a pedicel.

Peduncle, the main stalk or support to the inflorescence, more especially when this is limited to a solitary flower.

Pedunculate, furnished with a peduncle.

Pellucid, perfectly or partially transparent.

Peltate, where a support is inserted at some distance within the margin, and is not in the same plane as the flat surface which rests upon it.

Pendent, pendulous, inclined so that the apex is pointed vertically downwards.

Pentadelphous, having the stamens arranged in five sets or bundles.

Pentamerous, having the parts in fives or multiples of five.

Perennial, enduring for several years.

Perfoliate, applied to clasping leaves which have their basal lobes united, so that the axis about which they are placed appears to pass through them.

Perianth, the external floral whorl or whorls which surround the stamens and pistil, including the parts answering to calyx and corolla, when they are so much alike as not to be readily distinguishable.

Perigone, almost the same as *Perianth,* but more especially used when the floral envelopes are reduced to a single floral whorl, possessing a calyx-like character.

Perigynous, having the ovary free, but either the stamens or corolla adherent to the calyx.

Persistent, remaining beyond the period when similar parts in other plants become mature and fall; not falling off but remaining green as the leaves of evergreens.

Personate, a form of monopetalous bilabiate corolla, in which the orifice of the tube is closed by an inflated projection of the throat.

Petal, one of the foliaceous expansions of that part of the floral whorl, termed the corolla.

Petaline, of or belonging to a petal.

Petaloid, having a thin membranous character and coloured, thus assuming the more usual character of the petals of flowers.

Petiole, the stalk or support by which the blade, or limb of a leaf, is attached to the stem.

Pilose, hairy.

Pinnate, Pinnated, having the leaflets arranged on opposite sides of a common petiole; confined to pairs of leaflets that are equally or paripinnate; if terminated by an odd leaflet, they are unequally, or impari-pinnate.

Pinnatifid, having the lateral incisions (in a simple leaf) extending towards the axis; approaching the form termed pinnate.

Pistil, the female part of a flower, composed of the ovary with its ovules, and the stigma or stigmas, with sometimes (usually) an intervening style.

Pistillate, furnished with pistils; generally applied to unisexual flowers.

Pitcher-shaped, tubular, bulging below and contracted towards the orifice.

Pith, the central column of cellular tissue, in the stems and branches of exogenous plants.

Placenta, that part of the ovary which supports the ovules.

Plaited, plicate, or folded longitudinally.

Plicate, folded together in regularly disposed longitudinal plaits.

Plumose, applied to hairs invested with branches, arranged like the beard on a feather.

Pod, a two-valved seed-vessel, one-celled as in the legume or pod of the pea, and two-celled as in the silique of the wall-flower.

Pollen, the granular contents of an anther, either free and resembling dust, or variously agglutinated into waxy masses.

Pollen-mass, an agglutinated mass of pollen, such as occurs in Orchids.

Polyadelphous, having the stamens combined in several sets or bundles.

Polygamous, bearing on the same plant three descriptions of flowers; viz., hermaphrodite, male, and female.

Polypetalous, having two or more petals, and these perfectly distinct from each other.

Pome, a fleshy many-celled fruit, matured from an inferior ovary.

Pore, an aperture in the covering of any body, as in that of the anthers of heaths, which open by a hole or aperture, instead of the usual slit for the escape of the pollen.

Porose, having pores.

Pouch, a little bag; also the short silicules of some Crucifers, as in Alyssum.

Prickle, a more or less conical elevation of the substance of the bark, hard and sharp-pointed.

Prismatical, approaching the form of a prism, presenting angles disposed longitudinally.

Procumbent, lying upon or trailing along the ground.

Prostrate, procumbent.

Pubescence, elevated extensions of the cellular tissue of the epidermis, assuming the character of soft downy hairs.

Pubescent, furnished with pubescence.

Pulp, soft and juicy tissue.

Quadrangular, approximating to the form of a quadrangular prism.

Quaternate, Quaternary, having the parts arranged by fours.

Quinate, Quinary, having the parts arranged by fives, as the five petals of a buttercup, the five leaflets of a digitate leaf, etc.

Raceme, a form of inflorescence, where the flowers are furnished with pedicels arranged at intervals upon a common axis.

Racemose, arranged in racemes.

Radiant, Radiate, arranged like rays spreading from a common centre.

Radical, proceeding from a point

close to the summit or crown of the root.

Ray, the outer florets in the flower-head of Composites; the outer flowers, when differently formed from the inner, in umbels; also the branches of an umbel.

Receptacle, a part which bears or receives other parts, commonly applied to that which bears the flowers, as the expanded top of the peduncle of a dandelion, the inner surface of a fig, etc.

Recurved, bent backward.

Reflex, Reflexed, very much curved backwards.

Regular, uniform in structure or condition, as where subordinate parts of the same kind closely resemble each other, and are symmetrically arranged.

Remote, thinly set on the axis.

Reniform, kidney-shaped, that is, like a longitudinal section through a kidney.

Reticulate, resembling network.

Retuse, having a slight depression or sinus at the apex.

Rhizome, a prostrate or subterranean stem, from which roots are emitted, and scaly leaves or branches given off at the knots.

Rhomboid, Rhomboidal, rudely approximating to the form of a rhombus, that is to say, a quadrangular figure (not a square) whose sides are equal.

Rib, any strongly-marked nerve in the leaf, but more especially the central longitudinal one.

Ribbed, having one or more strongly marked nerves, proceeding from the base to the apex.

Ridge, an elevated line on the carpels of Umbellifers, of which some are primary and some secondary.

Rigid, almost or quite without flexibility.

Ringent, applied to bilabiate corollas whose lips are widely separate.

Root-stock, a subterranean or prostrate stem, which emits roots from its lower surface.

Rosulate, having the parts more or less laminated, and arranged in a whorl round an axis, in a manner somewhat resembling the disposition of the petals of a rose.

Rotate, wheel-shaped; having a monopetalous corolla, with a very short tube and spreading limb.

Rudimentary, either in an early state of development, or in an imperfectly-developed condition.

Rugose, having the surface covered with wrinkles.

Runcinate, having the large marginal incisions (in a leaf) directed in a curved manner towards the base.

Saccate, resembling a bag or sac.

Sagittate, pointed at the apex, and with the base prolonged backwards from the sides into two acute ears, like an arrow head.

Salver-shaped, hypocrateriform.

Samara, a compressed, few-seeded, coriaceous or membranaceous indehiscent pericarp, with a membranaceous expansion or wing at the end or edges, as in the fruit of the Sycamore.

Sapwood, the outermost layers in the trunks of exogenous trees.

Scabrous, harsh or rough to the touch, from the presence of stiff pubescence or scattered tubercles.

Scape, a long naked peduncle, rising from the crown of a root or a subterranean stem.

Scariose, Scarious, thin, dry, and membranous.

Scattered, without any apparent symmetry of arrangement.

Scorpioid, having the main axis of inflorescence curved in a circinate manner, like the tail of a scorpion.

Scurfy, bearing minute scales of membranous matter on the surface.

Secund, having the organs (generally applied to flowers) all turned to the same side of the axis round which they are arranged.

Seed, the fertilized ovule.

Semi, in composition, implies a partial or imperfect exhibition of the particular effect implied by the term with which it is compounded.

Sepal, one of the foliaceous expansions of that part of the floral whorl called the calyx.

Sepaline, having reference to sepals.

Serrated, having sharp marginal serratures pointed forward, like the teeth of a saw.

Sessile, applied when an organ is attached to its support without the intervention of some intermediate part : thus, *sessilifolius* when a leaf is without petiole, *sessiflorus* when a flower is without a pedicel.

Seta, any stiff bristly hair or straight slender prickle.

Setaceous, having the characters of setæ.

Setose, covered with setæ.

Sheath, a petiole, or a portion of it, embracing the stem to which it is attached ; a part rolled round a stem or other body.

Shrub, a woody plant which does not form a true trunk, like a tree, but has several stems rising from the roots.

Silicle, Silicula, formed like a silique, but about as broad as long or broader.

Siliculose, bearing silicles.

Silique, Siliqua, an elongated, dry, bivalvular, pod-like fruit, with a transverse internal membrane.

Siliquose, bearing siliqua.

Silky, having very long and fine hairs, with a glossy appearance like silk.

Simple, the opposite of compound ; without subordinate parts.

Sinuate, Sinuated, having alternate, rounded, rather large lobes and sinuses at the margin.

Sinus, the re-entering angle or depression between two projections or prominences.

Solitary, not closely associated with another object of the same description.

Spadix, the axis of a spiked, often fleshy inflorescence among monocotyledons, when the flowers are densely aggregated, usually accompanied by one or more spathes.

Spadiceous, of or belonging to a spadix.

Spathaceous, furnished with, or having the general appearance of a spathe.

Spathe, a foliaceous or membranaceous involucre, of one or few sheathing bracts, more or less enveloping the flowers in certain monocotyledons.

Spathulate, more or less rounded towards the summit and narrowed towards the base.

Spike, an inflorescence similar to the raceme, only that the flowers have no pedicels ; also, those forms in which spikelets are arranged in close and alternating series upon a common rachis, as in some grasses.

Spikelet, a small spike, of which several, aggregated round a common axis, constitute a compound spike ; more especially applied to the spiked arrangements of two or more flowers of grasses, which are variously disposed round a common axis.

Spine, a stiff, sharp-pointed process, containing some portions of woody tissue, degenerated branch, leaf, stipule, etc.

Spinous, bearing or covered with spines.

Spongy, having the cellular tissue copious, forming a sponge-like mass.

Spreading, having a gradual outward tendency, or bending from an axis.

Spur, a tubular expansion of some part of a flower.

Spurred, having a spur.

Stamen, that organ of the flower which contains the pollen.

Staminate, bearing stamens ; usually applied to unisexual flowers.

Standard, the dorsal petal in a papilionaceous flower.

Stellate, disposed in a radiating manner round a centre.

Stigma, that portion of a pistil,

generally its summit, by which the fertilizing influence of the pollen is conveyed to the ovules.

Stigmatic, of or belonging to the stigma.

Sting, a sharp, somewhat stiff hair, seated on a gland which secretes an acrid fluid.

Stipule, a foliaceous appendage, various in character, produced on each side the base of certain petioles.

Striated, marked with streaks or little furrows.

Style, the shaft which, in most flowers, is interposed between the stigma and ovary.

Sub, in composition, somewhat ; implying a near approach to the condition indicated by the term with which it is joined : thus, *subrotund* is roundish.

Subulate, awl-shaped.

Succulent, having abundant cellular tissue, replete with juices.

Suckers, tubercular processes on the stems of certain flowering parasites, by which they imbibe nourishment from the plants to which they attach themselves.

Superior, placed above another organ ; applied especially to indicate the position of the ovary with respect to the calyx.

Sword-shaped, straight and flat, with the point acute.

Symmetrical, when the parts of one series of organs agree with those of another in number, as in a flower which has five sepals, five petals, five or ten stamens, etc.

Syncarpous, bearing fruit composed of cohering carpels.

Syngenesious, having the stamens coherent by their anthers.

Tapering, gradually diminishing in diameter.

Tendril, a twisting, thread-like process, forming a modified condition of some appendage to the axis of vegetation.

Terete, nearly cylindrical, but somewhat tapering into a very elongated cone.

Terminal, situated at the extremity of some part.

Ternate, Ternary, arranged by threes about the same part.

Tetradynamous, having six stamens, four longer than the other two.

Thalamiflores, a group of dicotyledons, in which the several petals are distinct from the sepals, and the stamens are hypogynous.

Thorn, the same as spine.

Tetramerous, having the parts in fours or multiples of four.

Tooth, a small projection of the margin of some laminated part.

Toothed, furnished with teeth.

Triadelphous, having the stamens collected into three distinct sets or bundles.

Triandrous, having three stamens.

Triangular, a plane surface approaching a triangle in shape.

Trichotomous, subdivided, with the divisions in threes.

Tricoccous, composed of three cocci.

Trifid, divided into three subordinate parts, the incisions extending about half-way towards the base.

Trifoliate, Trifoliolate, having the leaflets disposed in threes at the extremities of their petioles.

Trigonous, triangular.

Tripartite, subdivided into three parts, much beyond the middle or nearly to the base.

Trimerous, having the parts in threes or multiples of three.

Tripetalous, consisting of three petals.

Triternate, having the petiole twice branched in a ternate manner, each partial petiole bearing three leaflets.

Truncate, terminating abruptly, as though shortened by the removal of the extremity.

Tube, the tubular portion formed by the cohesion of the subordinate parts composing a floral whorl.

Tuber, a fleshy, swollen, subterranean rhizome, like the potato.

Tubercle, a small wart-like excrescence.

Tubercled, Tuberculate, bearing or covered with tubercles.

Tuberous, resembling a tuber.

Tubular, hollow and cylindrical.

Tumid, swollen and inflated.

Turbinate, top-shaped.

Turgid, thick, as if swollen, but not inflated.

Twining, twisting in spiral folds round a support.

Umbel, a form of inflorescence in which all the pedicels start from the summit of a peduncle.

Umbellate, arranged in umbels.

Under-shrub, a plant only partially shrubby, the ends of the newly-formed branches continuing herbaceous, and dying away in winter.

Unilateral, disposed along one side.

Unisexual, applied to flowers which have either stamens alone or pistils alone.

Urceolate, shaped like a pitcher with a contracted mouth.

Unsymmetrical, having the parts neither equal nor proportional in number.

Utricle, a little bladder filled with air attached to the stems of certain aquatic plants.

Valve, distinct portions of certain organs which become detached by regular dehiscence along definite lines of suture, as the valves of anthers and the valves of carpels.

Valvate, united by the margins only, not overlapping.

Ventricose, bellied or swelling out on one side.

Vertical, having the axis perpendicular to the part from which it arises.

Verticillasters, short cymes in the axils of the leaves of labiate plants, giving the appearance of whorled flowers.

Viscid, glutinous and clammy.

Vitta, a narrow elongated receptacle of aromatic oil occurring, variously disposed, in the fruits of Umbellifers.

Waved, *Wavy*, having an alternately convex or concave surface or margin.

Waxy, resembling wax in texture.

Whorl, an arrangement in which the parts are set in a circle round an axis all on the same plane.

Whorled, disposed in whorls.

Wing, a membranous expansion ; also each of the two lateral petals in a papilionaceous or other irregular flower, when these differ from the rest.

Woolly, having long curled hairs, matted together like wool.

INDEX.

— ♦ —

2 E 2

THE END.

JOHN EDWARD TAYLOR, PRINTER,
LITTLE QUEEN STREET, LINCOLN'S INN FIELDS.

PLATE I.

A.—Anemone nemorosa, *Linnæus :* p. 6.
    1. Cluster of young carpels.
B.—Caltha palustris, *Linnæus :* p. 7.
    1. Cluster of young carpels.
    2. One of the carpels, separate.
    3. The full-grown carpels or follicles.
C.—Cardamine pratensis, *Linnæus :* p. 7.
    1. The stamens and pistil.
D.—Cheiranthus Cheiri, *Linnæus :* p. 8.
    1. The stamens and pistil.
    2. The ripe pod, with the valves separating.

Plate 1.

A

B

Anemone nemorosa.

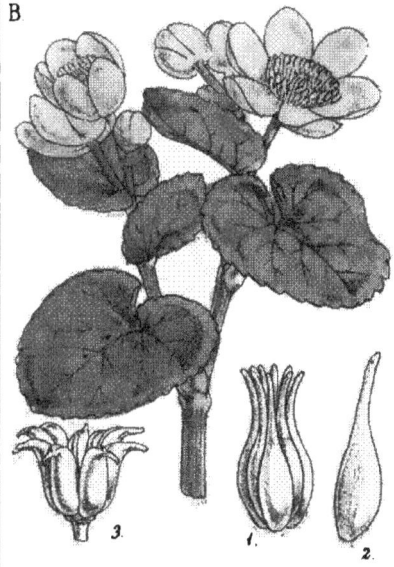

Caltha palustris.

C.

D

Cardamine pratensis.

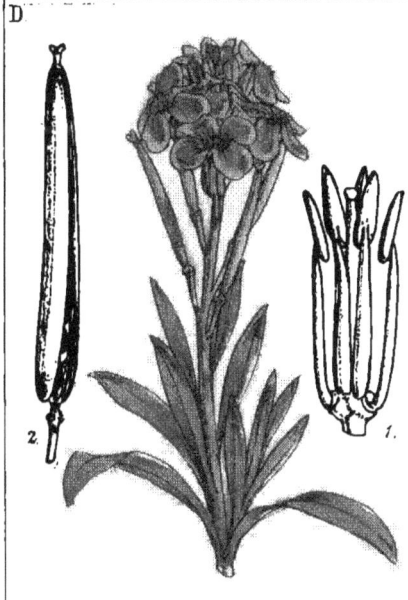

Cheiranthus Cheiri

W Fitch, del et lith.

Vincent Brooks Imp.

PLATE II.

A.—Viola odorata, *Linnæus:* p. 8.

    1. The stamens and pistil.

    2. One of the spurred stamens, separate.

    3. The pistil with its curved style.

B.—Acer Pseudo-platanus, *Linnæus:* p. 9.

    1. One of the flowers, separate.

    2. The winged pair of fruits or samara.

C.—Oxalis Acetosella, *Linnæus:* p. 10.

    1. The stamens and pistils.

    2. The pistil, with its five styles.

D.—Ribes rubrum, *Linnæus:* p. 11.

    1. A flower, separate.

    2. One of the berries.

Plate 2.

A.

2.

3.

1.

Viola odorata.

B.

2.

1.

Acer Pseudo-platanus

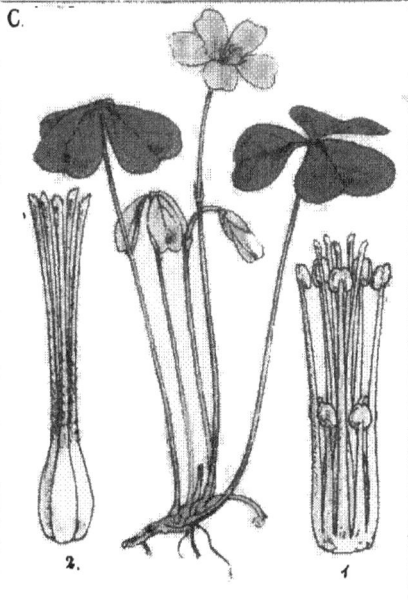

C.

2.

1.

Oxalis Acetosella.

D.

2.

1.

Ribes rubrum.

PLATE III.

A.—**Saxifraga granulata,** *Linnæus :* p. 11.

    1. Two of the clustered tubers.

    2. A flower, with the calyx and corolla removed.

    3. The calyx surrounding the ovary, with its two styles.

B.—**Leontodon Taraxacum,** *Linnæus :* p. 12.

    1. One of the florets, separate.

    2. One of the fruits or achenes, with the pappus expanded.

C.—**Bellis perennis,** *Linnæus :* p. 13.

    1. One of the ray florets, separate.

    2. One of the disk florets, separate.

D.—**Vinca minor,** *Linnæus :* p. 14.

    1. A portion of the corolla-tube, showing the attachment of the stamens to its inner surface.

    2. A flower, with the corolla and two of the calyx-segments removed, showing the ovaries, style, and pulley-shaped stigma.

Plate 3.

A

1.

3.

2.

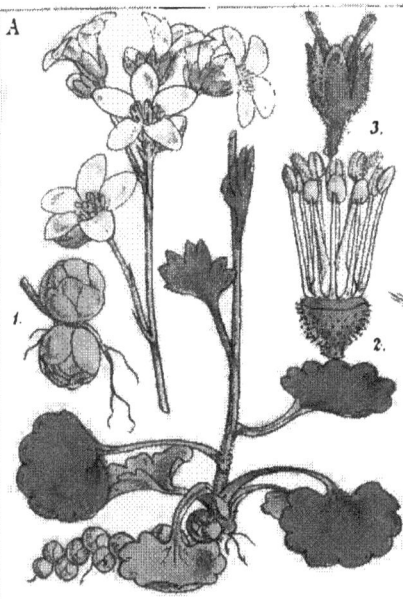

Saxifraga granulata.

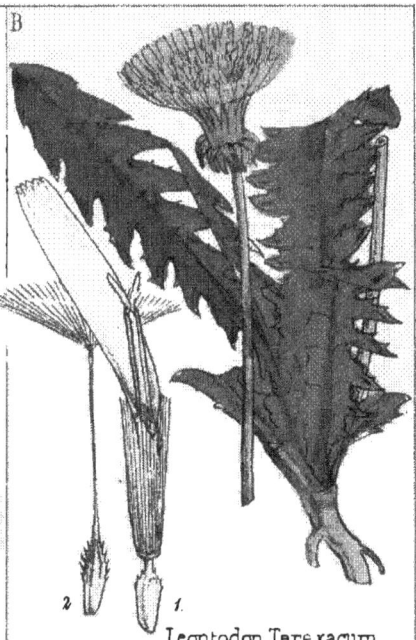

B

2

1.

Leontodon Taraxacum

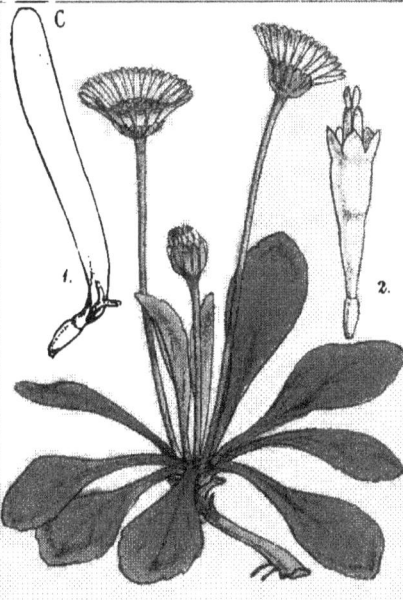

C

1.

2.

Bellis perennis.

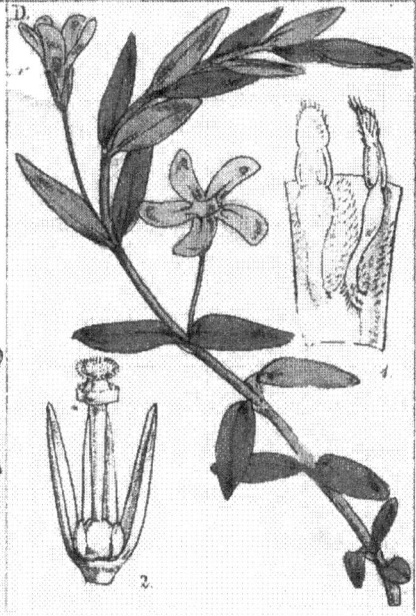

D.

2.

Vinca minor

W.Fitch.del et lith.

Vincent Brooks, Imp.

PLATE IV.

A.—**Menyanthes trifoliata,** *Linnæus :* p. 14.

    1. A corolla laid open, showing the stamens attached to the inner fringed surface.

    2. The pistil.

    3. One of the capsules.

B.—**Primula vulgaris,** *Hudson :* p. 1.

    1. The pistil.

C.—**Daphne Mezereum,** *Linnæus :* p. 15.

    1. One of the perianths laid open, showing the stamens attached to the inner surface.

    2. The pistil.

    3. A section of one of the berries.

D.—**Euphorbia amygdaloides,** *Linnæus :* p. 16.

    1. One of the small flower-heads, with crescent-shaped glands, several erect male flowers, and a recurved female flower.

    2. One of the male flowers separated, consisting of a scale and stamen only.

    3. The scale of the male flower.

    4. The tricoccous stalked ovary, with its three-cleft style, forming the female flower, which has no perianth.

Plate 4.

A.

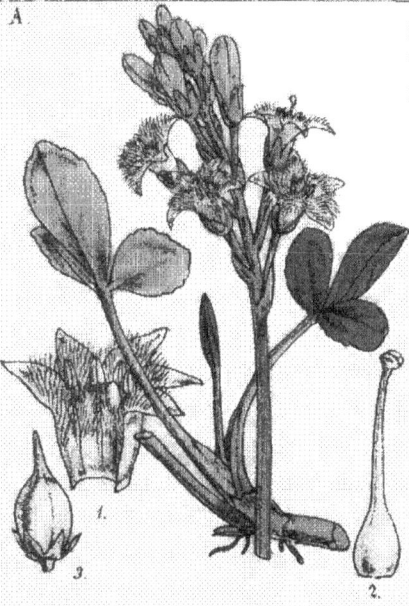

Menyanthes trifoliata.

B.

Primula vulgaris.

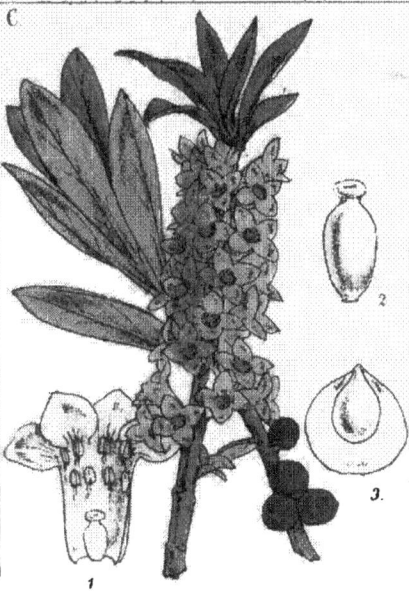

C.

Daphne Mezereum.

D.

Euphorbia amygdaloides

W. Fitch and auth.

Vincent Brooks imp.

PLATE V.

A.—**Ulmus montana,** *Smith :* p. 16.

    1. A perianth with stamens.

    2. The pistil.

    3. The winged seed.

B.—**Salix Caprea,** *Linnæus :* p. 17.

    1. A branch, with the male catkins.

    2. One of the male flowers, separate.

    3. A female catkin.

    4. One of the female flowers, separate.

    5. A seed.

C.—**Orchis maculata,** *Linnæus :* p. 19.

    (The name is misprinted "mascula" on the plate.)

    1. A flower, showing the three recurved sepals, two convergent petals, and three-lobed lip.

    2. The column and cells of the single perfect anther.

    3. One of the pollen-masses.

D.—**Cypripedium Calceolus,** *Linnæus :* p. 19.

    1. Front view of the column, with its abortive central stamen, and two lateral perfect ones.

    2. Back view of the same.

Plate 5

A.

B.

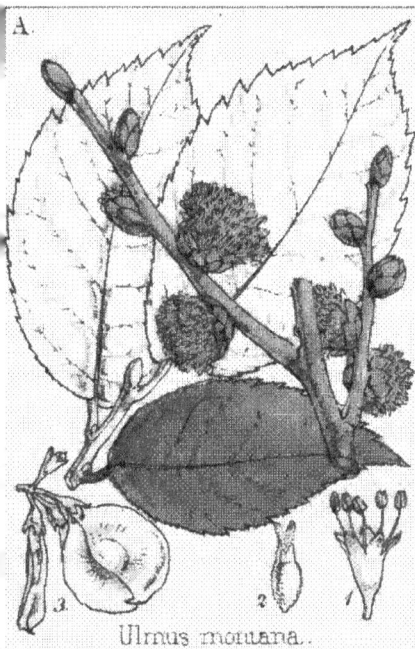

Ulmus montana.

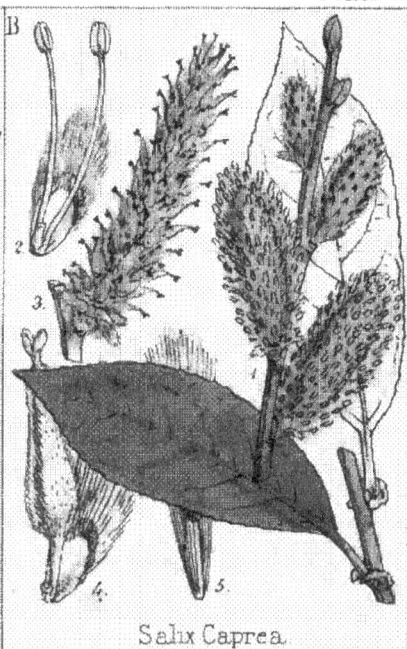

Salix Caprea.

D.

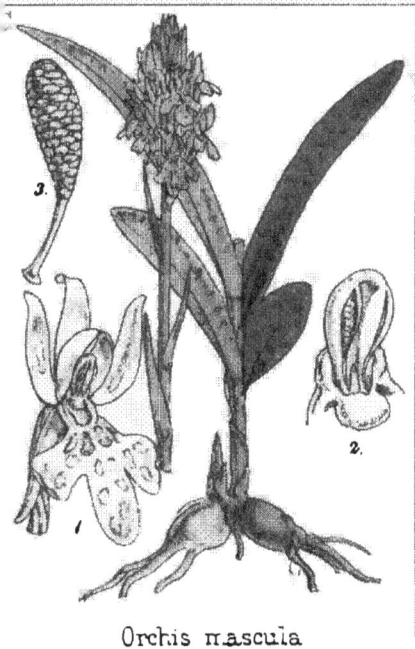

Orchis mascula

Cypripedium Calceolus.

W Fitch del. et. lith.

Vincent Brooks, Imp.

PLATE VI.

A.—Crocus vernus, *Willdenow :* p. 4.

    1. The base of the flower-tube, laid open in the upper part, showing the attachment of the stamens.

    2. One of the stamens.

    3. The three-cleft stigma.

B.—Galanthus nivalis, *Linnæus :* p. 3.

    1. A stamen.

    2. The pistil, showing the ovary at the base.

C.—Scilla verna, *Hudson :* p. 20.

    1. One of the flowers, separate.

    2. A segment of the star-shaped perianth, with its stamen.

    3. The pistil.

D.—Hyacinthus non-scriptus, *Linnæus :* p. 20.

    1. Portion of a flower, showing the pistil, and the insertion of the stamens.

    2. A transverse section of the ovary.

Plate 6.

A

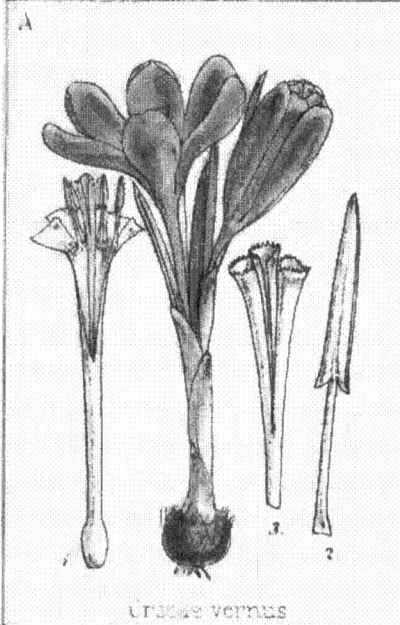

Crocus vernus

B

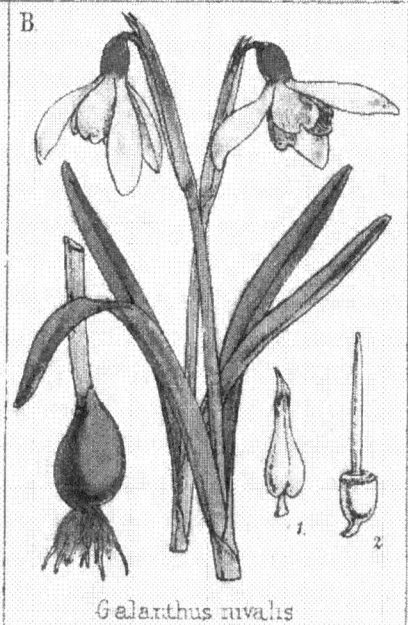

Galanthus nivalis

C

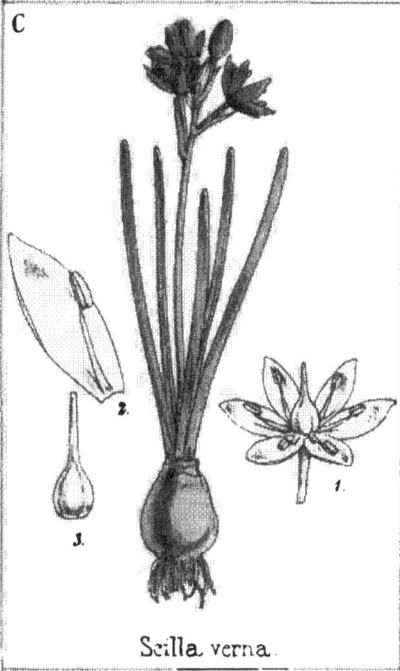

Scilla verna.

D

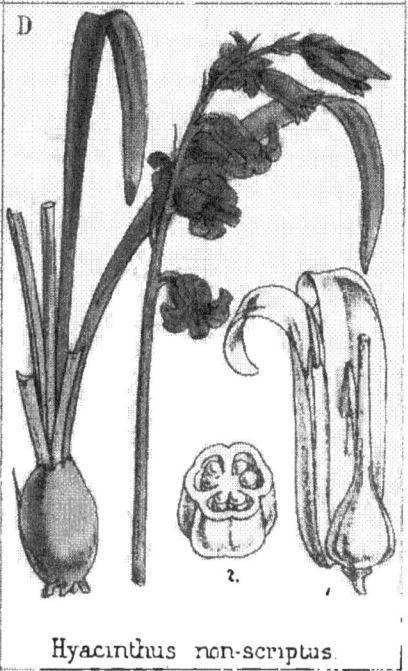

Hyacinthus non-scriptus.

PLATE VII.

A.—Berberis vulgaris, *Linnæus :* p. 80.

    1. A flower, separate.

    2. A petal, with its two glands.

    3. A stamen, showing the valves of its anthers.

    4. The pistil.

    5. One of the fruits.

B.—Nymphæa alba, *Linnæus :* p. 81.

    1, 2. Different forms of stamens.

    3. The pistil, with its sessile radiating stigma.

C.—Papaver Rhœas, *Linnæus :* p. 82.

    1. A stamen.

    2. The pistil, with its sessile radiating stigma.

    3. The ripe capsule, showing the apertures for the escape of the seeds.

D.—Reseda lutea, *Linnæus :* p. 82.

    1. One of the flowers, separate.

    2. One of the upper petals.

    3. Other forms of petals.

    4. The ovary, accompanied by a stamen.

Plate 7.

A.

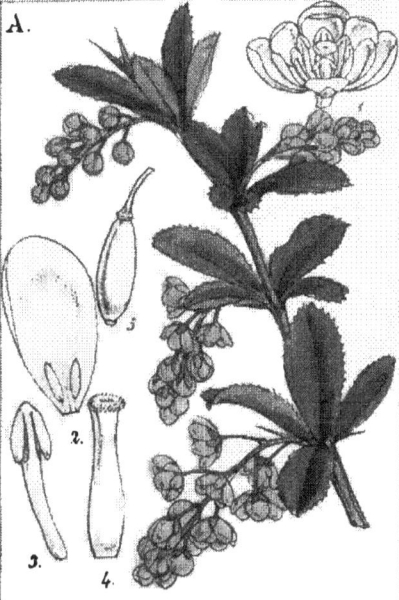

Berberis vulgaris

B.

Nymphæa alba

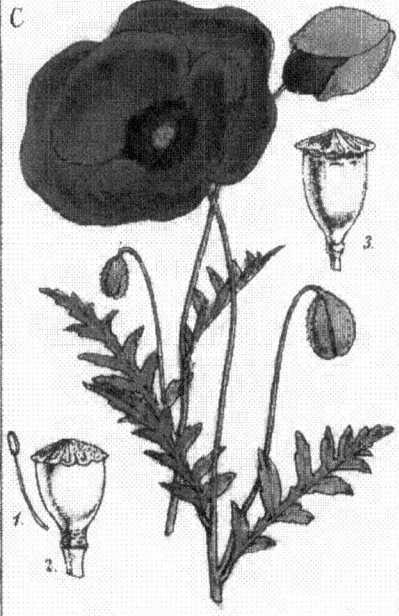

C.

Papaver Rhœas.

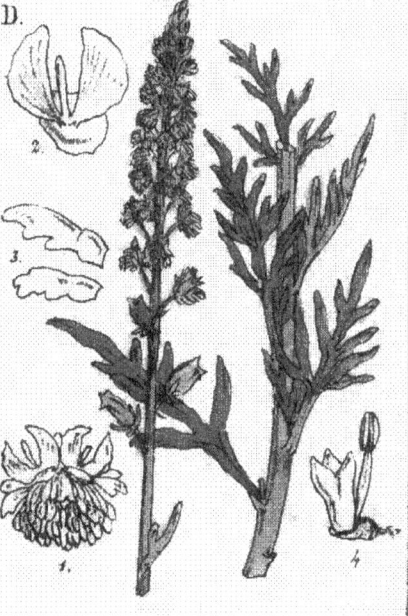

D.

Reseda lutea.

Whitau del et th

Vincent Brooks, Imp.

•

PLATE VIII.

A.—Helianthemum vulgare, *Gærtner:* p. 83.
    1. Back view of the calyx.
    2. The pistil.

B.—Drosera rotundifolia, *Linnæus:* p. 93.
    1. One of the gland-fringed leaves.
    2. The stamens and pistil.

C.—Polygala vulgaris, *Linnæus:* p. 83.
    1. A flower, seen from beneath.
    2. Side view of a flower.
    3. The pistil.

D.—Frankenia lævis, *Linnæus:* p. 84.
    1. A portion of the stem, showing the opposite leaves, and axillary clusters of smaller leaves.
    2. A flower.
    3. One of its petals.
    4. The pistil.

Plate 5

A.

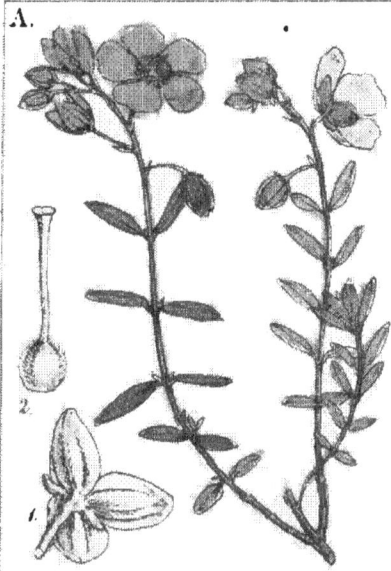

Helianthemum vulgare

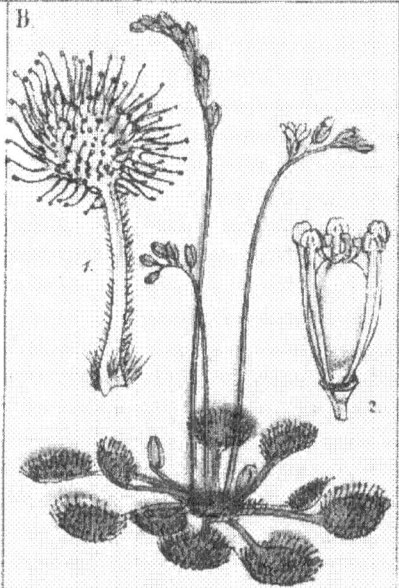

B.

Drosera rotundifolia.

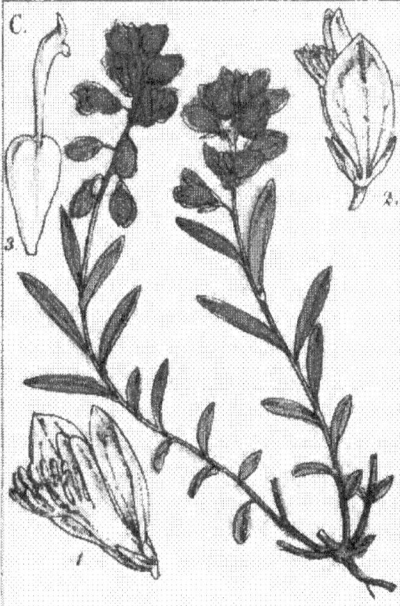

C.

Polygala vulgaris

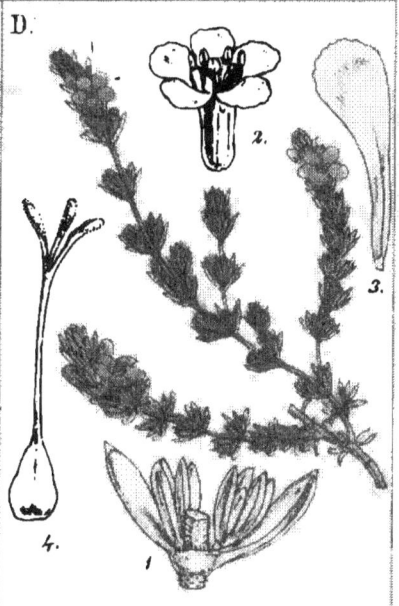

D.

Frankenia lævis.

Sowerby del. et lith.

Vincent Brooks, Imp.

PLATE IX.

A.—**Dianthus plumarius,** *Linnæus:* p. 84.

    1. One of the clawed petals.

    2. The pistil, with its two curved styles.

B.—**Lychnis Githago,** *Lamarck:* p. 85.

    1. A petal with adherent stamens.

    2. The pistil, with five styles.

C.—**Malva sylvestris,** *Linnæus:* p. 85.

    1. The disk-shaped fruit, composed of several contiguous carpels, and surrounded by the calyx.

    2. The staminal column with the stigmas protruding from the centre.

    3. One of the stamens, separate.

    4. The pistil separated from the column of stamens.

D.—**Tilia europæa,** *Linnæus:* p. 86.

    1. One of the flowers, separate.

Plate 9.

A

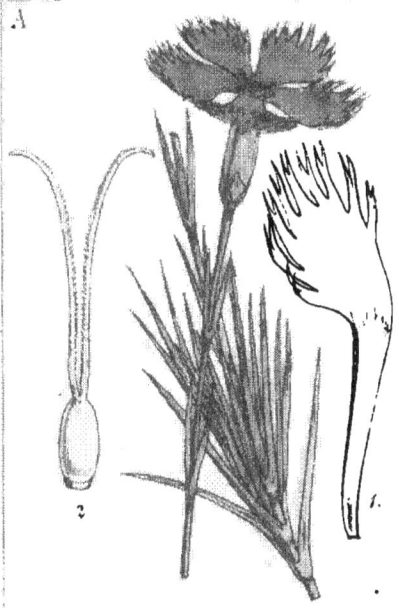

Dianthus plumarius

B

Lychnis Githago.

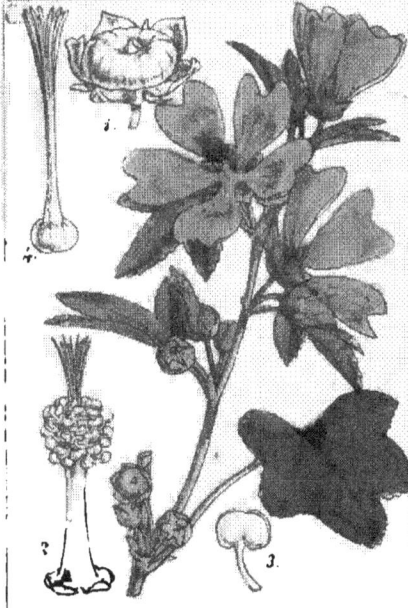

C

Malva sylvestris

D

Tilia europæa

PLATE X.

A.—**Hypericum pulchrum,** *Linnæus :* p. 86.

    1. One of the oblique petals.

    2. One of the parcels of stamens.

    3. The calyx surrounding the pistil, the petals and stamens being removed.

B.—**Geranium pratense,** *Linnæus :* p. 87.

    1. The stamens and pistil.

C.—**Linum usitatissimum,** *Linnæus :* p. 87.

    1. The stamens and pistil.

    2. The capsule surrounded by the calyx.

    3. One of the seeds.

D.—**Impatiens Noli-me-tangere,** *Linnæus :* p. 89.

    1. A flower with most of the petals removed.

    2. A young capsule.

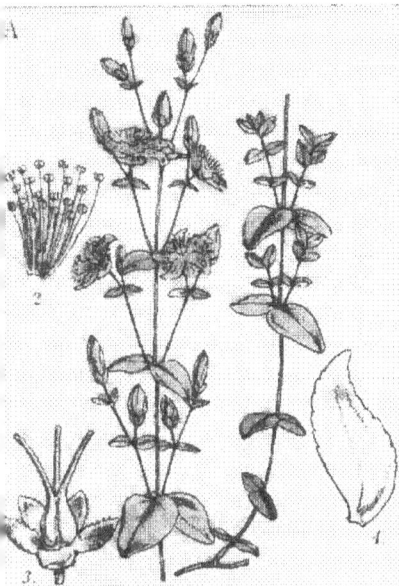

A. Hypericum pulchrum.

B. Geranium pratense.

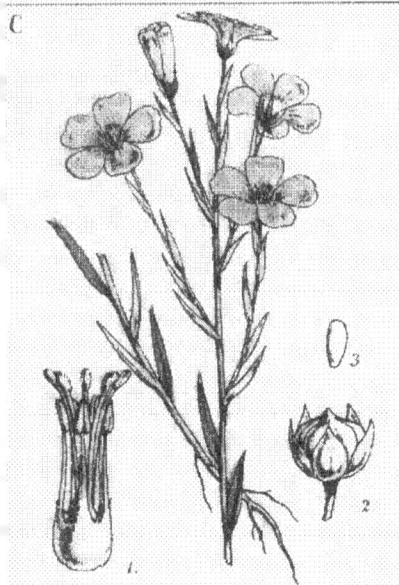

C. Linum usitatissimum

D. Impatiens Noli me tangere.

PLATE XI.

A.—Euonymus europæus, *Linnæus :* p. 90.
    1. A flower.
    2. A ripe fruit.

B.—Rhamnus catharticus, *Linnæus :* p. 90.
    1. A staminate or barren flower.
    2. A pistillate or fertile flower.
    3. The ripe fruit.

C.—Tamarix anglica, *Webb :* p. 88.
    1. Portion of a branchlet, showing the close scale-like leaves.
    2. A flower separated from the spike.
    3. The stamens and pistil.

D.—Rosa canina, *Linnæus :* p. 78.
    1. A vertical section of the flower, the petals being removed, showing the ovaries attached to the inside of the calyx-tube, with the stigmas just protruding from the orifice.
    2. A stamen.
    3. One of the ovaries.
    4. The ripe fruit.

Plate 11.

A

Euonymus europæus.

B

Rhamnus catharticus.

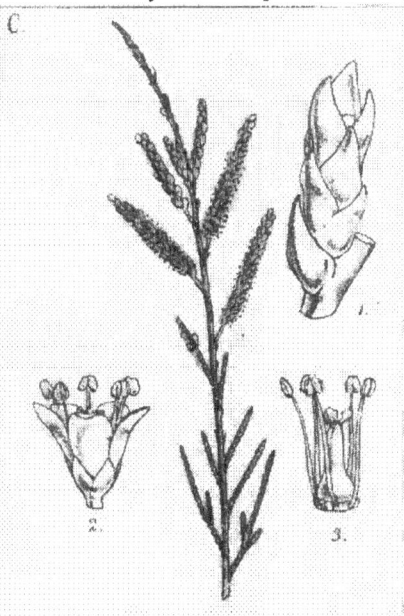

C

Tamarix anglica.

D

Rosa canina.

W. Fitch del et lith.

Vincent Brooks, Imp

PLATE XII.

A.—Lythrum Salicaria, *Linnæus:* p. 91.

    1. A vertical section of a flower.

    2. A capsule, bursting open.

B.—Hippuris vulgaris, *Linnæus:* p. 94.

    1. A flower, consisting of the ovary and style with one stamen.

    2. The stamen separate.

C.—Epilobium hirsutum, *Linnæus:* p. 94.

    1. The top of the ovary, with the calyx, stamens, and style.

    2. One of the seeds.

D.—Lathyrus pratensis, *Linnæus:* p. 91.

    1. The standard or vexillum forming the dorsal petal.

    2. The lateral petals or wings.

    3. The lower petal or keel.

    4. One of the flowers complete.

    5. The pod or legume.

E.—Bryonia dioica, *Jacquin:* p. 95.

    1. A male or staminate flower.

    2. A female or pistillate flower.

Plate 12

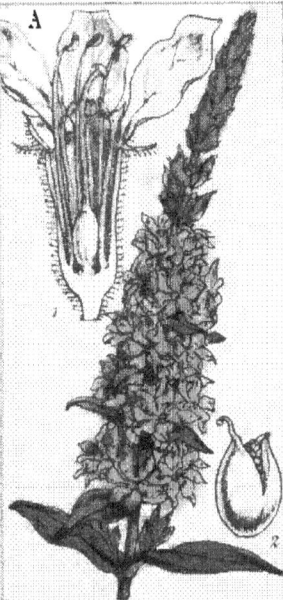

A. Lythrum Salicaria

B. Hippuris vulgaris

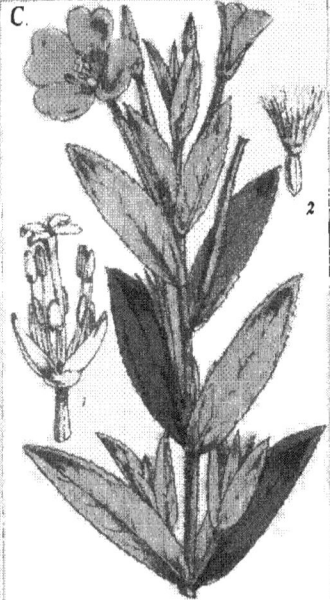

C. Epilobium hirsutum

D. Lathyrus pratensis

E. Bryonia dioica

W.Fitch.del et lith.   Vincent Brooks, Imp.

PLATE XIII.

A.—Œnanthe crocata, *Linnæus :* p. 96.

    1. A flower separated from the umbel.

    2. One of the cylindrical fruits.

    3. A transverse section of the fruit, showing the two car-
pels, and the ridges and vittæ.

B.—Pastinaca sativa, *Linnæus :* p. 97.

    1. A flower detached from the umbel.

    2. One of the flattened fruits.

    3. A transverse section of the fruit.

C.—Sedum acre, *Linnæus :* p. 92.

    1. A leaf, showing the spur behind the point of attach-
ment.

    2. The ripe follicles.

    3. A flower.

    4. The follicles or seed-vessels, before bursting.

D.—Cornus sanguinea, *Linnæus :* p. 98.

    1. One of the flowers.

    2. A fruit.

E.—Montia fontana, *Linnæus :* p. 92.

    1. The calyx or flower-cup.

    2, 3. Different views of the corolla and stamens.

    4. The pistil.

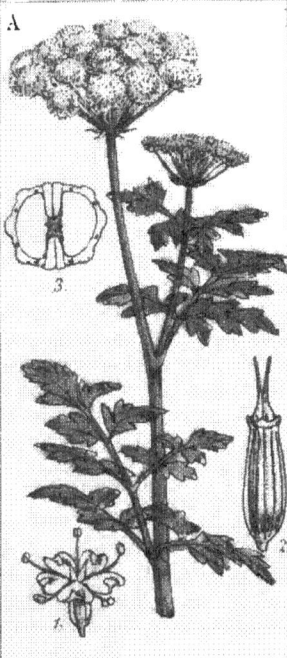

Plate 12

A

3.

1.

2.

Œnanthe crocata.

B

3.

2.

1.

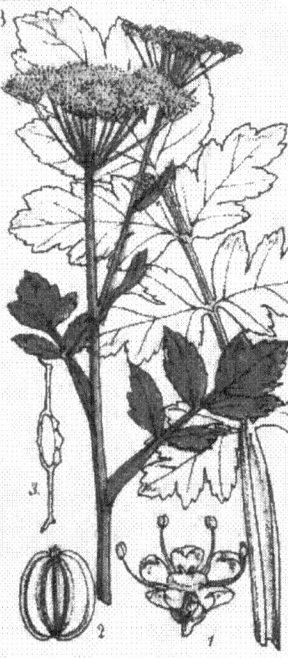

Pastinaca sativa.

C

1.

3.

4.

2.

Sedum acre.

D

1.

2.

Cornus sanguinea

E

1.

2.

3.

4.

Montia fontana

PLATE XIV.

A.—Lonicera Periclymenum, *Linnæus:* p. 98.

    1. One of the tubular two-lipped flowers.

    2. A cluster of the berries.

B.—Galium verum, *Linnæus:* p. 99.

    1. A flower complete.

    2. The pistil separated.

C.—Centranthus ruber, *De Candolle:* p. 99.

    1. One of the curious spurred flowers.

    2. The ripe fruit, with its feathery pappus unrolled.

D.—Dipsacus sylvestris, *Linnæus:* p. 100.

    1. A flower with its bract.

    2. A pistil.

Plate 14.

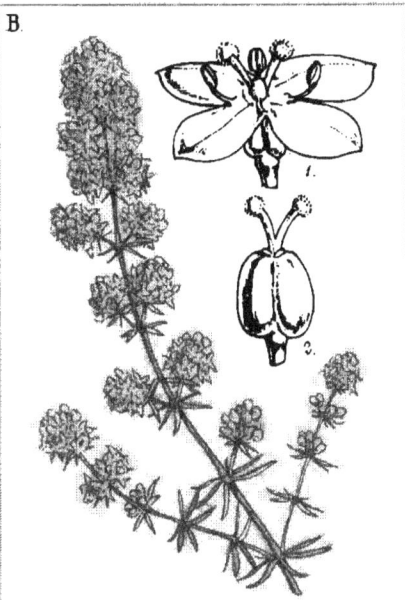

Lonicera Periclymenum.

Galium verum.

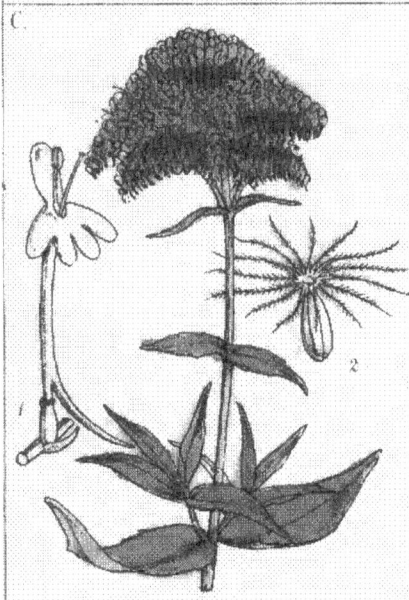

Centranthus ruber.

Dipsacus sylvestris.

W. Fitch del. et lith.

Vincent Brooks, Imp

PLATE XV.

A.—Carduus nutans, *Linnæus:* p. 101.

    1. One of the florets, separated from the capitule or head.

B.—Campanula rotundifolia, *Linnæus:* p. 102.

    1. The anthers and pistil, showing the inferior ovary.

C.—Erica cinerea, *Linnæus:* p. 103.

    1. One of the pitcher-shaped flowers.

    2. A stamen, showing the appendage at the base of the anther-cells, and the pores by which the pollen escapes.

    3. The pistil.

D.—Ligustrum vulgare, *Linnæus:* p. 104.

    1. A flower.

    2. A sprig bearing berries.

Plate 15.

A.

B.

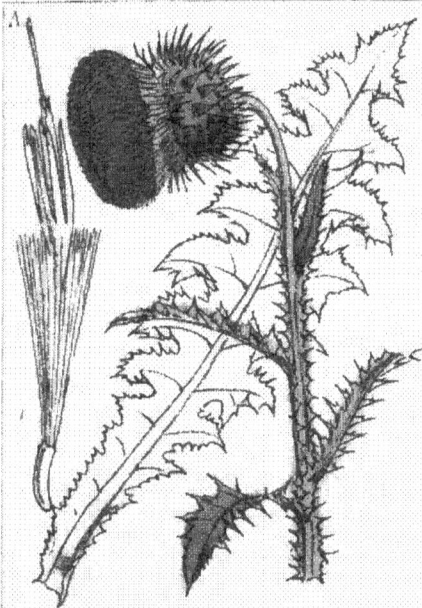

Carduus nutans.

Campanula rotundifolia

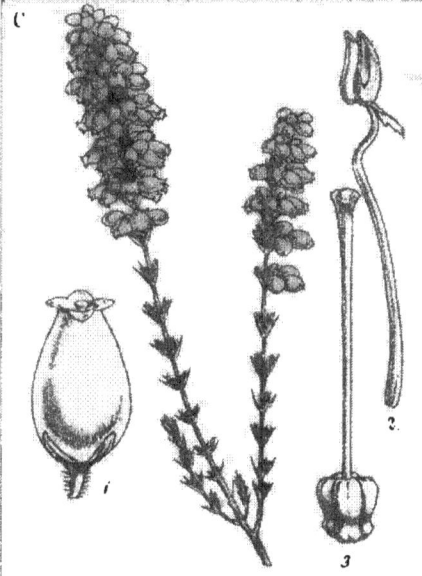

C

Erica cinerea.

D.

Ligustrum vulgare

W. Fitch, del et lith

Vincent Brooks, Imp

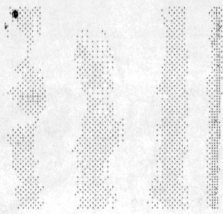

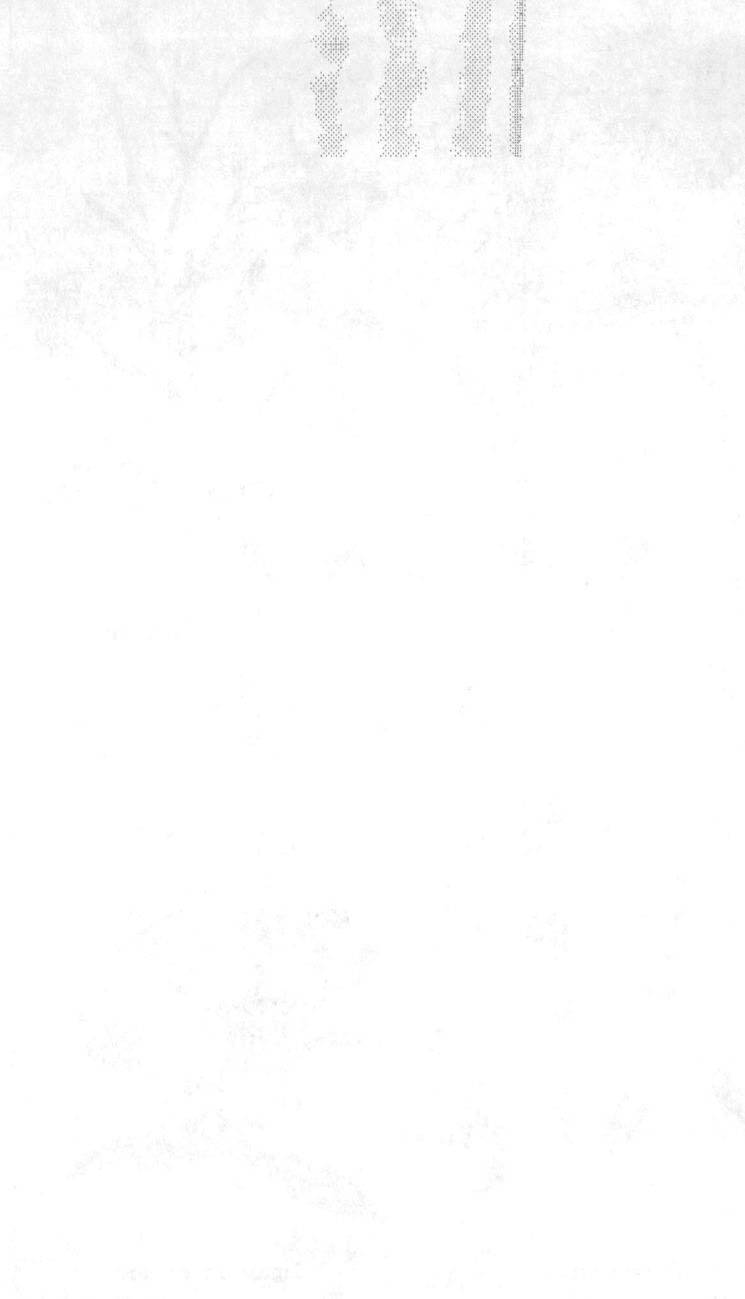

PLATE XVI.

A.—Villarsia nymphæoides, *Ventenat :* p. 104.

    1. A portion of the throat of the corolla, showing the attachment of the stamens.

    2. The pistil.

B.—Polemonium cœruleum, *Linnæus :* p. 105.

    1. A portion of the throat of the corolla, showing the insertion of the stamens.

    2. The pistil, surrounded by an annular disk.

    3. A capsule.

C.—Convolvulus arvensis, *Linnæus :* p. 106.

    1. One of the stamens.

    2. The pistil, surrounded by an annular disk.

D.—Solanum Dulcamara, *Linnæus :* p. 109.

    1. A flower, showing the connivent anthers.

    2. One of the anthers separate, showing the terminal pores.

    3. The pistil.

Plate 16

Villarsia nymphæoides.

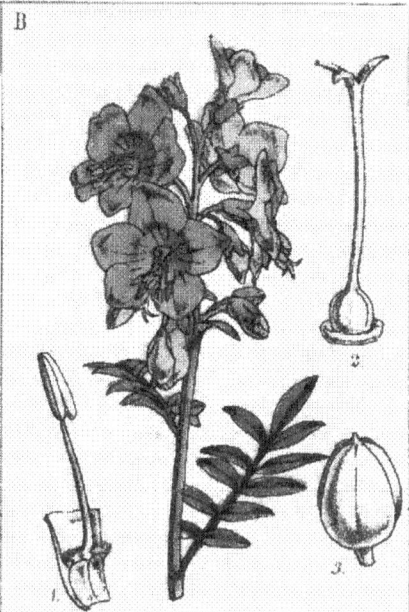

Polemonium cœruleum

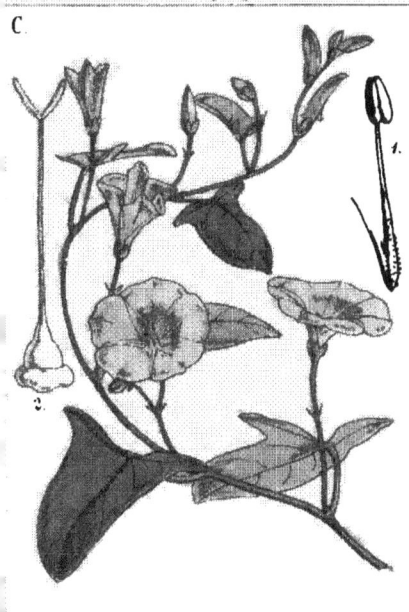

Convolvulus arvensis

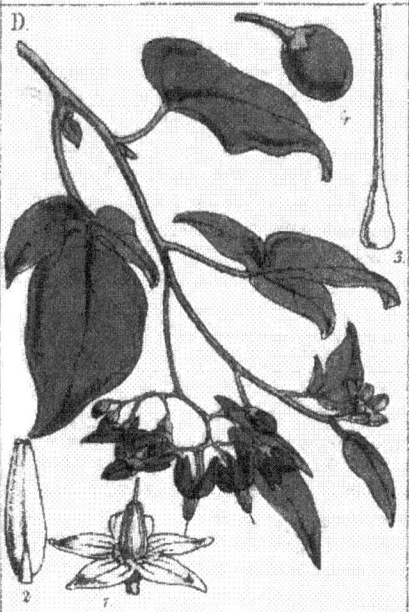

Solanum Dulcamara.

PLATE XVII.

A.—Digitalis purpurea, *Linnæus* : p. 111.

 1. The didynamous stamens, separated from the corolla.

 2. The pistil.

 3. A transverse section of the ovary.

B.—Orobanche minor, *Sutton* . p. 112.

 1. A flower separated from the spike, with its bract.

 2. The corolla laid open, showing the attachment of the stamens.

 3. One of the sepals.

 4. The pistil with a bract and sepals, the corolla removed.

C.—Verbena officinalis, *Linnæus* : p. 114.

 1. A flower, separate.

 2. The corolla laid open, showing the attachment of the stamens, and surrounding the pistil.

D.—Salvia pratensis, *Linnæus* : p. 113.

 1. A flower, separate.

 2. The stamens, showing the short filament, with the elongated connective bearing a perfect anther on its long arm and an abortive one on its short arm.

 3. Another view of the lower part of the connective.

 4. The pistil, showing the four-lobed ovary.

Plate 17.

A

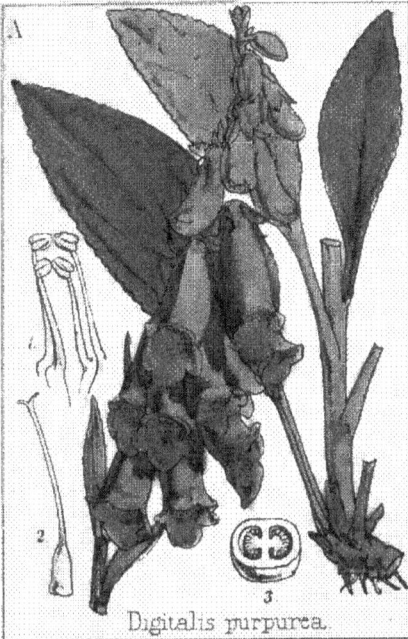

Digitalis purpurea.

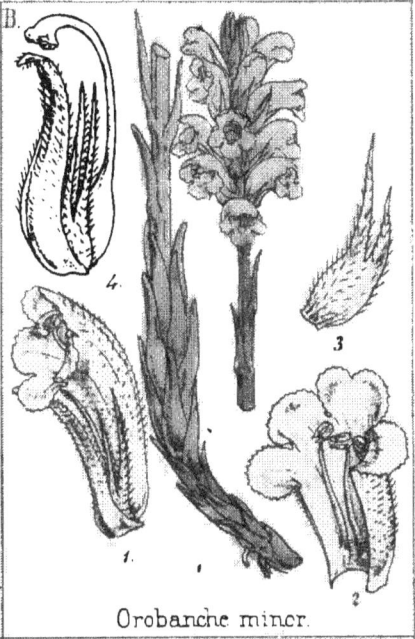

B

Orobanche minor.

C.

Verbena officinalis

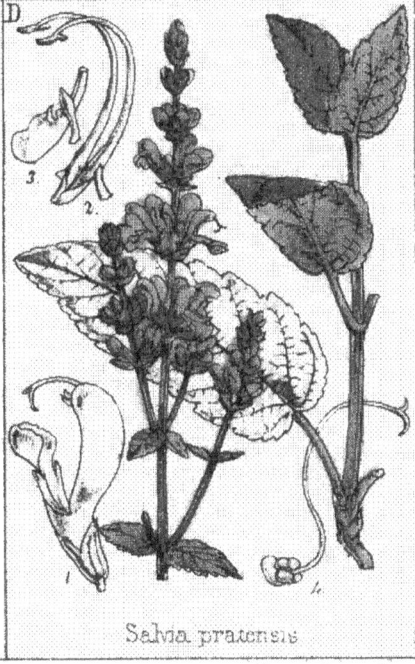

D

Salvia pratensis

PLATE XVIII.

A.—Myosotis palustris, *Withering :* p. 107.

    1. The corolla laid open, showing the stamens and the scaly appendages of the throat.

    2. The pistil with its four-lobed ovary.

B.—Pinguicula vulgaris, *Linnæus :* p. 114.

    1. A flower with the corolla removed.

    2. The two-valved one-celled capsule.

C.—Armeria maritima, *Willdenow :* p. 109.

    1. A flower removed from the head.

    2. One of the petals with a stamen attached to its base.

    3. The pistil, with five hairy styles.

D.—Plantago media, *Linnæus :* p. 110.

    1. A flower, separate.

    2. The pistil.

Plate 18

A.

Myosotis palustris.

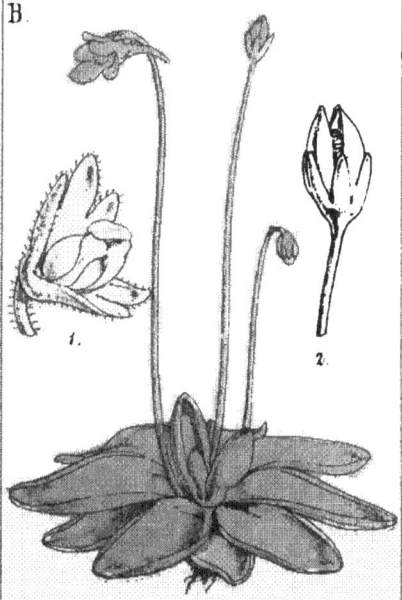

B.

Pinguicula vulgaris.

C.

Armeria maritima

D.

Plantago media.

W.Fich.deLet lith.   Vincent Brooks, Imp.

PLATE XIX.

A.—**Polygonum Bistorta,** *Linnæus :* p. 115.

    1. A flower, separate.

    2. The pistil with its three styles.

B.—**Aristolochia Clematitis,** *Linnæus :* p. 116.

    1. A flower complete.

    2. The pistil, showing the rayed stigma and epigynous stamens.

    3. A capsule.

C.—**Pinus sylvestris,** *Linnæus :* p. 117.

    1. One of the pairs of leaves, with its sheath.

    2. One of the male catkins, with its bract.

    3. An anther, with its two adnate cells, and scale-like connective.

    4. One of the scales of the female catkin, with its two naked ovules.

    5. Inner view of a similar scale when mature, with its two winged seeds.

D.—**Tamus communis,** *Linnæus :* p. 118.

    1. One of the male flowers with stamens.

    2. One of the female flowers with inferior ovary and three-branched style.

    3. The style, separate.

    4. A ripe berry.

Plate 19

A

B

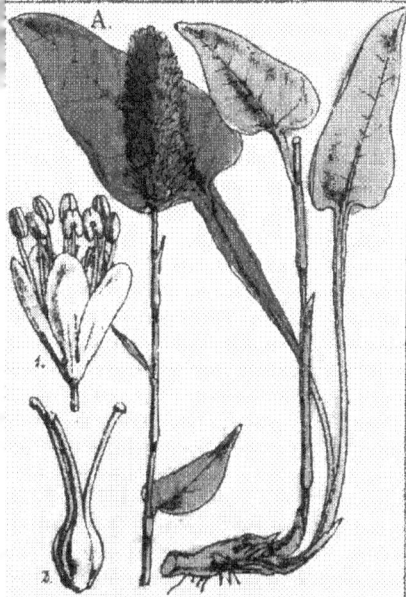

Polygonum Bistorta.

Aristolochia Clematitis.

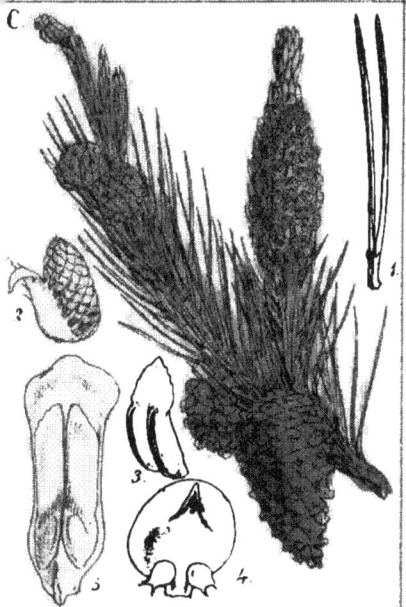

C

D

Pinus sylvestris.

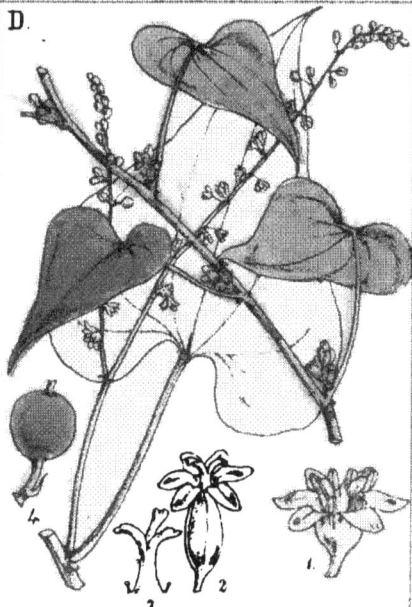

Tamus communis

PLATE XX.

A.—Paris quadrifolia, *Linnæus :* p. 119.

    1. One of the awl-shaped stamens.

    2. The pistil.

    3. A ripe berry, with the persistent perianth.

B.—Hydrocharis Morsus-ranæ, *Linnæus :* p. 119.

    1. The stamens removed from the flower.

    2. One of the stamens, separate.

    3. The pistils, with six two-cleft stigmas.

C.—Ophrys apifera, *Hudson :* p. 126.

    1. Front view of the lip.

    2. Side view of the lip.

    3. The column, showing the anther-case and pollen-masses.

    4. One of the pollen-masses separate.

D.—Iris Pseud-acorus, *Linnæus :* p. 125.

    1. A stamen.

E.—Convallaria majalis, *Linnæus :* p. 127.

    1. A flower with its pedicel and bract.

    2. Vertical section of a flower, showing three of the sta-
mens and the pistil.

    3. A ripe berry.

Plate 20.

A.

B.

3.

2.

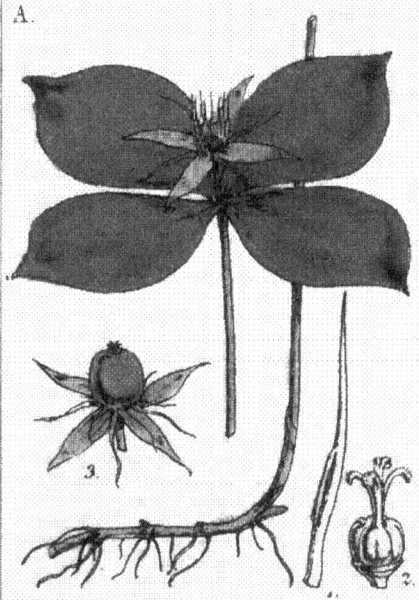

Paris quadrifolia.

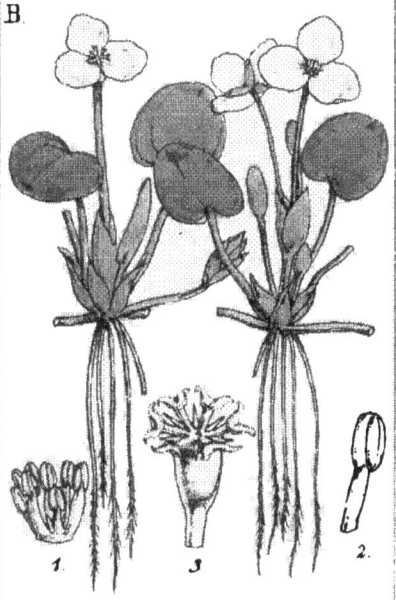

1.   3.   2.

Hydrocharis Morsus-ranæ.

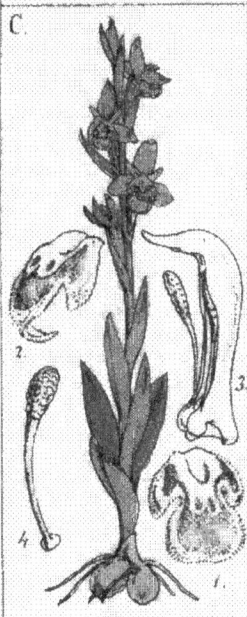

C.

2.

3.

4.

1.

D.

1.

E.

1.

2.

3.

Ophrys apifera.

Iris Pseud-acorus.

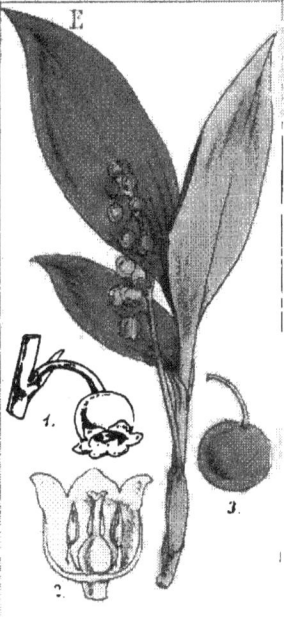

Convallaria majalis

W.Fitch del.et lith.   Vincent Brooks, imp.

PLATE XXI.

A.—Butomus umbellatus, *Linnæus :* p. 125.

    1. A flower with the perianth removed.

    2. One of the carpels.

    3. A transverse section of the carpels.

B.—Typha latifolia, *Linnæus :* p. 121.

    1. A male flower, separate.

    2. A female flower, separate.

C.—Acorus Calamus, *Linnæus :* p. 123.

    1. A flower with its green scales and stamens, and broad sessile stigma.

    2. One of the scales with its accompanying stamen, separate.

    3. An ovary.

D.—Potamogeton natans, *Linnæus :* p. 123.

    1. One of the flowers.

    2. A stamen, separate.

    3. One of the carpels.

E.—Lemna trisulca, *Linnæus :* p. 122.

    1. A frond with its branches and root.

    2. A flower.

F.—Juncus effusus, *Linnæus :* p. 124.

    1. A portion of the cylindrical stem.

    2. A flower.

    3. A capsule with the dry persistent perianth.

G.—Carex riparia, *Curtis :* p. 129.

    1. A flower of the male spikelet with its glume.

    2. A flower of the female spikelet with its glume.

Plate 21.

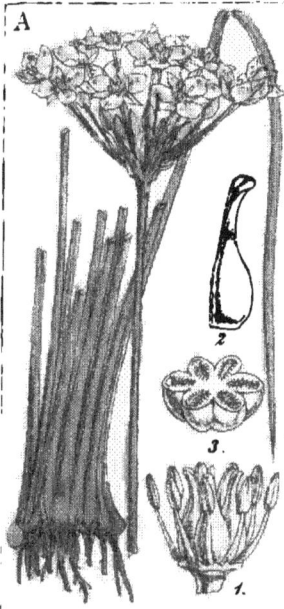

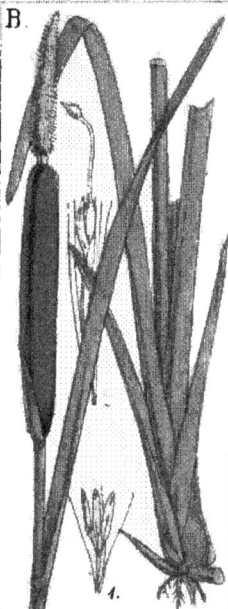

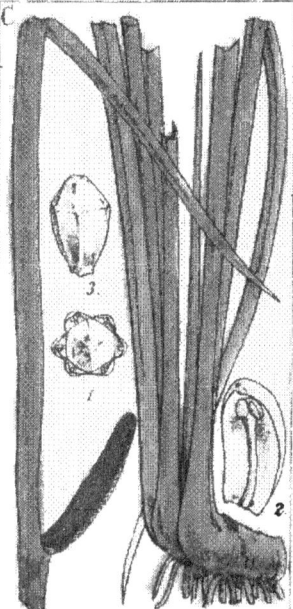

A — Butomus umbellatus

B — Typha latifolia.

C — Acorus Calamus

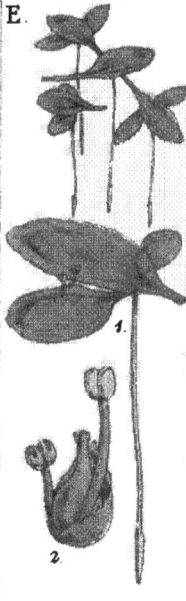

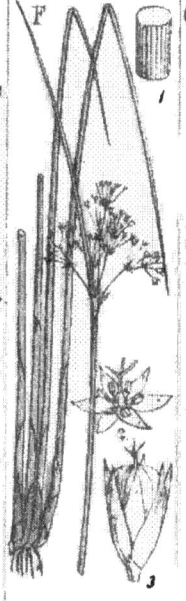

D — Potamogeton natans

E — Lemna trisulca

F — Juncus effusus

G — Carex riparia

Fitch, del et lith.

PLATE XXII.

A.—Arundo Phragmites, *Linnæus :* p. 131.

    1. One of the flowers, separate.

B.—Bromus mollis, *Linnæus :* p. 130.

    1. One of the flowers, separate.

    2. The pistil with its feathery styles.

C.—Parnassia palustris, *Linnæus :* p. 349.

    1. One of the fringed glands or nectaries.

    2. The pistil with its sessile stigmas.

D.—Ulex nanus, *Forster :* p. 349.

    1. A thorny branchlet with its flower.

    2. The standard or dorsal petal.

    3. The keel or lower combined petal.

E.—Hedera Helix, *Linnæus :* p. 350.

    1. A flower separated from the umbel.

Plate 22

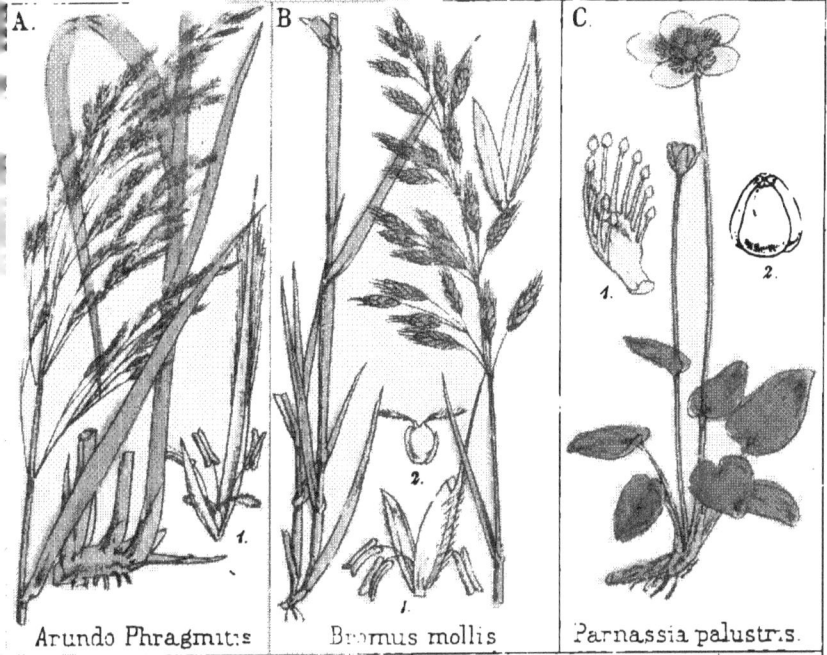

A.

B.

C.

Arundo Phragmitis

Bromus mollis

Parnassia palustris.

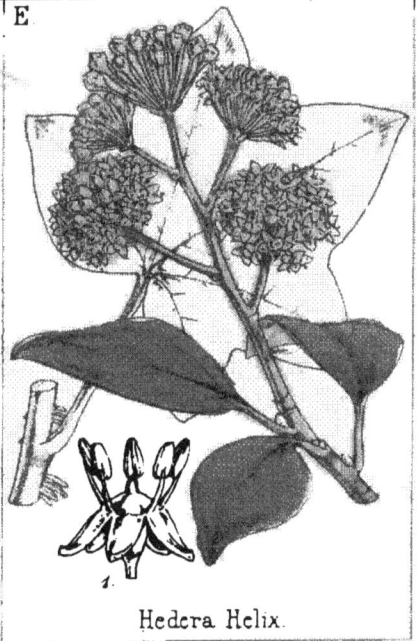

D.

E.

Ulex nanus.

Hedera Helix.

Vincent Brooks. imp

PLATE XXIII.

A.—Scabiosa succisa, *Linnæus :* p. 351.
 1. A flower separated from the head.
 2. The pistil.

B.—Arbutus Unedo, *Linnæus :* p. 352.
 1. A stamen, showing its pores and awns.
 2. The pistil.
 3. Some of the ripe berries.
 4. A transverse section of a berry.

C.—Gentiana Pneumonanthe, *Linnæus :* p. 353.
 1. A portion of the corolla, showing the attachment of the
  stamens to its inner surface.
 2. The pistil.

D.—Mentha Pulegium, *Linnæus :* p. 354.
 1. A flower separated from the verticillaster or whorl-like
  collection of flowers.
 2. The pistil.

E.—Chenopodium polyspermum, *Linnæus :* p. 354.
 1. A flower, separate.
 2. The fruit enclosed by the persistent perigone.

Plate 23

A.

1.

2.

Scabiosa succisa.

B.

1.

2.

3.

Arbutus Unedo.

C.

1.

2.

Gentiana Pneumonanthe

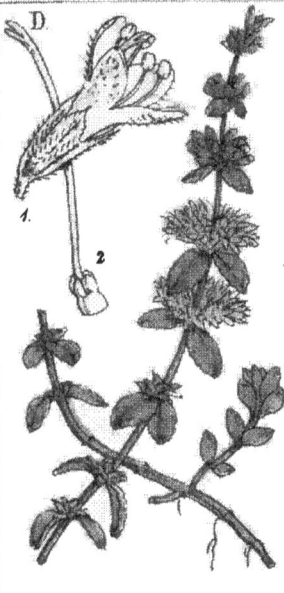

D.

1.

2.

Mentha Pulegium.

E.

1.

2.

Chenopodium polyspermum

W Fitch. del et lith.

Vincent Brooks, Imp

PLATE XXIV.

A.—Colchicum autumnale, *Linnæus :* p. 355.

    1. A stamen.

    2. The pistil invested by the base of the perianth-tube.

    3. The three-celled capsule.

B.—Eriocaulon septangulare, *Withering :* p. 356.

    1. A leaf.

    2. One of the male flowers.

    3. One of the female flowers.

    4. The pistil with its subulate stigmas.

C.—Ilex Aquifolium, *Linnæus :* p. 386.

    1. An abortive flower.

    2. A perfect flower.

    3. A ripe berry.

    4. A transverse section of the berry.

D.—Viscum album, *Linnæus :* p. 387.

    1. A cluster of male flowers.

    2. A cluster of female flowers.

    3. A ripe berry.

Plate 24

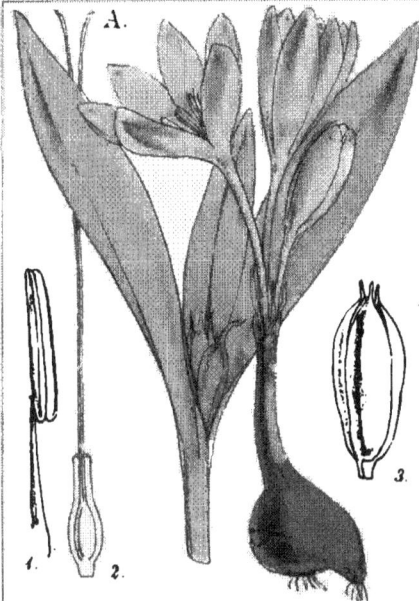

A. Colchicum autumnale.

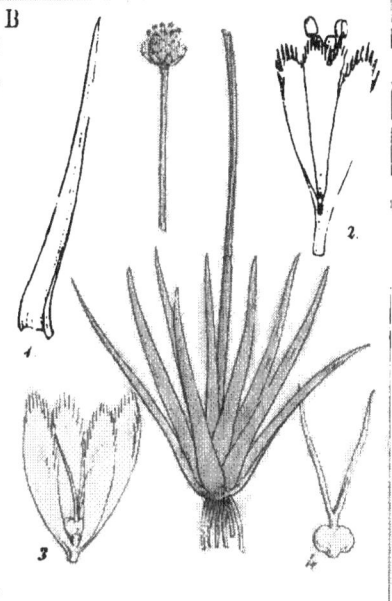

B. Eriocaulon septangulare.

C. Ilex Aquifolium

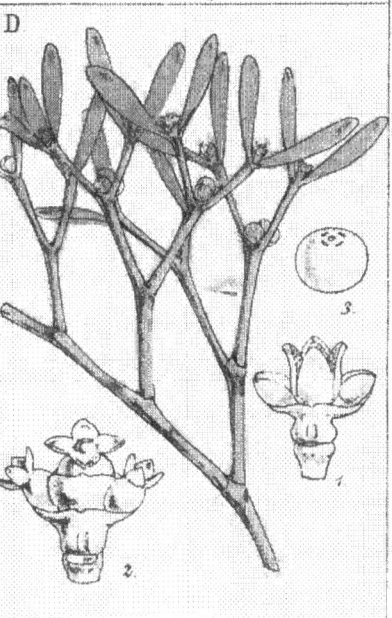

D. Viscum album.

W.Fitch, del et lith

Vincent Brooks, Imp.

www.ingramcontent.com/pod-product-compliance
Lightning Source LLC
Chambersburg PA
CBHW031932220326
41598CB00062BA/1671